Lecture Notes in Computer Science   8981

Commenced Publication in 1973
Founding and Former Series Editors:
Gerhard Goos, Juris Hartmanis, and Jan van Leeuwen

More information about this series at http://www.springer.com/series/7407

Editors
Maurizio Proietti
IASI-CNR
Rome
Italy

Hirohisa Seki
Nagoya Institute of Technology
Nagoya
Japan

ISSN 0302-9743   ISSN 1611-3349 (electronic)
Lecture Notes in Computer Science
ISBN 978-3-319-17821-9   ISBN 978-3-319-17822-6 (eBook)
DOI 10.1007/978-3-319-17822-6

Library of Congress Control Number: 2015937958

LNCS Sublibrary: SL1 – Theoretical Computer Science and General Issues

Springer Cham Heidelberg New York Dordrecht London

Printed on acid-free paper

Springer International Publishing AG Switzerland is part of Springer Science+Business Media
(www.springer.com)

Maurizio Proietti · Hirohisa Seki (Eds.)

Logic-Based Program Synthesis and Transformation

24th International Symposium, LOPSTR 2014
Canterbury, UK, September 9–11, 2014
Revised Selected Papers

 Springer

Preface

This volume contains a selection of the papers presented at LOPSTR 2014, the 24th International Symposium on Logic-Based Program Synthesis and Transformation held during September 9–11, 2014 at the University of Kent, Canterbury, UK. It was colocated with PPDP 2014, the 16th International ACM SIGPLAN Symposium on Principles and Practice of Declarative Programming.

Previous LOPSTR symposia were held in Madrid (2013 and 2002), Leuven (2012 and 1997), Odense (2011), Hagenberg (2010), Coimbra (2009), Valencia (2008), Lyngby (2007), Venice (2006 and 1999), London (2005 and 2000), Verona (2004), Uppsala (2003), Paphos (2001), Manchester (1998, 1992 and 1991), Stockholm (1996), Arnhem (1995), Pisa (1994), and Louvain-la- Neuve (1993). More information about the symposium can be found at: http://www.iasi.cnr.it/events/lopstr14/.

The aim of the LOPSTR series is to stimulate and promote international research and collaboration on logic-based program development. LOPSTR is open to contributions in all aspects of logic-based program development, all stages of the software life cycle, and issues of both programming-in-the-small and programming-in-the-large. LOPSTR traditionally solicits contributions, in any language paradigm, in the areas of synthesis, specification, transformation, analysis and verification, specialization, testing and certification, composition, program/model manipulation, optimization, transformational techniques in software engineering, inversion, applications, and tools. LOPSTR has a reputation for being a lively, friendly forum for presenting and discussing work in progress. Formal proceedings are produced only after the symposium so that authors can incorporate this feedback in the published papers.

In response to the call for papers, 34 contributions were submitted from 21 different countries. The Program Committee accepted 7 full papers for immediate inclusion in the formal proceedings, and 11 more papers presented at the symposium were accepted after a revision and another round of reviewing. Each submission was reviewed by at least 2 and on the average 3.0, Program Committee members or external referees. In addition to the 18 contributed papers, this volume includes the abstracts of the invited talks by two outstanding speakers: Roberto Giacobazzi (University of Verona, Italy), shared with PPDP and Viktor Kuncak (EPFL, Switzerland).

We would like to thank the Program Committee members, who worked diligently to produce high-quality reviews for the submitted papers, as well as all the external reviewers involved in the paper selection. We are very grateful to the LOPSTR 2014 General Co-chairs, Olaf Chitil and Andy King, and the local organizers for the great job they did in managing the symposium. Many thanks also to Olivier Danvy, the Program Committee Chair of PPDP, with whom we often interacted for coordinating the two events. We are grateful to Emanuele De Angelis and Fabrizio Smith, who helped us in maintaining the LOPSTR web site and editing these proceedings. We would also like to thank Andrei Voronkov for his excellent EasyChair system that automates many of the tasks involved in chairing a conference. Special thanks go to all the authors who submitted and presented their papers at LOPSTR 2014, without whom the symposium would have not been possible. Finally, Maurizio Proietti gratefully acknowledges financial support from the Italian National Group of Computing Science (GNCS-INDAM).

February 2015   Maurizio Proietti
                                                                  Hirohisa Seki

Organization

Program Committee

Slim Abdennadher	German University of Cairo, Egypt
Étienne André	Université Paris 13, France
Martin Brain	University of Oxford, UK
Wei-Ngan Chin	National University of Singapore, Singapore
Marco Comini	University of Udine, Italy
Włodek Drabent	IPI PAN Warszawa, Poland and Linköping University, Sweden
Fabio Fioravanti	University of Chieti-Pescara, Italy
Jürgen Giesl	RWTH Aachen, Germany
Miguel Gómez-Zamalloa	Complutense University of Madrid, Spain
Arnaud Gotlieb	SIMULA Research Laboratory, Norway
Gopal Gupta	University of Texas at Dallas, USA
Jacob Howe	City University London, UK
Zhenjiang Hu	National Institute of Informatics, Japan
Alexei Lisitsa	University of Liverpool, UK
Yanhong A. Liu	State University of New York at Stony Brook, USA
Jorge A. Navas	NASA Ames Research Center, USA
Naoki Nishida	Nagoya University, Japan
Corneliu Popeea	Technische Universität München, Germany
Maurizio Proietti	IASI-CNR, Rome, Italy (Co-chair)
Tom Schrijvers	Ghent University, Belgium
Hirohisa Seki	Nagoya Institute of Technology, Japan (Co-chair)
Jon Sneyers	Katholieke Universiteit, Leuven, Belgium
Fausto Spoto	University of Verona, Italy
Wim Vanhoof	University of Namur, Belgium
Germán Vidal	Universitat Politécnica de València, Spain

General Co-chairs

Olaf Chitil	University of Kent, UK
Andy King	University of Kent, UK

Organizing Committee

Emanuele De Angelis	IASI-CNR, Rome, Italy
Fabrizio Smith	IASI-CNR, Rome, Italy

Additional Reviewers

Bardin, Sebastien
Bucheli, Samuel
Cai, Zhouhong
Choppy, Christine
Cirstea, Horatiu
Di Gianantonio, Pietro
Emoto, Kento
Englebert, Vincent
Faber, Wolfgang
Fuhs, Carsten
Guo, Hai-Feng
Gutiérrez, Raúl
Haemmerlé, Rémy

Ieva, Carlo
Inuzuka, Nobuhiro
Ismail, Haythem
Kawabe, Yoshinobu
King, Andy
Komendantskaya,
   Ekaterina
Lenisa, Marina
Li, Jun
Lovato, Alberto
López-Fraguas, Francisco
   Javier
Marple, Kyle

Morihata, Akimasa
Narayanaswamy, Ganesh
Nishimura, Susumu
Pettorossi, Alberto
Salazar, Elmer
Ströder, Thomas
Tan, Tian Huat
Titolo, Laura
Yue, Tao
Zaki, Amira

Obscuring Code

Unveiling and Veiling Information in Programs[1]

Roberto Giacobazzi

University of Verona, Verona, Italy
roberto.giacobazzi@univr.it

Abstract. We survey the most recent developments in code obfuscation and protection from a programming languages perspective. Starting from known impossibility results on universal and general purpose code obfuscation, we show that provably secure obfuscation can be achieved by constraining the attack model. This corresponds to associate attacks with suitable forms of interpretation. In this context it is always possible to systematically making code obscure, making this interpretation failing in extracting (attacking) code. The code transformation can itself be specified as the specialization of a distorted interpreter.

---

[1] An extended version appears in the proceedings of the 16th International Symposium on Principles and Practice of Declarative Programming (PPDP 2014), September 8–10 2014, Canterbury, United Kingdom. ACM Press.

Synthesizing Functions from Relations in Leon

Viktor Kuncak[2], Etienne Kneuss, and Emmanouil Koukoutos

École Polytechnique Fédérale de Lausanne (EPFL),
Lausanne, Switzerland
viktor.kuncak@epfl.ch

Abstract. We present the synthesis functionality of the Leon system (leon.epfl.ch). Leon accepts a purely functional subset of Scala extended with a choice construct. We describe automated and manual synthesis and transformation techniques in Leon, which can eliminate the choice construct and thus transform input/output relation specifications into executable functions from inputs to outputs. The techniques employed include functional synthesis procedures for decidable theories such as term algebras and Presburger arithmetic, synthesis proof rules for decomposing specifications, as well as search-based techniques, such as counterexample-guided synthesis.

[2] This work is supported in part by the European Research Council (ERC) Project *Implicit Programming*.

Contents

Program Analysis and Transformation

Analyzing Array Manipulating Programs by Program Transformation

J. Robert M. Cornish[1], Graeme Gange[1], Jorge A. Navas[2], Peter Schachte[1],
Harald Søndergaard[1]([✉]), and Peter J. Stuckey[1]

[1] Department of Computing and Information Systems,
The University of Melbourne, Melbourne, VIC 3010, Australia
j.cornish@student.unimelb.edu.au
{gkgange,schachte,harald,pstuckey}@unimelb.edu.au
[2] NASA Ames Research Center, Moffett Field, Mountain View, CA 94035, USA
jorge.a.navaslaserna@nasa.gov

Abstract. We explore a transformational approach to the problem of
verifying simple array-manipulating programs. Traditionally, verification
of such programs requires intricate analysis machinery to reason with
universally quantified statements about symbolic array segments, such
as "every data item stored in the segment A[i] to A[j] is equal to the
corresponding item stored in the segment B[i] to B[j]." We define a simple
abstract machine which allows for set-valued variables and we show how
to translate programs with array operations to array-free code for this
machine. For the purpose of program analysis, the translated program
remains faithful to the semantics of array manipulation. Based on our
implementation in LLVM, we evaluate the approach with respect to its
ability to extract useful invariants and the cost in terms of code size.

1 Introduction

We revisit the problem of automated discovery of invariant properties in simple
array-manipulating programs. The problem is to extract interesting properties
of the contents of one-dimensional dynamic arrays (by dynamic we mean arrays
whose bounds are fixed at array variable creation time, but not necessarily at
compile time). We follow the *array partitioning* approach proposed by Gopan,
Reps, and Sagiv [9] and improved by Halbwachs and Péron [11]. This classical
approach uses two phases. In a first phase, a program analysis identifies all
(potential) symbolic *segments* by analyzing all array accesses in the program.
Each segment corresponds to an interval I_k of the array's full index domain, but
its bounds are symbolic, that is, bounds are *index expressions*. For example, the
analysis may identify three relevant segments $I_1 = [0, \ldots, i-1]$, $I_2 = [i]$, and
$I_3 = [i+1, \ldots, n-1]$. After this the original array A is considered partitioned
into segments A_{I_k} corresponding to the identified segments and each segment is
replaced with a *summary variable* a_k. In the second phase, the analysis aims at
discovering properties $\psi(a_k)$ on each summary variable a_k such that

$$\forall \ell \in I_k(\psi(a_k) \Rightarrow \psi(A[\ell])) \tag{1}$$

© Springer International Publishing Switzerland 2015
M. Proietti and H. Seki (Eds.): LOPSTR 2014, LNCS 8981, pp. 3–20, 2015.
DOI: 10.1007/978-3-319-17822-6_1

By partitioning arrays into segments, the analysis can produce stronger separate analysis for each segment rather than a single weaker combined result for the whole array. In particular, we can identify *singleton* segments (A_{I_2} in the example) and translate array writes to these as so-called strong updates. A *strong update* benefits from the fact that the old content of the segment is eliminated by the update, so the new content replaces the old. For a segment that may contain multiple elements, an assignment to an array cell may leave some content unchanged, so a *weak update* must be used, that is, we must use a lattice-theoretic "join" of the new result and the old result associated with ℓ.

Although very accurate, array partitioning methods have their drawbacks. Partitioning can be prohibitively expensive, with a worst-case complexity of $O(n!)$, where n is the number of program variables. Moreover, partitioning must be done before the array content analysis phase that aims at inferring invariants for the form (1), which could be less precise than doing both simultaneously [5]. To mitigate this problem, the index analysis, used to infer the relevant symbolic intervals, is run twice: once during the segmentation phase and again during the array content analysis, which needs it to separate the first fixed point iteration from the rest. In the more sophisticated approach of Halbwachs and Péron [11,16], the transfer functions are much more complex and a concept of "shift variables" is used, representing translation (in the geometric sense) of segments. This is not easily implemented using existing abstract interpretation libraries.

Contribution. We present a program transformation that allows scalar analyses techniques to be applied to array manipulating programs. As in previously proposed array analyses [9,11,16], we partition arrays into segments whose contents are treated as sets rather than sequences. To maintain the relationship among corresponding elements of different arrays, we abstract the state of all arrays within a segment to a set of vectors, one element per array. Thus we transform an array manipulating program into one that manipulates scalars and sets of vectors. A major challenge in this is to encode the disjunctive information carried by each array segment. We propose a technique that splits basic blocks. It has been implemented using the LLVM framework.

Importantly, a program transformation approach allows the separation of concerns: existing analyses based on any scalar abstract domains can be used directly to infer array content properties, even interprocedurally. While other approaches lift a scalar abstract domain to arrays by lifting each transfer function, our approach uses existing transfer functions unchanged, only requiring the addition of two simple transfer functions easily defined in terms of operations that already exist for most domains. The approach is also parametric in the granularity of array index sets, ranging from array smashing [2] to more precise (and expensive) instances. When we go beyond array smashing, the transformational approach inherits the exponential search cost present in the Halbwachs/Péron approach, as for some programs P, the transformed programs P' are exponentially larger than P. However, for simple array-based sort/search programs [9,11], a transformational approach is perfectly affordable, in particular as we can capitalize on code optimization support offered by the LLVM infrastructure.

Instructions	I	$\rightarrow$	$v_1 = constant \mid v_1 = \circ v_2 \mid v_1 = v_2 \diamond v_3 \mid$ A
Array assignments	A	$\rightarrow$	$v_1 = arr[v_2] \mid arr[v_1] = v_2$
Jumps	J	$\rightarrow$	If $(v_1 \bowtie v_2)$ $label_1$ $label_2 \mid$ Jmp $label \mid$ error $\mid$ end
Blocks	B	$\rightarrow$	$label$: I* J
Programs	P	$\rightarrow$	B$^+$

Fig. 1. A small control-flow language with array expressions

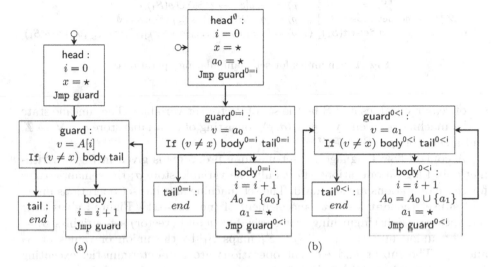

(a)   (b)

Fig. 2. (a) An array program fragment and (b) the corresponding set-machine program.

2   Source and Target Language

Our implementation uses LLVM IR as source and target language. However, as the intricacies of static single-assignment (SSA) form obscure, rather than clarify, the transformation, we base our presentation on a small traditional control flow language, whose syntax is given in Fig. 1. We shall usually shorten "basic block" to "block" and refer to a block's label as its identifier.

Source. Each block is a (possibly empty) sequence of instructions, followed by a (conditional) jump. Arithmetic unary and binary operators are denoted by $\circ$ and $\diamond$ respectively, and logical operators by $\bowtie$. We assume that there is a fixed set of arrays $\{A_1, \ldots, A_k\}$, which have global scope (and do not overlap in memory). The semantics is conventional and not discussed here. Figure 2(a) shows an example program in diagrammatic form.

Target. The abstract machine we consider operates on variables over two kinds of domains: standard scalar types, and sets of vectors of length k, where k is the number of arrays in the source. The scalar variables represent scalars of the source program, including index variables, as well as singleton array segments; sets of vectors represent non-singleton segments of all extant arrays. Let **V** be the

Instructions   I $\rightarrow$ $v_1 = constant \mid v_1 = \circ v_2 \mid v_1 = v_2 \diamond v_3 \mid$ S

Set operations   S $\rightarrow$ $S_1 = \mathsf{nondet\text{-}subset}(S_2) \mid S_1 = S_2 \cup S_3 \mid (v_1, \ldots, v_k) = \mathsf{nondet\text{-}elt}(S_1)$

Jumps   J $\rightarrow$ If $(v_1 \bowtie v_2)$ $label_1$ $label_2 \mid$ Jmp $label \mid error \mid end$

Blocks   B $\rightarrow$ $label$: I* J

Programs   P $\rightarrow$ B$^+$

Fig. 3. Control-flow language for the set machine.

$$\mathscr{S}[\![S_1 = S_2 \cup S_3]\!] \, \langle \sigma, \rho \rangle = \langle \sigma, \rho[S_1 \mapsto \rho(S_2) \cup \rho(S_3)] \rangle$$
$$\mathscr{S}[\![S_1 = \mathsf{nondet\text{-}subset}(S_2)]\!] \, \langle \sigma, \rho \rangle = \langle \sigma, \rho[S_1 \mapsto s] \rangle, s \subseteq \rho(S_2), s \neq \emptyset$$
$$\mathscr{S}[\![(v_1, \ldots, v_k) = \mathsf{nondet\text{-}elt}(S_1)]\!] \, \langle \sigma, \rho \rangle = \langle \sigma[v_1 \mapsto x_1, \ldots, v_k \mapsto x_k], \rho \rangle, (x_1, \ldots, x_k) \in \rho(S_1)$$

Fig. 4. Semantics for set manipulating operations.

set of scalar variables and **S** be the set of vector set variables. The runtime state of the machine is given by a pair $\langle \sigma, \rho \rangle$ consisting of a variable store $\sigma : \mathbf{V} \rightarrow \mathbb{Z}$, and a set store $\rho : \mathbf{S} \rightarrow \mathcal{P}(\mathbb{Z}^k)$.

A control flow language for set machine programs is given in Fig. 3. Arithmetic and logical operations affect only the variable store σ; the semantic rules for these operations are standard. The set machine also has set operations union ($\cup$), subset (nondet-subset) and element_of (nondet-elt). Figure 4 gives their semantic rules, distinguishing scalar variables v and (vector) set variables S.

The union update $S_1 = S_2 \cup S_3$ maps S_1 to the union of values of S_2 and S_3. The subset and element operations are non-deterministic: executing $S_1 = \mathtt{nondet\text{-}subset}(S_2)$ assigns to S_1 *some* non-empty subset of elements from S_2, but makes no guarantee as to which elements are selected. Similarly, the element operation $(v_1, \ldots, v_k) = \mathtt{nondet\text{-}elt}(S_1)$ nondeterministically selects some element of vector set S_1 to load into $v_1, \ldots, v_k$.

Translation. Figure 2(a)'s program scans an array for the first occurrence of value x, assumed to occur in A. The constraint $A[i] = x \wedge \forall \, k \in [0, i) \, (A[k] \neq x)$ is the desired invariant at tail. A corresponding array-free program is given in Fig. 2(b). The example illustrates some key features. Each contiguous array segment is represented by a set variable A_i. Each original block is duplicated for each feasible ordering of *interesting* variables. In the initial ordering $(0 = i)$ the only interesting segment is $A[0]$, represented as the singleton a_0; the read $v = A[i]$ is replaced by an assignment $v = a_0$. At guard$^{0<i}$, $A[i]$ is represented by a_1 so the read is replaced by $v = a_1$. When i is updated at body$^{0=i}$, the previous singleton a_0 becomes part of an "aggregate" segment $A[0, i-1]$. We then transform singleton a_0 to set A_0 and introduce a new singleton a_1 (representing $A[i]$ in the updated ordering). Similarly, when we update i in body$^{0<i}$, segments $A[0, i-1]$ and $A[i]$ are merged (yielding $A_0 = A_0 \cup \{a_1\}$), and a new singleton a_1 is introduced. Consider the resulting concrete set-machine states. At tail$^{0=i}$, we have $a_0 = x$, corresponding to $A[0] = x$ in the original program. At tail$^{0<i}$, we find $x \notin A_0$ and $a_1 = x$. These correspond, respectively, to array invariants $\forall \ell \in [0, i-1] \, . \, A[\ell] \neq x$ and $A[i] = x$ in the original program.

3   From Scalar to Set Machine Transfer Functions

We now show how to lift a scalar domain for use by set machines. Essentially, we use a scalar variable to approximate each component of each set; approximation of set-machine states can then be obtained by grouping states by values of the (original) scalar variables. Essentially, we approximate a set-machine state $\langle \sigma, \rho \rangle$ with set variables $\{S_1, \ldots, S_m\}$ by a set of scalar states $\{S_1^\diamond, \ldots, S_m^\diamond\}$ representing the possible results of selecting some element from each set:

$$\alpha^\diamond(\langle \sigma, \rho \rangle) = \{\sigma \cup \{S_1^\diamond \mapsto y_1, \ldots, S_m^\diamond \mapsto y_m\} \mid y_1 \in \rho(S_1), \ldots, y_m \in \rho(S_m)\}$$

Transfer functions for set operations then operate over the *universe* of possible states, rather than apply element-wise to each state.

Example 1. Consider a program with one scalar variable x, and one set variable S, with initial state $\langle \{x \mapsto 0\}, \{S \mapsto \{1, 2\}\}\rangle$. If we introduce a scalar variable y that selects a value from S via a we have two possible states:

$$\langle \{x \mapsto 0, y \mapsto 1\}, \{S \mapsto \{1, 2\}\}\rangle$$
$$\langle \{x \mapsto 0, y \mapsto 2\}, \{S \mapsto \{1, 2\}\}\rangle$$

If we represent S by scalar variable $S^\diamond$, we have initial states $\langle \{x \mapsto 0, S^\diamond \mapsto 1\}\rangle$ and $\langle \{x \mapsto 0, S^\diamond \mapsto 2\}\rangle$. When we wish to select a value for y, it is chosen nondeterministically from the possible values of $S^\diamond$, resulting in the states:

$$\langle \{x \mapsto 0, y \mapsto 1, S^\diamond \mapsto 1\}\rangle, \ \langle \{x \mapsto 0, y \mapsto 1, S^\diamond \mapsto 2\}\rangle$$
$$\langle \{x \mapsto 0, y \mapsto 2, S^\diamond \mapsto 1\}\rangle, \ \langle \{x \mapsto 0, y \mapsto 2, S^\diamond \mapsto 2\}\rangle$$

If we group states with equal values of x and y, we can see that these correspond to the final states of the original set-machine fragment.   □

Note that this is an (over-)approximation. We can only infer the set of values that *may* be elements of S—this representation cannot distinguish sets of elements which may occur together, nor the cardinality of S. For example, assume we have possible set-machine states $\langle \emptyset, \{S \mapsto \{1\}\}\rangle$ and $\langle \emptyset, \{S \mapsto \{2\}\}\rangle$. The scalar approximation is $\langle \{S^\diamond \mapsto 1\}, \{S^\diamond \mapsto 2\}\rangle$ which covers the feasible set-machine states, but also includes $\langle \emptyset, \{S \mapsto \{1, 2\}\}\rangle$, which is not feasible. More generally, if the set φ of concrete states allows sets $S \mapsto X_1, \ldots, S \mapsto X_k$, we have:

$$\forall X \ (X \subseteq X_1 \cup \ldots \cup X_k \Rightarrow (S \mapsto X) \in \gamma \circ \alpha(\varphi))$$

Consider a (not necessarily numeric) abstract domain $\mathcal{A}$, with meet ($\sqcap$), join ($\sqcup$) and rename operations, as well as a transfer function $\mathscr{F} : \mathsf{I} \to \mathcal{A} \to \mathcal{A}$ for the scalar fragment of the language. The rename operation constructs a new state where each variable x_i is replaced with y_i (then removes the existing bindings of x_i). Formally, the concrete semantics of rename is given by

$$\texttt{rename}(\sigma, [x_1, \ldots, x_k], [y_1, \ldots, y_k]) = \sigma \left[\begin{array}{c} y_1 \mapsto \sigma(x_1), \ldots, y_k \mapsto \sigma(x_k), \\ x_1 \mapsto \star, \ldots, x_k \mapsto \star \end{array} \right]$$

For each set variable S, we introduce k scalar variables $[s^1, \ldots, s^k]$ denoting the possible values of each vector in S. We then extend $\mathscr{F}$ to set operations as shown in Fig. 5.

$$\mathscr{F}[\![S_1 = \textsf{nondet-subset}(S_2)]\!]\ \varphi = \varphi \sqcap \mathbf{rename}(\varphi, [s_2^1, \ldots, s_2^k], [s_1^1, \ldots, s_1^k])$$

$$\mathscr{F}[\![S_1 = S_2 \cup S_3]\!]\ \varphi = \left(\begin{array}{c} \mathscr{F}[\![S_1 = \textsf{nondet-subset}(S_2)]\!]\ \varphi \\ \sqcup\ \mathscr{F}[\![S_1 = \textsf{nondet-subset}(S_3)]\!]\ \varphi \end{array} \right)$$

$$\mathscr{F}[\![(v_1, \ldots, v_k) = \textsf{nondet-elt}(S_1)]\!]\ \varphi = \varphi \sqcap \mathbf{rename}(\varphi, [s_1^1, \ldots, s_1^k], [v_1^1, \ldots, v_1^k])$$

Fig. 5. Extending the transfer function for scalar analysis to set operations

4   Orderings

The transformation relies on maintaining a single ordering of index variables at each transformed block. We now discuss such total orderings.

Our goal is to partition the array index space $(-\infty, \infty)$ into contiguous regions bounded by index variables. For index variables i and j, we need to be able to distinguish between the cases where $i < j$, $i = j$ and $i > j$. However, this is not enough; if we assign $A[i] = x$, but only know that $i < j$, we cannot distinguish between the cases $i = j - 1$ (every element between i and j is x) and $i < j - 1$ (there are additional elements with some other property). So, for index variables i and j, we choose to distinguish these five cases:

$$\boxed{i + 1 < j} \qquad \boxed{i + 1 = j} \qquad \boxed{i = j} \qquad \boxed{i = j + 1} \qquad \boxed{i > j + 1}$$

For convenience in expressing these orderings, we will introduce for each index variable i a new term i^+ denoting the value $i + 1$, and for a set of index variables **I** we will denote by $\mathbf{I}^+$ the augmented set $\mathbf{I} \cup \{v^+ \mid v \in \mathbf{I}\}$. We can then define a total ordering of a set of index variables **I** to be a sequence of sets $[B_1, \ldots, B_k]$, $B_s \subseteq \mathbf{I}^+$, such that the B_s's cover $\mathbf{I}^+$, are pairwise disjoint, and satisfy $i \in B_s \Leftrightarrow i^+ \in B_{s+1}$.

The meaning of the ordered list $\pi = [B_1, B_2, \ldots, B_k]$ is parameterised by the value of program variables involved, that is, it depends on a store σ. The meaning is: $[\![\pi]\!](\sigma) \equiv$

$$\bigwedge_{s,t \in [1..k]} (\forall e, e' \in B_s\ (\sigma(e) = \sigma(e')) \land \forall e \in B_s\ \forall e' \in B_t\ (s < t \to \sigma(e) < \sigma(e')))$$

An ordering π (plus virtual bounds $\{-\infty, \infty\}$) partitions the space of possible array indices into contiguous regions, given by $[\sigma(e), \sigma(e'))$ for $e \in B_i, e' \in B_{i+1}$. For any index variable i, a segment containing i^+ in the right bound is necessarily a singleton segment; all other segments are considered aggregate.

When a new index variable k enters scope, several possible orderings may result. Figure 6(c) gives a procedure for enumerating them. When an index variable k leaves scope, computing the resulting ordering consists simply of eliminating k and k^+ from π, and discarding any now-empty sets. Assignment of an index variable is handled as a removal followed by an introduction.[1]

---

[1] If the assigned index variable appears in the expression, we assign the index to a temporary variable, and replace the index with the temporary in the expression.

We can discard any ordering that arranges constants in infeasible ways, such as $4 < 3$. If we have performed some scalar analysis on the original program, we need only generate orderings which are consistent with the analysis results.

5   The Transformation

We now detail the transformation from an array manipulating program to a set-machine program, with respect to a fixed set of *interesting* segment bounds. Section 6 covers the selection of these bounds. Intuitively, the goal of the transformation is to partition the array into a collection of contiguous segments, such that each array operation uniquely corresponds to a singleton segment. Each singleton segment is represented by a tuple of scalars; each non-singleton segment is approximated by a set variable. There are two major obstacles to this. First, a program point does not typically admit a unique ordering of a given set of segment bounds; second, as variables are mutated in the program, the correspondence between concrete indices and symbolic bounds changes.

The transformation resolves this by replicating basic blocks to ensure that, at any program point, a unique partitioning of the array into segments is identifiable. Any time a segment-defining variable is modified, introduced or eliminated, we emit statements to distinguish the possible resulting partitions, and duplicate the remainder of the basic block for each case. For each partition, we also emit set operations to restore the correspondence between set variables and array segments, using nondet-elt and nondet-subset when a segment is subdivided, and ∪, when a boundary is removed, causing segments to be merged. This way every array read/write in the resulting program can be uniquely identified with a singleton segment. As singleton sets are represented by tuples of scalars, we can finally eliminate array operations, replacing them with scalar assignments.

In the following, we assume the existence of functions next_block, which allocates a fresh block identifier, and push_block, which takes an identifier, a sequence of statements and a branch, and adds the resulting block to the program. We also assume that there is a mutable global table T mapping block identifier and index variable ordering pairs $\langle id, \pi \rangle$ to ids, used to store previously computed partial transformations, and an immutable set $\mathcal{I}$ of segment bound variables and constants. The function get_block takes a block identifier, and returns the body of the corresponding block. The function vars returns the set of variables appearing lexically in the given expression. The function find_avar gives the variable name to which a given array and index will be translated, given an ordering.

Figure 6 gives the transformation. Procedure transform takes a block and transforms it, assuming a given total ordering π of the index variables. It is called once with the initial block of each function and an ordering containing only the constants in the index set. As there are finitely many $\langle id, \pi \rangle$ combinations, and each pair is constructed at most once, this process terminates.

The core of the transformation is done by a call to transform_body(B, π, id, ss). Here B is the portion of the current block to be transformed and π the current ordering. id and ss hold the identifier and body of the partially-transformed

```
% Check if the block has already been transformed
% under π. If not, transform it.
transform(id, π)
    if ((id, π) ∈ T)
        return T[(id, π)]
    id_t := next_block()
    T := T[(id, π) ↦ id_t]
    (stmts, br) := get_block(id)
    transform_body((stmts, br), π, id, [])
    return id_t
% Evaluate a branch.
transform_body(([], Jmp b), π, id, ss)
    id_b := transform(b, π)
    push_block(id, ss, Jmp id_b)

transform_body(([], If l then t else f), π, id, ss)
    if vars(l) ⊆ I
        dest := if eval(l, π) then t else f
        id_dest := transform(dest, π)
        push_block(id, ss, Jmp id_dest)
    else
        id_t := transform(t, π)
        id_f := transform(f, π)
        push_block(id, ss, If l then id_t else id_f)
% (Potentially) update an index.
transform_body(([x = expr|stmts], br), π, id, ss)
    if x ∈ I
        split_transform(x, (stmts, br), π, id, ss :: [x = expr])
    else
        transform_body((stmts, br), π, id, ss :: [x = expr])
% Transform an array read...
transform_body(([x = A[i]|stmts], br), π, id, ss)
    A_i := find_avar(π, A, i)
    transform_body((stmts, br), π, id, ss :: [x = A_i])
% or an array write.
transform_body(([A[i] = x|stmts], br), π, id, ss)
    A_i := find_avar(π, A, i)
    transform_body((stmts, br), π, id, ss :: [A_i = x])
```

(a) The top-level transformation process

```
split_transform(x, (stmts, br), π, id, ss)
    Π' := feasible_orders(π, x)
    split_rec(x, Π', (stmts, br), π, id, ss)

split_rec(x, [π'], (stmts, br), π, id, ss)
    asts := remap_avars(π, π')
    transform_body((stmts, br), π', id, ss :: asts)

split_rec(x, [π'|Π'], (stmts, br), π, id, ss)
    id_π' := next_block()
    id_Π' := next_block()
    cond := ord_cond(x, π')
    push_block(id, ss, If cond then id_π' else id_Π')
    asts := remap_avars(π, π')
    transform_body((stmts, br), π', id_π', asts)
    split_rec(x, Π', (stmts, br), π, id_Π', [])
```

(b) Fan-out of a block when an index variable is changed

```
feasible_orders(k, π) : insert(k, π, [])

insert(k, [], pre) : return {pre :: {k} :: {k⁺}}
insert(k, [S_i|S], pre)
    low := insert⁺(k, [S_i | S], pre :: {k})
    high := insert(k, S, pre :: S_i)
    if ∃ x . x⁺ ∈ S_i
        return low ∪ high
    else
        return low ∪ high ∪
            insert⁺(k, S, pre :: (S_i ∪ {k}))

insert⁺(k, [], pre) : return {pre :: {k⁺}}
insert⁺(k, [S_i | S], pre)
    if ∃ x . x⁺ ∈ S_i
        return {pre :: (S_i ∪ {k⁺}) :: S}
    else
        return {pre :: (S_i ∪ {k⁺}) :: S} ∪
            {pre :: {k⁺} :: S_i :: S}
```

(c) Enumerating the possible total orderings upon introducing a new index variable k

Fig. 6. Pseudo-code for stages of the transformation process.

block. As a block is processed, instructions not involving index or array variables are copied verbatim into the transformed block. During the process, we ensure that each (transformed) statement is reachable under exactly one index ordering π. Singleton segments under π are represented by scalar variables, and aggregate segments by set variables. Array reads and writes are replaced with accesses and assignments to the corresponding scalar or set variable, as determined by find_avar. Conditional branches whose conditions are determined by the current ordering are replaced by direct branches to the **then** or **else** part, as appropriate. Once no instructions remain to be transformed, the block id is emitted with body ss, together with the appropriate branch instruction.

Whenever an index variable is modified, the rest of the current block must be split, and the set variables must be updated accordingly. The rest of the block is then transformed under *each* possible new ordering π'. This is the job of split_transform shown in Fig. 6(b), while the job of feasible_orders in Fig. 6(c) is to determine the set of possible orders. The function ord_cond(x, π') generates logical expressions to determine whether the ordering π' holds, given that

$$\pi = [\{0\} < \{1\} < \{i\} < \{i^+\} < \{n\}]$$

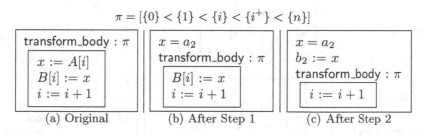

(a) Original   (b) After Step 1   (c) After Step 2

Fig. 7. Transformation of array reads and writes under ordering π. As the segment $[i, i^+]$ is a singleton, the array elements are represented as scalars.

π previously held. ord_cond checks the position of both x and x^+. If x is part of a larger equivalence class in π, ord_cond generates the corresponding equality; otherwise, it checks that x is greater than its left neighbour; similarly, it checks that x^+ is in its class or less than its right neighbour. Figure 6(b) shows the process of splitting a block upon introducing an index variable x.

5.1 Reading and Writing

Transformation of array reads and writes is simple, if the array index is in the set **I** of index variables. Figure 7(a–c) shows the step-by-step transformation of a block, under the specified ordering. After Step 1, reference $A[i]$ has been transformed to scalar a_2, since $\{i\}$ is a singleton. Similarly, Step 2 transforms $B[i]$ to b_2.

If the index of the read/write operation has been omitted, we must instead emit code to ensure the operation is dispatched to the correct set variable. The dispatch procedure is similar in nature to split_transform, as given in Fig. 6(c); essentially, we emit a series of branches to determine which (if any) of the current segments contains the read/write index. Once this has been determined, we apply the array operation to the appropriate segment. If the selected segment is a singleton, this is done exactly as in transform_body. For writes to an aggregate segment, we must first read some vector from the segment, substitute the element to be written, then merge the updated vector back into the segment.[2]

5.2 Index Manipulation

The updating of index variables is the most involved part of the transformation, as we must emit code not only to determine the updated ordering π', but also to ensure the array segment variables are matched to the corresponding bounds. Figure 8 illustrates this process, implemented by the procedure remap_avars, as it splits a block into three: one to test an index expression to determine what ordering applies, and one for each ordering. In the original code, ordering π applies, but following the assignment, either ordering π'_0 or π'_1 may apply. The test inserted by Step 2 distinguishes these cases, leaving only one ordering applicable to each of the $s_{\pi'_0}$ and $split_1$ blocks.

---

[2] Detailed pseudo-code for this is in Appendix A.

$$\pi = [\{0\} < \{1\} < \{i\} < \{i^+\} < \{n\}]$$
$$\pi_0' = [\{0\} < \{1\} < \{i\} < \{i^+\} < \{n\}]$$
$$\pi_1' = [\{0\} < \{1\} < \{i\} < \{i^+, n\}]$$

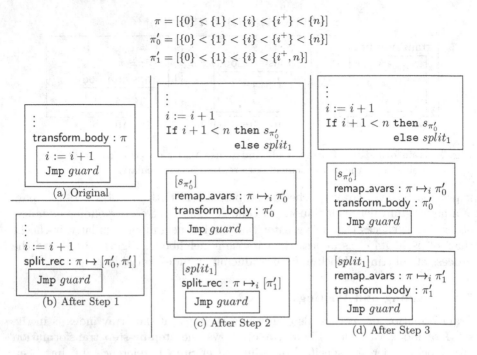

Fig. 8. Example of updating an index assignment. We assume an existing scalar analysis which has determined that, after $i = i + 1$, we have $1 < i < n$.

If we normalize index assignments such that for $k := E$, $k \notin E$, we can separate the updating of segment variables into two stages; first, computing intermediate segment variables A_i' after eliminating k from π, and then computing the new segment variables after introducing the updated value of k. Pseudo-code for these steps are given in Figs. 9(a) and 10(a). In practice, we can often eliminate many of these intermediate assignments, as segments not adjacent to the initial or updated values of k remain unchanged.

When we eliminate an index variable k from π, we merge segments that were bounded only by k or k^+. If k or k^+ appears alone at the very beginning or end of π, the segments are discarded entirely. If either appears alone between other variables in π, the segments on either side are merged to form a single segment. However, if k and k^+ are both equal to some other variables, the original segments are simply copied to the corresponding temporary variables. This is illustrated in Fig. 9(b).

The pseudo-code in Figs. 9 and 10 ignores the distinction between singleton and aggregate segments; the transformed operations differ slightly in the two cases. If we introduce a singleton segment into an aggregate segment, we select a single vector from the set $((a', b', c') = \mathtt{nondet\text{-}elt}(A))$; if an aggregate segment is introduced, we emit a subset operation $(A' = \mathtt{nondet\text{-}subset}(A))$.

The procedure for injecting k into π behaves similarly. If k is introduced at either end of π, we introduce new segments with indeterminate values. If k

$$\begin{array}{l}
\text{remap\_avars}(k, \pi, \pi') \\
\quad \text{eliminate}(k, \pi) :: \text{introduce}(k, \pi')
\end{array}$$

$$\begin{array}{l}
\text{eliminate}(k, \pi) \\
\quad \text{eliminate}(k, \pi, 0, 0, \emptyset)
\end{array}$$

$$\begin{array}{l}
\text{eliminate}(k, [], \_, i', E) \\
\quad \textbf{if}(i' = 0) \textbf{ return } [] \\
\quad \textbf{else return } [\text{emit\_merge}(A'_{i'-1}, E)]
\end{array}$$

$$\begin{array}{l}
\text{eliminate}(k, [\{c\} \mid \mathcal{S}], i, i', E) \\
\quad \textbf{where } c \in \{k, k^+\} \\
\quad \textbf{return } \text{eliminate}(k, \mathcal{S}, i+1, i', E \cup \{A_i\})
\end{array}$$

$$\begin{array}{l}
\text{eliminate}(k, [S_j \mid \mathcal{S}], i, i', E) \\
\quad \text{suff} := \text{eliminate}(k, \mathcal{S}, i+1, i'+1, \{A_i\}) \\
\quad \textbf{if}(i' = 0) \\
\quad\quad \% \text{ Ignore leading segment.} \\
\quad\quad \textbf{return } \text{suff} \\
\quad \textbf{else} \\
\quad\quad \textbf{return } \text{emit\_merge}(A'_{i'-1}, E) :: \text{suff}
\end{array}$$

$$\begin{array}{l}
\text{emit\_merge}(x, E) \\
\quad \textbf{return } [\text{x} = \bigcup E]
\end{array}$$

(a)

$$\pi = [\{k\} \overset{a_0}{<} \{k^+, n\} \overset{a_1}{<} \{n^+\}]$$
$$\pi_r = [\{n\} \overset{a'_0}{<} \{n^+\}\}]$$

$$\boxed{a'_0 := a_1}$$

$$\pi = [\{k, n\} \overset{a_0}{<} \{k^+, n^+\}]$$
$$\pi_r = [\{n\} \overset{a'_0}{<} \{n^+\}\}]$$

$$\boxed{a'_0 := a_0}$$

$$\pi = [\{i\} \overset{a_0}{<} \{i^+, k\} \overset{a_1}{<} \{k^+\} \overset{A_2}{<} \{n\} \overset{a_3}{<} \{n^+\}]$$
$$\pi_r = [\{i\} \overset{a'_0}{<} \{i^+\} \overset{A'_1}{<} \{n\} \overset{a'_2}{<} \{n^+\}]$$

$$\boxed{\begin{array}{l} a'_0 := a_0 \\ A'_1 := \{a_1\} \cup A_2 \\ a'_2 := a_3 \end{array}}$$

(b)

Fig. 9. (a) Algorithm for generating instructions to keep segment variables updated; (b) resulting assignments when k is eliminated from various orderings, also showing the remaining order π_r and scalar or set variables corresponding to each segment.

is introduced somewhere within an existing segment, we introduce new child segments—each of which is a subset of the original segment.

5.3 Control Flow

When transforming control flow, there are three cases we must consider:

1. Unconditional jumps
2. Conditional jumps involving some non-index variables
3. Conditional jumps involving *only* index variables

In cases (1) and (2), the transformation process operates as normal; we recursively transform the jump targets, and construct the corresponding jump with the transformed identifiers. However, when we have a conditional jump If $i \bowtie j$ then t else f where i and j are both index terms, the relationship between i and j is statically determined by the current ordering π. As a result, we can simply evaluate the condition $i \bowtie j$ under the ordering π, and use an unconditional branch to the corresponding block. This is illustrated in Fig. 11.

$$\text{introduce}(k, \pi)$$
$$\quad \text{introduce}(k, \pi, 0, 0)$$

$$\text{introduce}(k, [], \_, i')$$
$$\quad \textbf{return } []$$

$$\text{introduce}(k, [\{c\} \mid \mathcal{S}], i, i')$$
$$\qquad \textbf{where } c \in \{k, k^+\}$$
$$\quad \text{suff} := \text{introduce}(k, \mathcal{S}, i+1, i')$$
$$\quad \textbf{if}(i' = 0)$$
$$\qquad \textbf{return } [A_i = \star] :: \text{suff}$$
$$\quad \textbf{else}$$
$$\qquad \textbf{return } [A_i = \text{nondet-subset}(A_i')] :: \text{suff}$$

$$\text{introduce}(k, [S_j \mid \mathcal{S}], i, i')$$
$$\quad \text{suff} := \text{introduce}(k, \mathcal{S}, i+1, i'+1)$$
$$\quad \textbf{if}(i' = 0)$$
$$\qquad \% \text{ Ignore leading segment.}$$
$$\qquad \textbf{return suff}$$
$$\quad \textbf{else}$$
$$\qquad \textbf{return } [A_i = A_{i'}'] :: \text{suff}$$

(a)

$$\pi_p = [\{n\} < \{n^+\}]$$
$$\pi = [\{k\} < \{k^+, n\} < \{n^+\}]$$

$$a_0 := \star$$
$$a_1 := a_0'$$

$$\pi_p = [\{n\} < \{n^+\}]$$
$$\pi = [\{k, n\} < \{k^+, n^+\}]$$

$$a_0 := a_0'$$

$$\pi_p = [\{i\} < \{i^+\} < \{n\} < \{n^+\}]$$
$$\pi = [\{i\} < \{i^+, k\} < \{k^+\} < \{n\} < \{n^+\}]$$

$$a_0 := a_0'$$
$$(a_1) = \text{nondet-elt}(A_1')$$
$$A_2 = \text{nondet-subset}(A_1')$$
$$a_3 := a_2'$$

(b)

Fig. 10. (a) Generating instructions for (re-)introducing a variable k into a given ordering, and (b) the resulting assignments when k is introduced into various orderings. Note the difference between introducing singleton and aggregate segments.

$$\pi = [\{0\} < \{1\} < \{i\} < \{i^+\} < n]$$

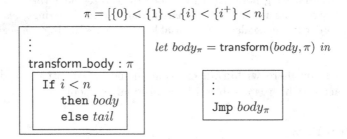

$let\ body_\pi = \text{transform}(body, \pi)\ in$

Fig. 11. Transforming a jump, conditional on index variables only, under ordering π

6   Selecting segment Bounds

Until now we have assumed a pre-selected set of *interesting* segment bounds. The selection of segment boundaries involves a trade-off: we can improve the precision of the analysis by introducing additional segment bounds, but the transformed program grows exponentially as the number of segment bounds increases. As do [11], we can run a data-flow analysis to find the set of variables that may (possibly indirectly) be used as, or to compute, array indices. Formally, we collect the set $\mathbb{I}$ of variables and constants i occurring in these contexts:

$$A[i] \quad \text{where } A \text{ is an array} \tag{2}$$

$$i' = \mathsf{op}(i) \quad \text{where } i' \in \mathbb{I} \tag{3}$$

$$i = \mathsf{op}(i') \quad \text{where } i' \in \mathbb{I} \text{ and } i' \text{ is not a constant} \tag{4}$$

Any variable which does not satisfy these conditions can safely be discarded as a possible segment bound. For the experiments in Sect. 7 we used all elements of $\mathbb{I}$ as segment bounds (so $\mathbf{I} = \mathbb{I}$), which yields an analysis corresponding roughly to the approaches of [9,11]. We could, however, discard some subset of $\mathbb{I}$ to yield a smaller, but less precise, approximation of the original program. The cases (3) and (4) are needed because of possible aliasing; this is particularly critical in an SSA-based language, as SSA essentially replaces mutation with aliasing. It is worth noting that these dependencies extend to aliases introduced prior to the relevant array operation, as in the snippet "$i := x; \dots A[i] := k; \dots y := x + 1;$"

7 Experimental Evaluation

We have implemented our method using the LLVM framework, in two distinct transformation phases. In a first pass, transformation is done as described above, but without great regard for the size of the transformed program. At the same time, we also use a (polyhedral) scalar analysis of the original program (treating arrays as unknown value sources) to detect any block whose total ordering is infeasible. In the second pass, we prune these unreachable blocks away. As can be gleaned from Table 1, these measures reduce the complexity of the transformed program significantly.

To extract array properties from the corresponding invariants discovered in a transformed program, we require users to specify, at transformation time, the range of array segments that are of interest. In our implementation, this is described by a strict index inequality that must apply to segments in the range. For example, specifying $0 < n$ indicates that we are interested in invariants of the form $\forall \ell \ (0 \le \ell < n \Rightarrow \psi(A_1[\ell], \dots, A_k[\ell]))$, where $A_1, \dots, A_k$ are the arrays in scope and ψ is some property. At the end of the transformation we use a newly created block to join all copies of the original exit block whose total ordering is consistent with the given range. The various scalar representations for each array segment, as well as other variables in scope in each copy, are merged together in phi nodes inside this final block. Properties discovered about the segment phi nodes then translate directly to properties about the corresponding array segments in the original program.

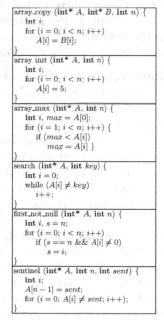

```
array_copy (int* A, int* B, int n) {
    int i;
    for (i = 0; i < n; i++)
        A[i] = B[i];
}

array_init (int* A, int n) {
    int i;
    for (i = 0; i < n; i++)
        A[i] = 5;
}

array_max (int* A, int n) {
    int i, max = A[0];
    for (i = 1; i < n; i++) {
        if (max < A[i])
            max = A[i] }
}

search (int* A, int key) {
    int i = 0;
    while (A[i] ≠ key)
        i++;
}

first_not_null (int* A, int n) {
    int i, s = n;
    for (i = 0; i < n; i++)
        if (s == n && A[i] ≠ 0)
            s = i;
}

sentinel (int* A, int n, int sent) {
    int i;
    A[n − 1] = sent;
    for (i = 0; A[i] ≠ sent; i++);
}
```

Fig. 12. Simple test programs

Table 1. Sizes of transformed test programs and analysis time for `polka` and `uva`

Program	Original		Transformed		Post-processed		Running time (s)		
	blocks	insts.	blocks	insts.	blocks	insts.	transf.	polka	uva
array_copy	5	12	274	898	33	149	0.80	67.07	0.18
array_init	5	11	274	644	33	115	0.94	19.08	0.22
array_max	7	19	220	562	51	139	0.95	110.87	0.45
search	5	10	90	167	27	69	0.49	2.05	2.75
first_not_null	8	17	1057	2217	216	694	3.73	2378.35	4.85
sentinel	5	13	1001	1936	294	765	3.07	1773.01	4.97

We have tested our method by running first the `polka` polyhedra domain [12] on the output of our transformation when applied to the programs given in Fig. 12. The interesting invariants that we infer are as follows (each property holds at the end of the corresponding function):

$$\text{array\_copy} \quad : \forall \ell \ (0 \leq \ell < n \Rightarrow A[\ell] = B[\ell])$$
$$\text{array\_init} \quad : \forall \ell \ (0 \leq \ell < n \Rightarrow A[\ell] = 5)$$
$$\text{array\_max} \quad : \forall \ell \ (0 \leq \ell < n \Rightarrow A[\ell] \leq max)$$
$$\text{search} \quad : \forall \ell \ (0 \leq \ell < i \Rightarrow A[\ell] \neq key)$$
$$\text{first\_not\_null} : \forall \ell \ (0 \leq \ell < s \Rightarrow A[\ell] = 0)$$
$$\text{sentinel} \quad : \forall \ell \ (0 \leq \ell < i \Rightarrow A[\ell] \neq sent)$$

Figure 12 shows test programs from related papers [9, 11]. Table 1 lists sizes of the original, transformed, and post-processed transformed versions (columns Original, Transformed, and Post-processed respectively), as well as the time to perform the transformation (column transf). Column `polka` shows the analysis time in seconds for running the polka polyhedra domain. `uva` is explained below.

Enhancing an existing analyzer. As a separate experiment we use IKOS [3], an abstract interpretation-based static analyzer developed at NASA. IKOS has been used successfully to prove absence of buffer overflows in Unmanned Aircraft Systems flight control software written in C. The latest (unreleased) version of IKOS provides an uninitialized variable analysis that aims at proving that no variable can be used without being previously defined, otherwise the execution of the program might result in undefined behaviour.

Currently IKOS is not sufficiently precise for array analysis. (Figure 13), IKOS cannot deduce that $A[5]$ is definitely uninitialized at line 4. However, using the transformational approach, IKOS proves that $A[5]$ is definitely uninitialized. The problem is far from trivial; as Regehr [17] notes, `gcc` and `clang` (with -*Wuninitialized*) do not even raise warnings for this example, but stay completely silent.

```
int array_init_unsafe (void) {
1:     int A[6], i;
2:     for (i = 0; i < 5; i++)
3:         A[i] = 1;
4:     return A[5];
}
```

Fig. 13. Regehr's example [17]

We ran IKOS on the transformed version of array_init_unsafe. IKOS successfully reported a definite error at line 4 in 0.22 s. Conversely, transformation enabled IKOS to show that no array element was left undefined in the case of array_init. Finally we ran IKOS on the rest of the programs in Fig. 12. For the purpose of the uninitialized variable analysis we added loops to force each array to be treated as initialized, when appropriate. For the transformed version of array_copy, IKOS proved that A is definitely initialized after the execution of the loop. For the rest of the programs IKOS proved that the initialized array A is still initialized after the loops. Column uva in Table 1 shows the analysis time in seconds of the uninitialized variable analysis implemented in IKOS.

Note that the polka analysis does not eliminate out-of-scope variables. Our program transformation introduces many variables, and since polka incurs a super-linear per-variable cost, the overall time penalty is considerable. We expect to be able to greatly reduce the cost by utilising a projection operation and improving the fixed-point finding algorithm.

8 Related Work

Amongst work on automated reasoning about array-manipulating code, we can distinguish work on *analysis* from work that focuses on *verification*. Our paper is concerned with the analysis problem, that is, how to use static analysis for automated generation of (inductive) code invariants. As mentioned in Sect. 1, we follow the tradition of applying abstract interpretation [4] to the array content analysis problem [5,9,11,16]. Alternative array analysis methods include Gulwani, McCloskey and Tiwari's lifting technique [10] (requiring the user to specify *templates* that describe when quantifiers should be introduced), Kovács and Voronkov's theorem-prover based method [13], Dillig, Dillig and Aiken's fluid updates [7] (supporting points-to and value analysis but excluding relational analyses), and incomplete approaches based on dynamic analysis [8,15].

Unlike previous work, we apply abstract interpretation to a transformed program in which array reads and writes have been translated away; any standard analysis, relational or not, can be applied to the resulting program, with negligible additional implementation cost.

There is a sizeable body of work that considers the *verification* problem for array-processing programs. Here the aim is to establish that given assertions hold at given program points. While abstract interpretation may serve this purpose (given a well-chosen abstract domain), more direct approaches are goal-directed, using assertions actively, to drive reasoning, rather than passively, as checkpoints. Many alternative techniques have been suggested for the verification of (sometimes restricted) array programs, including lazy abstraction [1], template-based methods [14], and, more closely related to the present paper, techniques that employ translation, for example to Horn clauses [6].

9   Conclusion

We have described a new abstract machine that supports set-valued variables and shown how array manipulating programs can be translated to array-free code for this machine. By compiling array programs for this machine, we are able to discover non-trivial universally quantified loop invariants, simply by analysing the transformed program using off-the-shelf scalar analysers. As an example of how this allows an existing analysis to be lifted to array programs in a straightforward manner, we have extended an uninitialised-variable analysis; Fig. 13 showed the usefulness of this approach. The indisputable price for the ease of implementation is a potentially excessive size of the transformed program. However, much array-processing code tends to make simple array traversals and access, and the transformational approach is viable for more than just small programs. Future work includes performing the transformation lazily, to avoid generating unneeded blocks. This should significantly speed up the analysis.

Acknowledgements. This work was supported through ARC grant DP140102194.

A   Array Operations with Non-segment Variables

Figure 6(a) assumes that the index variable of every read or write is included in the set of segment bounds. Figure 14 gives a revised version of transform_body which handles writes to indices that are not included in the set of segment bounds. When we transform a write to an index in the set of segment bounds (determined by the predicate is_idx), the transformation is as usual. Otherwise, we emit code to walk through the current set of segments, and apply the write operation to the appropriate one. The dispatch process is similar to the operation of split_transform, except that all leaves jump back to the continuation of the basic block after the write, rather than continuing under the modified ordering.

```
transform_body((([A[i] = x|stmts], br), π, id, ss)
    if is_idx(i)
        A_i := find_avar(π, A, i)
        transform_body((stmts, br), π, id, ss :: [A_i = x])
    else
        id' := next_block()
        transform_body((stmts, br), π, id', [])
        dispatch_write(A[i] = x, ε, π, id, ss, id')

dispatch_write(A[i] = x, sv, [], id, ss, id')
    push_block(id, ss, Jmp id')

dispatch_write(A[i] = x, s_<, [p, ...|π], id, ss, id')
    id_≥ := next_block()
    id_= := next_block()
    s_= := next_svar(s_<)
    s_> := next_svar(s_=)
    if s_< = ε
        push_block(id, If i < p then id' else id_≥, ss)
    else
        id_< := next_block()
        push_block(id, If i < p then id_< else id_≥, ss)
        push_block(id_<, Jmp id',
            [(v_1, ..., v_A, ..., v_k) ∈ s_<,
            s_< = s_< ∪ {(v_1, ..., x, ..., v_k)}])
        push_block(id_>, If i = p then id_= else id_>, id_=, id_>)
        (..., v_A, ...) := s_=
        push_block(id_=, Jmp id', [v_A = x])
        dispatch_write(A[i] = x, s_>, π, id_>, [], id')
```

Fig. 14. Revised pseudo-code for transforming array writes, allowing for omitted indices. Array reads are transformed similarly.

References

1. Alberti, F., Bruttomesso, R., Ghilardi, S., Ranise, S., Sharygina, N.: Lazy abstraction with interpolants for arrays. In: Bjørner, N., Voronkov, A. (eds.) LPAR-18 2012. LNCS, vol. 7180, pp. 46–61. Springer, Heidelberg (2012)
2. Blanchet, B., Cousot, P., Cousot, R., Feret, J., Mauborgne, L., Miné, A., Monniaux, D., Rival, X.: Design and implementation of a special-purpose static program analyzer for safety-critical real-time embedded software. In: Mogensen, T.Æ., Schmidt, D.A., Sudborough, I. (eds.) The Essence of Computation. LNCS, vol. 2566, pp. 85–108. Springer, Heidelberg (2002)
3. Brat, G., Navas, J.A., Shi, N., Venet, A.: IKOS: a framework for static analysis based on abstract interpretation. In: Giannakopoulou, D., Salaün, G. (eds.) SEFM 2014. LNCS, vol. 8702, pp. 271–277. Springer, Heidelberg (2014)

4. Cousot, P., Cousot, R.: Abstract interpretation: a unified lattice model for static analysis of programs by construction or approximation of fixpoints. In: POPL 1977, pp. 238–252. ACM Press (1977)
5. Cousot, P., Cousot, R., Logozzo, F.: A parametric segmentation functor for fully automatic and scalable array content analysis. In: POPL 2011, pp. 105–118. ACM Press (2011)
6. De Angelis, E., Fioravanti, F., Pettorossi, A., Proietti, M.: Verifying array programs by transforming verification conditions. In: McMillan, K.L., Rival, X. (eds.) VMCAI 2014. LNCS, vol. 8318, pp. 182–202. Springer, Heidelberg (2014)
7. Dillig, I., Dillig, T., Aiken, A.: Fluid updates: beyond strong vs. weak updates. In: Gordon, A.D. (ed.) ESOP 2010. LNCS, vol. 6012, pp. 246–266. Springer, Heidelberg (2010)
8. Ernst, M.D., Perkins, J.H., Guo, P.J., McCamant, S., Pacheco, C., Tschantz, M.S., Xiao, C.: The Daikon system for dynamic detection of likely invariants. Sci. Comput. Program. **69**(1–3), 35–45 (2007)
9. Gopan, D., Reps, T., Sagiv, M.: A framework for numeric analysis of array operations. In: POPL 2005, pp. 338–350. ACM Press (2005)
10. Gulwani, S., McCloskey, B., Tiwari, A.: Lifting abstract interpreters to quantified logical domains. In: POPL 2008, pp. 235–246. ACM Press (2008)
11. Halbwachs, N., Péron, M.: Discovering properties about arrays in simple programs. In: PLDI 2008, pp. 339–348. ACM Press (2008)
12. Jeannet, B., Miné, A.: APRON: a library of numerical abstract domains for static analysis. In: Bouajjani, A., Maler, O. (eds.) CAV 2009. LNCS, vol. 5643, pp. 661–667. Springer, Heidelberg (2009)
13. Kovács, L., Voronkov, A.: Finding loop invariants for programs over arrays using a theorem prover. In: Chechik, M., Wirsing, M. (eds.) FASE 2009. LNCS, vol. 5503, pp. 470–485. Springer, Heidelberg (2009)
14. Larraz, D., Rodríguez-Carbonell, E., Rubio, A.: SMT-based array invariant generation. In: Giacobazzi, R., Berdine, J., Mastroeni, I. (eds.) VMCAI 2013. LNCS, vol. 7737, pp. 169–188. Springer, Heidelberg (2013)
15. Nguyen, T.V., Kapur, D., Weimer, W., Forrest, S.: Using dynamic analysis to discover polynomial and array invariants. In: Proceedings of the 34th International Conference on Software Engineering, pp. 760–770. IEEE (2012)
16. Péron, M.: Contributions à l'analyse statique de programmes manipulant des tableaux. Ph.D. thesis, Université de Grenoble (2010)
17. Regehr, J.: Uninitialized variables. Web blog, http://blog.regehr.org/archives/519, Accessed 18 June 2014

Analysing and Compiling Coroutines with Abstract Conjunctive Partial Deduction

Danny De Schreye, Vincent Nys[(✉)], and Colin Nicholson

Department of Computer Science, KU Leuven,
Celestijnenlaan 200A, 3001 Heverlee, Belgium
{danny.deschreye,vincent.nys}@cs.kuleuven.be

Abstract. We provide an approach to formally analyze the computational behavior of coroutines in Logic Programs and to compile these computations into new programs, not requiring any support for coroutines. The problem was already studied near to 30 years ago, in an analysis and transformation technique called Compiling Control. However, this technique had a strong ad hoc flavor: the completeness of the analysis was not well understood and its symbolic evaluation was also very ad hoc. We show how Abstract Conjunctive Partial Deduction, introduced by Leuschel in 2004, provides an appropriate setting to redefine Compiling Control. Leuschel's framework is more general than the original formulation, it is provably correct, and it can easily be applied for simple examples. We also show that the Abstract Conjunctive Partial Deduction framework needs some further extension to be able to deal with more complex examples.

1 Introduction

The work reported on in this paper is an initial step in a new project, in which we aim to formally analyze and automatically compile certain types of coroutining computations. Coroutines are a powerful means of supporting complex computation flows. They can be very useful for improving the efficiency of declaratively written programs, in particular for generate-and-test based programs. On the other hand, obtaining a deep understanding of the computation flows underlying the coroutines is notoriously difficult.

In this paper we restrict our attention to pure, definite Logic Programs. In this context, the problem was already studied nearly 30 years ago. Bruynooghe et al. (1986) and Bruynooghe et al. (1989) present an analysis and transformation technique for coroutines, called Compiling Control (CC for short). The purpose of the CC transformation is the following: transform a given program, P, into a program P', so that computation with P' under the standard selection rule mimics the computation with P under a non-standard selection rule. In particular, given a coroutining selection rule for a given Logic Program, the transformed program will execute the coroutining if it is evaluated under the standard selection rule of Prolog.

© Springer International Publishing Switzerland 2015
M. Proietti and H. Seki (Eds.): LOPSTR 2014, LNCS 8981, pp. 21–38, 2015.
DOI: 10.1007/978-3-319-17822-6_2

To achieve this, CC consists of two phases: an analysis phase and a synthesis phase. The analysis phase analyzes the computations of a program for a given query pattern and under a (non-standard) selection rule. The query pattern is expressed in terms of a combination of type, mode and aliasing information. The selection rule is instantiation-based, meaning that different choices in atom selection need to be based on different instantiations in these atoms. The analysis results in what is called a "trace tree", which is a finite upper part of a symbolic execution tree that one can construct for the given query pattern, selection rule and program. In the synthesis phase, a finite number of clauses are generated, so that each clause synthesizes the computation in some branch of the trace tree and such that all computations in the trace tree have been synthesized by some clause. The technique was implemented, formalized and proven correct, under certain fairly technical conditions.

Unfortunately, the CC transformation has a rather ad hoc flavor. It was very hard to show that the analysis phase of the transformation was complete, in the sense that a sufficiently large part of the computation had been analyzed to be able to capture all concrete computations that could possibly occur at run time. Even the very idea of a "symbolic execution" had an ad hoc flavor. It seemed that it should be possible to see this as an instance of a more general framework for analysis of computations.

Fortunately, since the development of CC a number of important advances have been achieved in analysis and transformation:

- General frameworks for abstract interpretation (e.g. Bruynooghe 1991) were developed. It is clear that abstract interpretation has the potential to provide a better setting for developing the CC analysis.
- Partial deduction of Logic Programs was developed (e.g. Gallagher 1986). Partial deduction seems very similar to CC, but the exact relationship was never identified. When John Lloyd and John Shepherdson formalized the issues of correctness and completeness of partial deduction in Lloyd and Shepherdson (1991), this provided a new framework for thinking about a complete analysis of a computational behavior and it was clear that some variant of this could improve the CC analysis.
- Conjunctive partial deduction (see De Schreye et al. 1999) seems even closer to CC. In an analysis for a CC transformation, one really does not want to split up the conjunctions of atoms into separate ones and then analyze the computations for these atoms separately. It is crucial that one can analyze the computation for certain atoms in conjunction (which is how conjunctive partial deduction generalizes partial deduction), so that their behavior under the non-standard selection rule may be observed.
- Finally, abstract (conjunctive) partial deduction (Leuschel 2004) brings all these features together. It provides an extension of (conjunctive) partial deduction in which the analysis is based on abstract interpretation, rather than on concrete evaluation.

In this paper we will demonstrate – mostly on the basis of examples – that abstract conjunctive partial deduction (ACPD for short) is indeed a suitable framework to redefine CC in such a way that the flaws of the original approach are overcome. We show that for simple problems in the CC context, ACPD can, in principle, produce the transformation automatically. We also show that for more complex CC transformations, ACPD is still not powerful enough. We suggest an extension to ACPD that allows us to solve the problem and illustrate with an example that this extension is very promising.

After the preliminaries, in Sect. 3, we introduce a fairly refined abstract domain, including type, mode and aliasing information, and we show, by means of an example, how ACPD allows us to analyze a coroutine and compile the transformed program. In Sect. 4 we propose a more complex example and show why it is out of scope for ACPD. We introduce an additional abstraction in our domain and illustrate that this abstraction solves the problem. This abstraction, however, does not respect the requirements of the formalization of ACPD in Leuschel (2004). We end with a discussion.

2   Preliminaries

We assume that the reader is familiar with the basics of Logic Programming (Lloyd 1987). We also assume knowledge of the basics of abstract interpretation (Bruynooghe 1991) and of partial deduction (Lloyd and Shepherdson 1991).

In this paper, names of variables will start with a capital. Names of constants will start with a lower case character. Given a logic program P, Con_p, Var_p, Fun_p and $Pred_p$ respectively denote the sets of all constants, variables, functors and predicate symbols in the language underlying P. $Term_p$ will denote the set of all terms constructable from Con_p, Var_p and Fun_p. $Atom_p$ denotes the set of all atoms which can be constructed from $Pred_p$ and $Term_p$. We will often need to refer to conjunctions of atoms of $Atom_p$ and we denote the set of all such conjunctions as $ConAtom_p$.

We will introduce an abstract domain in the following section. The abstract domain will be based on a set of abstract constant symbols, $ACon_p$. Based on these, there is a corresponding set of abstract terms, $ATerm_p$, which consists of the terms that can be constructed from $ACon_p$ and Fun_p. $AAtom_p$ will denote the set of abstract atoms, being the atoms which can be constructed from $ATerm_p$ and $Pred_p$. Finally, $AConAtom_p$ denotes the set of conjunctions of elements of $AAtom_p$.

3   An Example of a CC Transformation, Using ACPD

In this section, we provide the intuitions behind our approach by means of a simple example. We use permutation sort as an illustration. The intention is to transform this program so that calls to $perm/2$ and $ord/1$ are interleaved.

Example 1 (Permutation sort).

$$sort(X,Y) \longleftarrow perm(X,Y), ord(Y).$$
$$perm([],[]).$$
$$perm([X|Y],[U|V]) \longleftarrow$$
$$\quad del(U,[X|Y],W), perm(W,V).$$

$$del(X,[X|Y],Y).$$
$$del(X,[Y|U],[Y|V]) \longleftarrow del(X,U,V).$$
$$ord([]).$$
$$ord([X]).$$
$$ord([X,Y|Z]) \longleftarrow$$
$$\quad X \leq Y, ord([Y|Z]).$$

We now introduce the abstract domain. This domain consists of two types of new constant symbols: g and $a_i, i \in \mathbb{N}$. The symbol g denotes any ground term in the concrete language. The basic intuition for the symbols a_i is that they are intended to represent variables of the concrete domain. However, as we want the abstract domain to be closed under substitution (if an abstract term denotes some concrete term, then it should also denote all of its instances), an abstract term a_i will actually represent any term of the concrete language.

The subscript i in a term a_i is used to represent aliasing. If an abstract term, abstract atom or abstract conjunction of atoms contains a_i several times (with the same subscript), the denoted concrete terms, atoms or conjunctions of atoms contain the *same* term in all positions corresponding to those occupied by a_i. For instance, the abstract conjunction $perm(g, a_1), ord(a_1)$ denotes the concrete conjunctions $\{perm(t_1, t_2), ord(t_2) | t_1, t_2 \in Term_p \text{ and } t_1 \text{ is ground}\}$.

In addition to g and a_i, we will include all concrete constants in the abstract domain, so $Con_p \subseteq ACon_p$. This is not essential for the approach: we could develop a sound and effective ACPD for the CC transformation based on the abstract constants g and $a_i, i \in \mathbb{N}$, alone. However, including Con_p in $ACon_p$ makes the analysis more precise: some redundant paths in the analysis are avoided.

Definition 1 (Abstract domain). *The abstract domain consists of:*

– $ACon_p = Con_p \cup \{g\} \cup \{a_i | i \in \mathbb{N}\}$.
– $ATerm_p$, $AAtom_p$ and $AConAtom_p$ *are defined as the sets of the terms, atoms and conjunctions of atoms constructable from* $ACon_p$, Fun_p *and* $Pred_p$.

Next, we define the semantics of the abstract domain, through a concretization function γ. With slight abuse of notation, we use the same symbol γ to denote the concretization functions on $ATerm_p$, $AAtom_p$ and $AConAtom_p$.

In order to formalize the semantics of the aliasing, we need two auxiliary concepts: the subterm selection sequence and the aliasing context.

Definition 2 (Subterm selection sequence). *Let t be a term, atom or conjunction of atoms (either concrete or abstract).*

– $i \in \mathbb{N}_0$ *is a subterm selection sequence for t, if $t = f(t_1, ..., t_n)$ and $i \leq n$. The subterm of t selected by i is t_i.*
– $i_1.i_2.....i_n$ *is a subterm selection sequence for t, if $t = f(t_1, ..., t_n)$, $i_1 \leq n$, $i_1 \in \mathbb{N}_0$ and $i_2.....i_n$ is a subterm selection sequence for t_{i_1}. With an inductively defined notation, we denote by $t_{i_1.i_2.....i_k}$ the subterm of $t_{i_1....i_{k-1}}$ selected by i_k, with $1 < k \leq n$. We also refer to $t_{i_1.i_2.....i_n}$ as the subterm of t selected by $i_1.i_2.....i_n$.*

Note that, in this definition, we assume that a conjunction of atoms A_1, A_2,...,A_n is denoted as $\wedge(A_1, A_2, ..., A_n)$.

Example 2 (Subterm selection sequence). Let $t = f(g(h(X), 5), f(h(a), Y))$*, then* $t_{1.1.1} = X$*,* $t_{2.1.1} = a$*.*

Definition 3 (Aliasing context). *Let* t *be an abstract term, atom or conjunction of atoms. The aliasing context of* t*, denoted* $AC(t)$*, is the finite set of pairs* (sss_1, sss_2) *of subterm selection sequences of* t*, such that* $t_{sss_1} = t_{sss_2} = a_i$ *for some* $i \in \mathbb{N}$*.*

Example 3 (Aliasing context). Let $t = p(f(a_2, g), a_1, a_2, g(h(a_1)))$*, then* $AC(t) = \{(1.1, 3), (2, 4.1.1)\}$*.*

Definition 4 (Concretization function). *The concretization function* γ : $ATerm_p \cup AAtom_p \cup ACon Atom_p \rightarrow 2^{Term_p} \cup 2^{Atom_p} \cup 2^{Con Atom_p}$ *is defined as:*

- $\gamma(c) = \{c\}$*, for any* $c \in Con_p$
- $\gamma(g) = \{t \in Term_p | t \text{ is ground}\}$
- $\gamma(a_i) = Term_p$*,* $i \in \mathbb{N}$
- $\gamma(f(at_1, ..., at_n)) = \{f(t_1, ..., t_n) | t_i \in \gamma(at_i), i = 1...n$*, and let* t *denote* $f(t_1, ..., t_n)$*, then for all* $(sss_1, sss_2) \in AC(f(at_1, ..., at_n)) : t_{sss_1} = t_{sss_2}\}$

Example 4 (Concretization function). $\gamma(p(f(a_2, g), a_1, a_2, g(h(a_1)))) = \{p(f(t_1, t_2), t_3, t_1, g(h(t_3))) | t_1, t_3 \in Term_p, t_2 \text{ ground term of } Term_p\}$

The abstract domain introduced above is infinitely large. There are two causes for this. Terms can be nested unboundedly deep, therefore infinitely many different terms exist. In addition, there are infinitely many $a_i, i \in \mathbb{N}$, symbols.

If so desired, the abstract domain can be refined, so that it becomes finite. This is done by using depth-k abstraction and by defining an equivalence relation on $\{a_i | i \in \mathbb{N}\}$. For the purpose of this paper, the infinite size of the abstract domain is not a problem.

Let us return to the permutation sort example. ACPD requires a top-level abstract atom (or conjunction) to start the transformation. Let $sort(g, a_1)$ be this atom. In the context of the $\mathcal{A}$-coveredness condition of partial deduction, our initial set $\mathcal{A}$ is $\{sort(g, a_1)\}$.

Below, we construct a finite number of finite, abstract partial deduction derivation trees for abstract (conjunctions of) atoms. The construction of these trees assumes an "abstract unification" and an "abstract unfold" operation. Their formal definitions can be found in Annex (2014). For now, we only show their effects in abstract partial derivation trees.

Next, we need an "oracle" that decides on the selection rule applied in the abstract derivation trees. This oracle mainly has two functions:

- to decide whether an obtained goal should be unfolded further, or whether it should be kept residual (to be split and added to $\mathcal{A}$),
- to decide which atom of the current goal should be selected for unfolding.

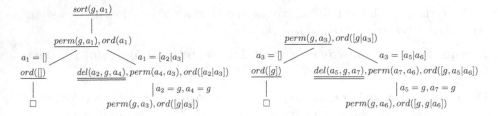

Fig. 1. Abstract tree for $sort(g, a_1)$ **Fig. 2.** Abstract tree for $perm(g, a_3)$, $ord([g|a_3])$

In fact, we will use a third type of decision that the oracle may make: it may decide to "fully evaluate" a selected atom. This type of decision is not commonly supported in partial deduction. What it means is that we decide not to transform a certain predicate of the original program, but merely keep its original definition in the transformed program. In partial deduction, this can be done by never selecting these atoms, including them in $\mathcal{A}$ and including their original definition in the transformed program.

In our setting, however, we want to know the effect that solving the atom has on the remainder of the goal. Therefore, we will assume that an abstract interpretation over our abstract domain computes the abstract bindings that solving the atom results in. These are applied to the remainder of the goal. Note that this cannot easily be done in standard partial deduction, as fully evaluating an atom during (concrete) partial deduction may not terminate. In Vidal (2011), a similar functionality is integrated in a hybrid approach to conjunctive partial deduction.

For now, we simply assume the existence of the oracle. Figures 1, 2 and 3 show the abstract partial derivation trees that ACPD may build for permutation sort and top level $\mathcal{A} = \{sort(g, a_1)\}$.

In these figures, in each goal, the atom selected for abstract unfolding is underlined. If an atom is underlined twice, this expresses that the atom was selected for full abstract interpretation.

Both unfolding and full abstract evaluation may create bindings. Our abstract unification only collects bindings made on the a_i terms. Bindings created on g terms are not relevant.

In the left branch of the tree in Fig. 1 we see the effect of including the concrete constants in the abstract domain. As a result, the binding for a_1 is $[]$, instead of g. If we had not included Con_p in $ACon_p$, then $ord(g)$ would have required a full analysis, using the three clauses for $ord/1$.

A goal with no underlined atom indicates that the oracle selects no atom and decides to keep the conjunction residual. After the construction of the tree in Fig. 1, ACPD adds the abstract conjunction $perm(g, a_3), ord([g|a_3])$ to $\mathcal{A}$. ACPD starts a new tree for this atom. This tree is shown in Fig. 2.

The tree is quite similar to the one in Fig. 1. The main difference is that, in the residual leaf, the ord atom now has a list argument with two g elements.

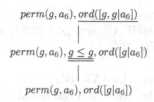

$$perm(g, a_6), ord([g, g|a_6])$$
$$|$$
$$perm(g, a_6), \underline{g \leq g}, ord([g|a_6])$$
$$|$$
$$perm(g, a_6), ord([g|a_6])$$

Fig. 3. Abstract tree for $perm(g, a_6), ord([g, g|a_6])$

This pattern does not yet exist in the current $\mathcal{A}$ and is therefore added to $\mathcal{A}$. A third abstract tree is computed for $perm(g, a_6), ord([g, g|a_6])$, shown in Fig. 3.

In Fig. 3, the residual leaf $perm(g, a_6), ord([g|a_6])$ is a renaming of the conjunction $perm(g, a_3), ord([g|a_3])$, which is already contained in $\mathcal{A}$. Therefore, ACPD terminates the analysis, concluding $\mathcal{A}$-coveredness for $\mathcal{A} = \{sort(g, a_1), \wedge(perm(g, a_3), ord([g|a_3])), \wedge(perm(g, a_6), ord([g, g|a_6]))\}$.

In standard (concrete) conjunctive partial deduction, the analysis phase would now be completed. In ACPD, however, we need an additional step. In the abstract partial derivation trees, we have not collected the concrete bindings that unfolding would produce. These are required to generate the resolvents. Therefore, we need an additional step, constructing essentially the same three trees again, but now using concrete terms and concrete unification.

We only show one of these concrete derivation trees in Fig. 4. It corresponds to the tree in Fig. 2. We define the root of a concrete derivation tree corresponding to an abstract tree as follows.

Definition 5 (Concrete conjunctions in the root). *Let acon $\in \mathcal{A}$, then the conjunction in the root of the corresponding concrete tree, denoted as $c(acon)$, is obtained by replacing any g or a_i symbol in acon by a fresh free variable, ensuring that multiple occurrences of a_i, with the same subscript i, are replaced by identical variables.*

When unfolding the concrete tree, every abstract unfolding of the abstract tree is mimicked, using the same clauses, over the concrete domain.

The step of full abstract interpretation of the $del(a_5, g, a_7)$ atom in Fig. 2 has no counterpart in Fig. 4. The atom $del(U, [X_1|X_2], W)$ is kept residual and the $del/3$ clauses are added to the transformed program.

$$perm(X, Y), ord([Z|Y])$$

$X = [], Y = []$ ⟋　⟍ $X = [X_1|X_2], Y = [U|V]$

$ord([Z])$ 　　 $del(U, [X_1|X_2], W), perm(W, V), ord([Z, U|V])$
$|$
$\square$

Fig. 4. Concrete tree corresponding to Fig. 2

More specifically, using a renaming $p_1(X, Y, Z)$ for $\wedge(perm(X, Y), ord([Z|Y]))$ and $p_2(W, V, Z, U)$ for $\wedge(perm(W, V), ord([Z, U|V]))$, we synthesize the following resolvents from the tree in Fig. 4:

$p_1([], [], Z) \longleftarrow .$

$p_1([X_1|X_2], [U|V], Z) \longleftarrow del(U, [X_1|X_2], W), p_2(W, V, Z, U).$

From the counterparts of the trees in Figs. 1 and 3, we obtain the following additional resultants:

$sort([], []).$

$sort([X_1|X_2], [Y_1|Y_2]) \longleftarrow del(Y_1, [X_1|X_2], Z), p_1(Z, Y_2, Y_1).$

$p_2(U, V, W, X)W \leq X, p_1(U, V, X).$

This transformation inherits correctness results from ACPD. In particular, $\mathcal{A}$-closedness and independence guarantee the completeness and correctness of the analysis. In addition, the transformation preserves all computed answers (in both directions) and finite failure of the transformed program implies finite failure of the original.

4   A More Complex Example, Introducing the *multi* Abstraction

In Sect. 3 we have shown that ACPD is indeed sufficient to formally revisit CC for a simple example. However, for more complex examples, ACPD still lacks expressivity. Consider the following prime number generator.

Example 5 (Prime numbers)

```
primes(N,P) ← integers(2,I),sift(I,P),len(P,N).
integers(N,[]).
integers(N,[N|I]) ← M is N+1, integers(M,I).
sift([N|Is],[N|Ps]) ← filter(N,Is,F), sift(F,Ps).
sift([],[]).
divides(N,M) ← X is M mod N, X is 0.
not_divide(N,M) ← X is M mod N, X > 0.
filter(N,[M|I],F) ← divides(N,M), filter(N,I,F).
filter(N,[M|I],[M|F]) ← not_divide(N,M), filter(N,I,F).
filter(N,[],[]).
len([],0).
len([H|T],N) ← M is N - 1, len(T,M).
```

The program is intended to be called with a goal $primes(N, P)$, with N a positive integer and P a free variable. The $integers/2$ predicate generates growing lists of integer numbers. $filter/3$ represents the removal of all multiples of a single integer N from a list. $sift/2$ recursively filters out multiples of an initial list element which is prime.

The complete ACPD style analysis is available in Annex (2014). We only present some relevant parts.

The top level goal for the abstract analysis is $primes(g, a_1)$, so that the initial set $\mathcal{A}$ is $\{primes(g, a_1)\}$. A first abstract derivation tree describes the initialization for the computation. It contains a branch leading to an empty goal (success branch) and a branch with the leaf: $\wedge(integers(g, a_3), filter(g, a_3, a_5),$ $sift(a_5, a_4), len(a_4, g))$, which is added to $\mathcal{A}$.

Next, we construct an abstract derivation tree for the latter conjunction. This gives a successful branch with an empty conjunction in the leaf, a branch ending in a renamed version of the above conjunction, and a third branch, with the following leaf, which is added to $\mathcal{A}$: $\wedge(integers(g, a_4), filter(g, a_4, a_5), filter$ $(g, a_5, a_7), sift(a_7, a_6), len(a_6, g))$.

At this point it becomes clear that an analysis following only the steps shown in Sect. 3 will not terminate. The two abstract conjunctions, most recently added to $\mathcal{A}$, are identical – up to renaming of a_i's – except that the latter conjunction contains two atoms $filter(g, a_i, a_j)$, instead of just one. A further analysis, building additional derivation trees, will result in the construction of continuously growing conjunctions, with continuously increasing numbers of $filter/3$ atoms.

We could solve this by cutting the goal into two smaller conjunctions and adding these to $\mathcal{A}$. However, all these atoms are generators or testers in the coroutine and depend on eachother. By splitting the conjunction, we would no longer be able to analyze the coroutine.

One of the restrictions imposed by ACPD is that for any abstract conjunction of atoms, $acon \in AConAtom_p$, there exists a concrete conjunction, $con \in ConAtom_p$, such that: for all $con_i \in \gamma(acon)$: con_i is an instance of con. In practice, this means that an abstract conjunction is not allowed to represent a set of concrete conjunctions whose elements have a distinct number of conjuncts. However, in order to solve the problem observed in our example, we need the ability to represent a set of conjunctions, with a growing number of atoms, by an abstract atom. Therefore, we need to extend ACPD.

We extend our abstract domain and introduce a new abstraction, $multi/4$, which makes it possible to represent growing conjunctions, with a number of copies of a single abstract atom.

To define this abstraction is rather difficult. This is because we do not only want to be able to represent a conjunction of multiple, identically instantiated atoms, but also their aliasing with the context in which they occur, as well as the aliasing between consecutive atoms in the conjunction.

We first introduce a parameterized naming scheme for a_i constants and apply this to abstract atoms.

Definition 6 (Parameterized naming and parameterized abstract atom). *Let $A \in AAtom_p$. By $Id(A)$, we denote a unique identifier associated with A.*

Let $a_j \in ACon_p, j \in \mathbb{N}$, such that a_j occurs in A, then the parameterized naming of a_j is the symbol $a_{Id(A),i,j}$.

Let $A \in AAtom_p$. The parameterized atom for A, $p(A)$, is obtained by replacing every a_j occurring in A by its parameterized naming, $a_{Id(A),i,j}$.

The new abstraction $multi/4$ will depend on the context (the abstract conjunction) in which it occurs. This context may contain abstract constants, a_j. It may also contain parameterized namings of abstract constants, $a_{Id(A),i,j}$. This is due to the fact that a $multi/4$ abstraction will typically contain parameterized namings and that an abstract conjunction will be allowed to contain multiple $multi/4$ abstractions. Therefore, the context of one $multi/4$ abstraction may contain another $multi/4$ abstraction.

Definition 7 (Context). *A context is an abstract conjunction and is denoted as C. Given a context C, we denote $a(C) = \{a_j \in ACon_p | a_j \text{ occurs in } C\}$, we denote $pa(C) = \{a_{Id(A),i,j} | a_{Id(A),i,j} \text{ occurs in } C\}$.*

Definition 8 (multi abstraction). *A multi abstraction is a construct of the form $multi(p(A), First, Consecutive, Last)$, where:*

- *$p(A)$ is the parameterized atom for some $A \in AAtom_p$.*
- *$First$ is a conjunction of equalities $a_{Id(A),1,j} = b_j$, where $b_j \in a(C) \cup pa(C)$ and all left-hand sides of the equalities are distinct.*
- *$Consecutive$ is a conjunction of equalities $a_{Id(A),i+1,j} = a_{Id(A),i,j'}$, where $j, j' \in \mathbb{N}$ and all left-hand sides of the equalities are distinct.*
- *$Last$ is a conjunction of equalities $a_{Id(A),k,j} = b_j$, where $b_j \in a(C) \cup pa(C)$ and all left-hand sides of the equalities are distinct.*

Example 6 (multi/4 abstraction). We return to the primes example, with the two abstract conjunctions already added to $\mathcal{A}$. We can rename the indices of the a_j constants in one of these conjunctions in order to make the contexts in which the $filter(g, a_i, a_j)$ atoms occur identical for both conjunctions, e.g.:
$\wedge(integers(g, a_3), filter(g, a_3, a_5), sift(a_5, a_4), len(a_4, g))$ and $\wedge(integers(g, a_3), filter(g, a_3, a_6), filter(g, a_6, a_5), sift(a_5, a_4), len(a_4, g))$.

Now we can generalize these two abstract conjunctions using the $multi/4$ abstraction: Let $A = filter(g, a_3, a_6)$. Then, the abstract conjunction is:
$\wedge(integers(g, a_3), multi(filter(g, a_{Id(A),i,3}, a_{Id(A),i,6}), \wedge(a_{Id(A),1,3} = a_3),$
$\wedge(a_{Id(A),i+1,3} = a_{Id(A),i,6}), \wedge(a_{Id(A),k,6} = a_5)), sift(a_5, a_4), len(a_4, g))$
Here, expressions such as $\wedge(a_{Id(A),1,3} = a_3)$ represent conjunctions with only one conjunct.

Conversely, abstract conjunctions containing $multi/4$ abstractions, such as the one above, represent infinitely many abstract conjunctions without the $multi/4$ abstraction. In the example, these contain either one or multiple $filter(g, a_i, a_j)$ atoms.

In what follows, we will omit the $Id(A)$ subscript in the parameterized namings $a_{Id(A),i,j}$ and just refer to $a_{i,j}$ instead. The $Id(A)$ subscript is only relevant for abstract conjunctions containing multiple $multi/4$ abstractions, a case which we will not consider for the moment.

In order to describe the abstract conjunctions represented by an abstract conjunction containing a $multi/4$ abstraction, we need the ability to map parameterized namings back to ordinary a_j constants. This requires the following concepts.

Definition 9 (concrete index assignment mapping). *Let $n \in \mathbb{N}$. The concrete index assignment mapping, $R(i, n)$, is a mapping defined on any syntactic construct, S, containing parameterized namings $a_{i,j}$. $R(i, n)$ replaces every occurrence of a parameterized naming $a_{i,j}$ in S by the parameterized naming $a_{n,j}$.*

Example 7 (concrete index assignment mapping). $R(i, 1)(filter(g, a_{i,3}, a_{i,6})) = filter(g, a_{1,3}, a_{1,6})$. $R(i, k)(filter(g, a_{i,3}, a_{i,6})) = filter(g, a_{k,3}, a_{k,6})$.

Definition 10 (double-index mapping). *The double-index mapping, ψ, is a mapping defined on any syntactic construct, S, containing parameterized namings $a_{i,j}$. ψ replaces every occurrence of a parameterized naming $a_{i,j}$ in S by a_{i_j}, where i_j denotes a fresh element of $\mathbb{N}$, not occurring in any a_i yet.*

Example 8 (double-index mapping). $\psi(filter(g, a_{i,3}, a_{i,6})) = filter(g, a_{i_3}, a_{i_6})$, with i_3, i_6 fresh elements of $\mathbb{N}$.

Definition 11 (substitution corresponding to equality constraints). *Let Constraint be a conjunction of equality constraints, $a_{i,j} = b_j$, with $a_{i,j}$ parameterized namings, and such that all left-hand sides of equalities are mutually distinct. The substitution corresponding to Constraint is the substitution $\Theta_{Constraint} = \{\psi(a_{i,j})/\psi(b_j) | a_{i,j} = b_j \in Constraint\}$.*

Note that this definition is meant to deal with the conjunctions of equalities in the *First*, *Consecutive* and *Last* arguments of the *multi/4* abstraction.

Example 9 (substitutions corresponding to equality constraints). For the conjunctions of equality constraints in Example 6, the corresponding substitutions are: $\Theta_{First} = \{a_{1_3}/a_3\}$, $\Theta_{Consecutive} = \{a_{(i+1)_3}/a_{i_6}\}$, $\Theta_{Last} = \{a_{k_6}/a_5\}$.

With these notions, we can now describe the abstract conjunctions represented by a *multi/4* abstraction.

Definition 12 (Abstract conjunctions represented by multi/4). *The abstract conjunctions represented by $multi(p(A), First, Consecutive, Last)$ are:*

- $\psi(R(i, 1)(p(A)))\Theta_{First} \circ \Theta_{R(k,1)(Last)}$, *and*
- $\psi(R(i, 1)(p(A)))\Theta_{First} \wedge \psi(R(i, 2)(p(A)))\Theta_{R(i,1)(Consecutive)} \wedge$
  $\ldots \wedge \psi(R(i, k)(p(A)))\Theta_{R(i,k-1)(Consecutive)} \circ \Theta_{Last}$, *with $k > 1$.*

Example 10 (Abstract conjunctions represented by multi/4). For the multi/4 abstraction in Example 6, $multi(filter(g, a_{i,3}, a_{i,6}), \wedge(a_{1,3} = a_3), \wedge(a_{i+1,3} = a_{i,6}), \wedge(a_{k,6} = a_5))$, the represented abstract conjunctions are:

- $filter(g, a_3, a_5)$, *and*
- $filter(g, a_3, a_{1_6}) \wedge filter(g, a_{1_6}, a_{2_3}) \wedge \ldots \wedge filter(g, a_{(k-1)_6}, a_5), k > 1$.

Next, we need to define the abstract unfolding of a *multi/4* abstraction. Unfolding a *multi/4* abstraction makes a case split. Either the *multi/4* abstraction represents only one abstract atom, or it represents more than one. In both cases the bindings with the context and, in the latter case, the bindings between consecutive atoms, need to be respected.

Definition 13 (Abstract unfold of $multi/4$**).** *Abstract unfold of multi produces a branching in the abstract derivation tree. An abstract atom* $multi(p(A),$ *$First, Consecutive, Last)$ is replaced in one branch by*
$\psi(R(i,1)(p(A)))\Theta_{First} \circ \Theta_{R(k,1)(Last)}$ *and in a second branch by*
$\psi(R(i,1)(p(A)))\Theta_{First} \wedge multi(p(A), NewFirst, Consecutive, Last)$, *where*
$NewFirst = \wedge\{a_{1,j} = a_{1_{j'}} | a_{(i+1),j} = a_{i,j'} \in Consecutive\}$.

Example 11 (Abstract unfold of $multi/4$*). Again returning to Example 6, abstract unfold of* $multi(filter(g, a_{i,3}, a_{i,6}), \wedge(a_{1,3} = a_3), \wedge(a_{i+1,3} = a_{i,6}), \wedge(a_{k,6} = a_5))$ *produces in one branch* $filter(g, a_3, a_5)$ *and in the other branch* $filter(g, a_3, a_{1_6}) \wedge$ *$multi(filter(g, a_{i,3}, a_{i_6}), \wedge(a_{1,3} = a_{1,6}), \wedge(a_{i+1,3} = a_{i,6}), \wedge(a_{k,6} = a_5))$.*

A few comments on this definition are in order. First, the definition of $NewFirst$ may seem strange, because both sides of the equalities have a "1" index. However, note that on the left-hand side of the equality, it is in a parameterized naming, $a_{1,j}$, referring to the first atom represented by the $multi/4$, while on the right-hand side, it is in an abstract atom $a_{1_{j'}}$, referring to an atom that was just moved outside of the $multi/4$. Second, it is important to remember that the abstract constants a_{1_j} are produced by a $\psi(a_{1,j})$ call and that their index 1_j needs to be a fresh index, not yet occurring in the expressions. This is particularly important in cases where we perform several abstract unfoldings of $multi/4$ in sequence. At each unfold, new fresh subscripts need to be introduced.

Finally, we need to define abstract generalization with $multi/4$, allowing us to replace conjunctions of identically instantiated and similarly aliased abstract atoms by a $multi$ construct.

Definition 14 (Abstract generalization with $multi/4$**).** *Let* $A \in AAtom_p$. *Let* $A_1, ..., A_k \in AAtom_p$ *and let* $\bigwedge_{l=1,k} A_l$ *occur in a context of abstract atoms* C. *Let* $a(C)$ *and* $pa(C)$ *respectively be the abstract constants and the parameterized namings occurring in* C. *Let* $r_l, l = 1, k$, *be renamings of* A, *such that* $r_l(A) = A_l$. *In particular, for any* a_i *occurring in* A, $r_l(a_i)$ *occurs at the same subterm selection sequence position in* A_l.

$Gen(\bigwedge_{l=1,k} A_l) = multi(p(A), First, Consecutive, Last)$ is the abstract generalization with $multi/4$ *of* $\bigwedge_{l=1,k} A_l$ *in* C *if:*

- *for any* $b_j \in a(C) \cup pa(C)$, $a_{1,j} = b_j \in First$ *if and only if* $r_1(a_j) = b_j$
- $a_{i+1,j} = a_{i,j'} \in Consecutive$ *if and only if* $r_{i+1}(a_j) = r_i(a_{j'})$
- *for any* $b_j \in a(C) \cup pa(C)$, $a_{k,j} = b_j \in Last$ *if and only if* $r_k(a_j) = b_j$

We can extend the above definition to allow generalizations $Gen(\bigwedge_{l=1,k} A_l \wedge multi(p(A), First, Consecutive, Last)) = multi(p(A), First', Consecutive, Last)$ and generalizations $Gen(multi(p(A), First, Consecutive, Last), \wedge \bigwedge_{l=1,k} A_l) = multi(p(A), First, Consecutive, Last')$. We omit the details for these generalizations. We illustrate abstract generalization with $multi/4$ in our running example below.

Let us return to the prime numbers example. Observing the growing number of $filter/3$ atoms in our last conjunction (w.r.t. the conjunction already present in $\mathcal{A}$), we perform the generalization: $Gen(\wedge(filter(g, a_4, a_5), filter(g, a_5, a_7)))$ $= multi(filter(g, a_{1,i,4}, a_{1,i,5}), \wedge(a_{1,1,4} = a_4), \wedge(a_{1,i+1,4} = a_{1,i,5}), \wedge(a_{1,k,5} = a_7))$. Here, we include the $Id(A)$ again, because we will have multiple $multi/4$ abstractions. We arbitrarily select $Id(A)$ to be 1.

Then we construct a new abstract derivation tree for this conjunction, including – among others – an abstract unfold of $multi/4$ and abstract generalizations with $multi/4$. In Fig. 5, we show this abstract tree.

After abstract unfolding of $integers(g, a_1)$, the tree contains an abstract unfolding of $multi(filter(g, a_{1,i,1}, a_{1,i,2}), \wedge(a_{1,1,1} = [g|a_4]), \wedge(a_{1,i+1,1} = a_{1,i,2})$, $\wedge(a_{1,k,2} = a_2))$. This unfolding can lead to one instance of $filter/3$ or several. If there is only one filter, a full evaluation of $divides(g, g)$ eventually leads to an empty goal. A full evaluation of $does\_not\_divide(g, g)$, on the other hand, leads to a new generalization which produces a renaming of the root of this tree.

Eventually, the analysis ends up with a final set $\mathcal{A}$:

$\{ \wedge (primes(g, a_1)),$

$\wedge (integers(g, a_1), multi(filter(g, a_{1,i,1}, a_{1,i,2}), \wedge(a_{1,1,1} = a_1),$

$\wedge (a_{1,i+1,1} = a_{1,i,2}), \wedge(a_{1,k,2} = a_2)), sift(a_2, a_3), len(a_3, g)),$

$\wedge (multi(filter(g, a_{1,i,1}, a_{1,i,2}), \wedge(a_{1,1,1} = [ ]), \wedge(a_{1,i+1,1} = a_{1,i,2}),$

$\wedge (a_{1,k,2} = a_2)), sift(a_2, a_3), len(a_3, g)),$

$\wedge (integers(g, a_4), multi(filter(g, a_{2,i,4}, a_{2,i,6}), \wedge(a_{2,1,4} - a_4),$

$\wedge (a_{2,i+1,4} = a_{2,i,6}), \wedge(a_{2,k,6} = a_6)), multi(filter(g, a_{1,i,1}, a_{1,i,2}),$

$\wedge (a_{1,1,1} = [g|a_{2,k,6}]), \wedge(a_{1,i+1,1} = a_{1,i,2}), \wedge(a_{1,k,2} = a_2)), sift(a_2, a_3),$

$len(a_3, g))\}$

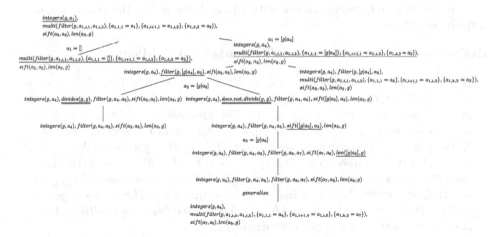

Fig. 5. Abstract unfolding of $integers(g, a_1), multi(filter(g, a_{1,i,1}, a_{1,i,2}), \wedge(a_{1,1,1} = a_1), \wedge(a_{1,i+1,1} = a_{1,i,2}), \wedge(a_{1,k,2} = a_2)), sift(a_2, a_3), len(a_3, g)$

All non-empty leaves in the abstract derivation trees for these atoms are (renamings of) elements of $\mathcal{A}$. This shows $\mathcal{A}$-coveredness and the abstract phase of the analysis terminates.

Similar to what was observed for permutation sort in Sect. 3, we still need an extra analysis to collect the concrete bindings, so that the resultants can be generated. Special care is required for the $multi/4$ abstraction. There are three issues: how to represent $multi/4$ in the concrete domain, how to deal with the concrete counterparts of abstract generalization with $multi/4$ and abstract unfolding of $multi/4$.

Definition 5, in Sect. 3, defined the concrete counterparts of the conjunctions in $\mathcal{A}$. We extend it to $multi(A)$:

Definition 15 (Concrete conjunction for $multi(A, First, Consecutive,$ $Last)$**).** Let $A \in AAtom_p$, then $c(multi(p(A), First, Consecutive, Last)) = multi$ $([c(A)|T])$, with T a fresh variable.

It may seem strange that in the concrete analysis phase we omit the three arguments $First$, $Consecutive$ and $Last$. These arguments are needed in the abstract analysis to correctly capture the data flow and to correctly model the unfolding under the coroutining selection rule. In the concrete analysis phase, as we are completely mimicking the unfolding in the corresponding abstract trees, we are still performing the correct selection. Moreover, the only point of the concrete analysis phase is to collect the bindings produced by unfolding the concrete clauses. The extra arguments are not needed for this purpose.

Example 12 $c(multi(filter(g, a_{1,1,1}, a_{1,1,2}), \wedge(a_{1,1,1} = a_1), \wedge(a_{1,i+1,1} = a_{1,i,2}), \wedge$ $(a_{1,k,2} = a_2))) = multi([filter(X, I1, F1)|T])$

For the abstract generalization with $multi/4$, we define the concrete counterpart as follows.

Definition 16 (Concrete generalization). *Let* $A \in AAtom$.

- *If the abstract generalization with* $multi/4$ *is of the type* $Gen(\bigwedge_{i=1,n} A) = multi(A, First, Consecutive, Last)$, *then the corresponding node in the concrete derivation contains* $c(\bigwedge_{i=1,n} A)$. *The concrete generalization is defined as* $ConGen(c(\bigwedge_{i=1,n} A)) = multi(c([A, \ldots, A]))$, *with* n *members in the list.*
- *If the abstract generalization with* $multi/4$ *is of the type* $Gen((\bigwedge_{i=1,n} A) \wedge multi(A, First, Consecutive, Last)) = multi(A, First', Consecutive, Last)$, *then the corresponding node in the concrete derivation contains* $c(\bigwedge_{i=1,n} A) \wedge multi(List)$, *where* $List$ *is a list of at least one* $c(A)$. *The concrete generalization is defined as* $ConGen(c(\bigwedge_{i=1,n} A) \wedge multi(List)) = multi([c(A), \ldots, c(A)|List])$ *with* n *new members added to* $List$.
- *The third case,* $Gen(multi(A, First, Consecutive, Last) \wedge (\bigwedge_{i=1,n} A)) = multi(A, First, Consecutive, Last')$, *is treated similarly to the previous case, but the concrete atoms are appended to the existing list.*

Example 13 (Concrete   generalization). Let integers(A, B), filter(C, B, D), filter(E, D, F), sift(F, G), len(G, H) occur in a concrete conjunction in a concrete derivation tree, where abstract generalization with multi/4 is performed on the corresponding abstract conjunction. Then, as a next step in the concrete derivation tree, this conjunction is replaced by integers(A, B), multi$([$filter(C, B, D), filter$(E, D, F)])$, sift(F, G), len(G, H).

Note that this "generalization" actually does not generalize anything. It only brings the information in a form that can be generalized.

The actual generalization happens implicitly in the move to the construction of the next concrete derivation tree. If our conjunction is a leaf of the concrete derivation tree, then the corresponding abstract conjunction is added to the set $\mathcal{A}$. Let $\wedge(integers(g, a_4), multi(filter(g, a_{1,i,4}, a_{1,i,5}), \wedge(a_{1,1,4} = a_4), \wedge(a_{1,i+1,4} = a_{1,i,5}), \wedge(a_{1,k,5} = a_7)), sift(a_7, a_6), len(a_6, g))$, for instance, be the corresponding abstract conjunction that is added to $\mathcal{A}$. Then, a new concrete tree is built for a concrete conjunction corresponding to this abstract one.

In this example, the root of that concrete tree is:

$$\wedge(integers(A, B), multi([filter(C, B, D)|T]), sift(E, F), len(F, G))$$

Finally, we still need to define the counterpart of abstract unfold of $multi/4$ in the concrete tree. To do this, we add the following definition of $multi/1$ to the original program P.

```
multi([H]) ← H.
multi([H|T]) ← H, multi(T).
```

It should be clear that concrete unfolding of concrete $multi/1$ atoms with the above definition for $multi/1$ gives us the desired counterpart of the case split performed in abstract unfold of $multi/1$ if we apply the same bindings used in the abstract unfold.

With the concepts above, we construct a concrete derivation tree, mimicking the steps in the abstract derivation tree – but over the concrete domain – for every conjunction in the set $\mathcal{A}$. Collecting all the resultants from these concrete trees, we get the transformed program. A working Prolog program can be found in Annex (2014). Transformations of permutation sort, graph coloring and lucky numbers are available from the same source.

5   Discussion

In this paper, we have presented an approach to formally analyze the computations, for logic programs, performed under coroutining selection rules, and to compile such computations into new logic programs. On the basis of an example, we have shown that simple coroutines, in which the execution of a single, atomic generator is interleaved with a single, atomic tester, can be successfully analyzed and compiled within the framework of ACPD (Leuschel 2004). These "simple" coroutines essentially correspond to the *strongly regular* logic programs of Vidal (2011), based on Hruza and Stepanek (2003).

To achieve this, we defined an expressive abstract domain, capturing modes, types and aliasing. In the paper, we have focused on the intuitions, more than on the full formalization, as space restrictions would not allow both. However, we have developed the formal definitions for the ordering on the abstract domain, abstract unification, abstract unfold and others. Because the approach – for simple coroutines – fits fully within the ACDP framework, it inherits the correctness results from ACPD.

We have proposed an extension to our abstract domain: the *multi*/4-abstraction. A *multi*/4 atom can represent (sets of) conjunctions of one or more concrete atoms. We have defined abstract unfold and abstract generalisation operations for this abstraction. We have shown, in an example, that this abstraction and these operations allow us to extend ACPD, enabling it to perform a complete analysis, and to compile the more complex coroutines.

On a more general level, our work provides a new, rational reconstruction of the CC-transformation (Bruynooghe et al. 1986), avoiding ad hoc features of the CC approach. In addition, the work presents a new application for ACPD.

As a rule, coroutining improves the efficiency of declarative programs by testing partial solutions as quickly as possible. In addition, a program may become more flexible when the transformation is applied. For instance, a generate-and-test based program for the graph coloring problem which was transformed in the course of this research was originally meant to be called with a ground list of nations and a list of free variables of the correct length. A transformed variant of this program can be run in the same way, but the top-level predicate can also be called with a ground list of nations and a free variable. This is because SLD resolution sends the original program down an infinite branch of the search tree. The transformed program checks results earlier and, as a result, infers that both top-level arguments must be lists of the same size. In this scenario, compiling control transforms an infinite computation into a finite one.

The CC-transformation raised challenges for a number of researchers and a range of compediting transformation and synthesis techniques. A first reformulation of the CC-transformation was proposed in the context of the "programs-as-proofs" paradigm, in Wiggins (1990). It was shown that CC-transformations, to a limited extent, could be formalized in a proof-theoretic program synthesis context.

In Boulanger et al. (1993), CC-transformation was revisited on the basis of a combination of abstract interpretation and constraint processing. This improved the formalization of the technique, but it did not clarify the relation with partial deduction.

The seminal survey paper on Unfold/Fold transformation, Pettorossi and Proietti (1994), showed that basic CC-transformations are well in the scope of Unfold/Fold transformation. In later works (e.g. Pettorossi and Proietti 2002), the same authors introduced list-introduction into the Unfold/Fold framework, whose function is very similar to that of the *multi*/4 abstraction in our approach. Also related to our work are Puebla et al. (1997), providing alternative transformations to improve the efficiency of dynamic scheduling, and Vidal (2011) and

Vidal (2012), which also provide a hybrid form of partial deduction, combining abstract and concrete levels.

As an alternative approach to the one proposed in this paper, one could also apply the first Futamura projection. Given a meta-interpreter implementing a dynamic selection strategy, one could attempt to transform a program P by partially evaluating P and the meta-interpreter. This would require an appropriate analysis, for instance abstract partial deduction.

There are a number of issues that are open for future research. First, we aim to investigate the generality of the $multi/4$ abstraction. Although it seems to work well in a number of examples, we will study more complex ones. We also want to revisit the ACPD framework, in order to extend it to the new abstraction we aim to support. This will involve a new formalization of ACPD, capable of supporting analysis and compilation of coroutines in full generality. This will also formally establish the correctness results for the more general cases, such as the one presented in Sect. 4. Obviously, we also want to have a full implementation of these concepts and to show that the analysis and compilation can be fully automated.

Acknowledgements. We thank the anonymous reviewers for their very useful suggestions.

References

Annex.: Definitions, concepts and elaboration of an example (2014). https://perswww. kuleuven.be/~u0055408/tag/lopstr14.html

Doulanger, D., Bruynooghe, M., De Schreye, D.: Compiling control revisited. a new approach based upon abstract interpretation for constraint logic programs. In: LPE, pp. 39–51 (1993)

Bruynooghe, M.: A practical framework for the abstract interpretation of logic programs. J. Logic Program. **10**(2), 91–124 (1991)

Bruynooghe, M., De Schreye, D., Krekels, B.: Compiling control. In: Proceedings of the 1986 Symposium on Logic Programming. IEEE Society Press, Salt Lake City (1986)

Bruynooghe, M., De Schreye, D., Krekels, B.: Compiling control. J. Logic Program. **6**(1), 135–162 (1989)

De Schreye, D., Glück, R., Jørgensen, J., Leuschel, M., Martens, B., Sørensen, M.H.: Conjunctive partial deduction: foundations, control, algorithms, and experiments. J. Logic Program. **41**(2), 231–277 (1999)

Gallagher, J.P.: Transforming logic programs by specialising interpreters. In: ECAI, pp. 313–326 (1986)

Hruza, J., Stepanek, P.: Speedup of logic programs by binarization and partial deduction. arXiv preprint arXiv:cs/0312026 (2003)

Leuschel, M.: A framework for the integration of partial evaluation and abstract interpretation of logic programs. ACM Trans. Program. Lang. Syst. (TOPLAS) **26**(3), 413–463 (2004)

Lloyd, J.: Foundations of Logic Programming. Springer-Verlag, Berlin (1987)

Lloyd, J.W., Shepherdson, J.C.: Partial evaluation in logic programming. J. Logic Program. **11**(3), 217–242 (1991)

Pettorossi, A., Proietti, M.: Transformation of logic programs: foundations and techniques. J. Logic Program. **19**, 261–320 (1994)

Pettorossi, A., Proietti, M.: The list introduction strategy for the derivation of logic programs. Formal Aspects Comput. **13**(3–5), 233–251 (2002)

Puebla, G., de la Banda, M.J.G., Marriott, K., Stuckey, P.J.: Optimization of logic programs with dynamic scheduling.In: ICLP, vol. 97, pp. 93–107 (1997)

Vidal, G.: A hybrid approach to conjunctive partial evaluation of logic programs. In: Alpuente, M. (ed.) LOPSTR 2010. LNCS, vol. 6564, pp. 200–214. Springer, Heidelberg (2011)

Vidal, G.: Annotation of logic programs for independent and-parallelism by partial evaluation. Theor. Pract. Logic Program. **12**(4–5), 583–600 (2012)

Wiggins, G.A.: The improvement of prolog program efficiency by compiling control: a proof-theoretic view. Department of Artificial Intelligence, University of Edinburgh (1990)

Constraint Handling Rules

Confluence Modulo Equivalence in Constraint Handling Rules

Henning Christiansen[✉] and Maja H. Kirkeby

Research group PLIS: Programming, Logic and Intelligent Systems Department
of Communication, Business and Information Technologies, Roskilde University,
Roskilde, Denmark
{henning,majaht}@ruc.dk

Abstract. Previous results on confluence for Constraint Handling Rules,
CHR, are generalized to take into account user-defined state equivalence
relations. This allows a much larger class of programs to enjoy the advantages of confluence, which include various optimization techniques and
simplified correctness proofs. A new operational semantics for CHR is
introduced that significantly reduces notational overhead and allows to
consider confluence for programs with extra-logical and incomplete built-
in predicates. Proofs of confluence are demonstrated for programs with
redundant data representation, e.g., sets-as-lists, for dynamic programming algorithms with pruning as well as a Union-Find program, which
are not covered by previous confluence notions for CHR.

1 Introduction

A rewrite system is confluent if all derivations from a common initial state end in
the same final state. Confluence, like termination, is often a desirable property,
and proof of confluence is a typical ingredient of a correctness proof. For a
programming language based on rewriting such as Constraint Handling Rules,
CHR [8,9], it ensures correctness of parallel implementations and application
order optimizations.

Previous studies of confluence for CHR programs are based on Newman's
lemma. This lemma concerns confluence defined in terms of alternative derivations ending in the exact same state, which excludes a large class of interesting
CHR programs. However, the literature on confluence in general rewriting systems has, since the early 1970s, offered a more general notion of confluence modulo an equivalence relation. This means that alternative derivations only need
to end in states that are equivalent with respect to some equivalence relation
(and not necessarily identical). In this paper, we show how confluence modulo
equivalence can be applied in a CHR context, and we demonstrate interesting
programs covered by this notion that are not confluent by any previous definition of confluence for CHR. The use of redundant data representations is one

M.H. Kirkeby—The second author's contribution has received funding from the
European Union Seventh Framework Programme (FP7/2007-2013) under grant
agreement no 318337, ENTRA - Whole-Systems Energy Transparency.

© Springer International Publishing Switzerland 2015
M. Proietti and H. Seki (Eds.): LOPSTR 2014, LNCS 8981, pp. 41–58, 2015.
DOI: 10.1007/978-3-319-17822-6_3

example of what becomes within reach, and programs that search for one best among multitudes of alternative solutions is another.

Example 1. The following CHR program, consisting of a single rule, collects a number of separate items into a (multi-) set represented as a list of items.

```
set(L), item(A) <=> set([A|L]).
```

This rule will apply repeatedly, replacing constraints matched by the left hand side by those indicated to the right. The query

```
?- item(a), item(b), set([]).
```

may lead to two different final states, $\{\texttt{set([a,b])}\}$ and $\{\texttt{set([b,a])}\}$, both representing the same set. This can be formalized by a state equivalence relation $\approx$ that implies $\{\texttt{set}(L)\} \approx \{\texttt{set}(L')\}$, whenever L is a permutation of L'. The program is not confluent in the classical sense as the end states are not identical, but it will be shown to be confluent modulo $\approx$.

Our generalization is based on a new operational semantics that permits extralogical and incomplete predicates (e.g., Prolog's `var/2` and `is/2`), which is out of the scope of previous approaches. It also leads to a noticeable reduction of notational overhead due to a simpler structure of states.

It is shown that previous results for CHR confluence, based upon critical pairs, to a large extent can be generalized for confluence modulo equivalence. We introduce additional mechanisms to handle the extra complexity caused by the equivalence relation. We do not present any (semi-) automatic approach to confluence proofs, as this would need a formal language for specifying equivalences, which has not been considered at present.

Section 2 reviews previous work on confluence, in general and for CHR. Sections 3 and 4 give preliminaries and our operational semantics. Section 5 considers how to prove confluence modulo equivalence for CHR. Section 6 shows confluence modulo equivalence for a CHR version of the Viterbi algorithm; it represents a wider class of dynamic programming algorithms with pruning, also outside the scope of earlier proposals. Section 7 shows confluence modulo equivalence for the Union-Find algorithm, which has become a standard test case for confluence in CHR; it is not confluent in any previously proposed way (except with contrived side-conditions). Section 8 comments on related work in more detail, and the final section provides a summary and a conclusion.

2   Background

A binary *relation* $\rightarrow$ on a set A is a subset of $A \times A$, where $x \rightarrow y$ denotes membership of $\rightarrow$. A *rewrite system* is a pair $\langle A, \rightarrow \rangle$; it is *terminating* if there is no infinite chain $a_0 \rightarrow a_1 \rightarrow \cdots$. The *reflexive transitive closure* of $\rightarrow$ is denoted $\overset{*}{\rightarrow}$. The *inverse relation* $\leftarrow$ is defined by $\{(y,x) \mid x \rightarrow y\}$. An *equivalence (relation)* $\approx$ is a binary relation on A that is reflexive, transitive and symmetric.

A rewrite system $\langle A, \to \rangle$ is *confluent* if and only if $y \xleftarrow{*} x \xrightarrow{*} y' \Rightarrow \exists z.\ y \xrightarrow{*} z \xleftarrow{*} y'$, and is *locally confluent* if and only if $y \leftarrow x \to y' \Rightarrow \exists z.\ y \xrightarrow{*} z \xleftarrow{*} z'$. In 1942, Newman showed his fundamental lemma [13]: *A terminating rewrite system is confluent if and only if it is locally confluent.* An elegant proof of Newman's lemma was provided by Huet [11] in 1980.

The more general notion of *confluence modulo equivalence* was introduced in 1972 by Aho et al. [3] in the context of the Church-Rosser property.

Definition 1 (Confluence Modulo Equivalence). *A relation* $\to$ *is confluent modulo an equivalence* $\approx$ *if and only if*

$$\forall x, y, x', y'.\quad y \xleftarrow{*} x \approx x' \xrightarrow{*} y' \quad \Rightarrow \quad \exists z, z'.\ y \xrightarrow{*} z \approx z' \xleftarrow{*} y'.$$

This shown as a diagram in Fig. 1a. In 1974, Sethi [17] showed that confluence modulo equivalence for a bounded rewrite system is equivalent to the following properties, α and β, also shown in Fig. 1b.

Definition 2 (α & β). *A relation* $\to$ *has the* α *property and the* β *property if and only if it satisfy the* α *condition and the* β *condition, respectively:*

$$\alpha: \quad \forall x, y, y'.\quad y \leftarrow x \to y' \quad \Longrightarrow \quad \exists z, z'.\ y \xrightarrow{*} z \approx z' \xleftarrow{*} y'$$

$$\beta: \quad \forall x, x', y.\quad x \approx x' \to y \quad \Longrightarrow \quad \exists z, z'.\ x' \xrightarrow{*} z' \approx z \xleftarrow{*} y$$

In 1980, Huet [11] generalized this result to any terminating system.

Definition 3 (Local Confl. Mod. Equivalence). *A rewrite system is* locally confluent modulo an equivalence $\approx$ *if and only if it has the* α *and* β *properties.*

Theorem 1. *Let* $\to$ *be a terminating relation. For any equivalence* $\approx$, $\to$ *is confluent modulo* $\approx$ *if and only if* $\to$ *is locally confluent modulo* $\approx$.

The known results on confluence for CHR are based on Newman's lemma. Abdennadher *et al.* [2] in 1996 seem to be the first to consider this, and they showed that confluence (without equivalence) for CHR is decidable and can be checked by examining a finite set of states formed by a combination of heads of rules. A refinement, called observational confluence was introduced in 2007 by Duck *et al.* [6], in which only states that satisfy a given invariant are considered.

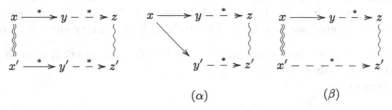

(α)   (β)

(a) Confluence modulo $\approx$.   (b) Local Confluence modulo $\approx$.

Fig. 1. Diagrams for the fundamental notions. A dotted arrow (single wave line) indicates an inferred step (inferred equivalence).

3 Preliminaries

We assume standard notions of first-order logic such as predicates, atoms and terms. For any expression E, $vars(E)$ refers to the set of variables that occurs in E. A substitution is a mapping from a finite set of variables to terms, which also may be viewed as a set of first-order equations. For substitution σ and expression E, $E\sigma$ (or $E \cdot \sigma$) denotes the expression that arises when σ is applied to E; composition of two substitutions σ, τ is denoted $\sigma \circ \tau$. Special substitutions $failure$, $error$ are assumed, the first one representing falsity and the second one runtime errors.

Two disjoint sets of *(user) constraints* and *built-in* predicates are assumed. For the built-ins, we use a semantics that is more in line with implemented CHR systems than previous approaches and also allows extra-logical devices such as Prolog's `var/1` and incomplete ones such as `is/2`. While [2,5,6] collect built-ins in a separate store and determine their satisfiability by a magic solver that mirrors a first-order semantics, we execute a built-in right away. Thereby, it serves as a test, possibly giving rise to a substitution that is immediately applied to the state.

An evaluation procedure Exe for built-ins b is assumed, such that $Exe(b)$ is either a (possibly identity) substitution to a subset of $vars(b)$ or one of $failure$ and $error$. It extends to sequences of built-ins as follows.

$$Exe((b_1, b_2)) = \begin{cases} Exe(b_1) & \text{when } Exe(b_1) \in \{failure, error\}, \\ Exe(b_2 \cdot Exe(b_1)) & \text{when otherwise } Exe(b_2 \cdot Exe(b_1)) \\ & \in \{failure, error\}, \\ Exe(b_1) \circ Exe(b_2 \cdot Exe(b_1)) & \text{otherwise} \end{cases}$$

A subset of built-in predicates are the *logical* ones, whose meaning is given by a first-order theory $\mathcal{B}$. For a logical atom b with $Exe(b) \neq error$, the following conditions must hold.

- Partial correctness: $\mathcal{B} \models \forall_{vars(b)}(b \leftrightarrow \exists_{vars(Exe(b)) \setminus vars(b)} Exe(b))$.
- Instantiation monotonicity: $Exe(b \cdot \sigma) \neq error$ for all substitutions σ.

A logical predicate p is *complete* whenever, for any atom b with predicate symbol p, we have $Exe(b) \neq error$; later we define completeness with respect to a state invariant. Any built-in predicate which is not logical is called *extra-logical*. The following predicates are examples of built-ins; ϵ is the empty substitution.

1. $Exe(t = t') = \sigma$ where σ is a most general unifier of t and t'; if no such unifier exists, the result is *failure*.
2. $Exe(\texttt{true})$ is ϵ.
3. $Exe(\texttt{fail})$ is *failure*.
4. $Exe(t \texttt{ is } t') = Exe(t = v)$ whenever t' is a ground term that can be interpreted as an arithmetic expression e with the value v; if no such e exists, the result is *error*.
5. $Exe(\texttt{var}(t))$ is ϵ if t is a variable and *failure* otherwise.

6. $Exe(\mathbf{ground}(t))$ is ϵ when t is ground and *failure* otherwise.
7. $Exe(t == t')$ is ϵ when t and t' are identical and *failure* otherwise.
8. $Exe(t \mathbin{\backslash =} t')$ is ϵ when t and t' are non-unifiable and *failure* otherwise.

The first three predicates are logical and complete; "is" is logical but not complete without an invariant that grounds its second arguments (considered later). The remaining ones are extra-logical.

The practice in previous semantics [2,5,6] of conjoining built-ins and testing them by satisfiability leads to ignorance of runtime errors and incompleteness.

To represent the propagation history, we introduce *indices:* An *indexed set* S is a set of items of the form $x{:}i$ where i belongs to some index set and each such i is unique in S. When clear from context, we may identify an indexed set S with its cleaned version $\{x \mid x{:}i \in S\}$. Similarly, the item x may identify the indexed version $x{:}i$. We extend this to any structure built from indexed items.

4   Constraint Handling Rules

We define an abstract syntax of CHR together with an operational semantics suitable for considering confluence. We use the *generalized simpagation* form as a common representation for the rules of CHR. Guards may unify variables that occur in rule bodies, but not variables that occur in the matched constraints. In accordance with the standard behaviour of implemented CHR systems, failure and runtime errors are treated the same way in the evaluation of a guard, but distinguished when occurring in a query or rule body, cf. Definitions 4 and 8, below.

Definition 4. *A rule r is of the form*

$$H_1 \setminus H_2 <=> g \mid C,$$

where H_1 and H_2 are sequences of constraints, forming the head of r, g is the guard being a sequence of built-ins, and C is a sequences of constraints and built-ins called the body of r. Any of H_1 and H_2, but not both, may be empty. A program is a finite set of rules.

For any fresh variant of rule r with notation as above, an application instance r'' is given as follows.

1. *Let r' be a structure of the form*
   $H_1\tau \setminus H_2\tau <=> C\tau\sigma$
   where τ is a substitution for the variables of H_1, H_2, $Exe(g\tau) = \sigma$, $\sigma \notin \{failure, error\}$, and it holds that $(H_1 \setminus H_2)\tau = (H_1 \setminus H_2)\tau\sigma$,
2. *r'' is a copy of r' in which each atom in its head and body is given a unique index, where the indices used for the body are new and unused.*

The substitution $g\tau$ is referred to the as the guard of r''. The application record for r'' is a structure of the form

$$r @ i_1, \ldots, i_n$$

where $i_1, \ldots, i_n$ is the sequence of indices of H_1, H_2 in the order they occur.

A rule is a *simplification* when H_1 is empty, a *propagation* when H_2 is empty; in both cases, the backslash is left out, and for a propagation, the arrow symbol is written ==> instead. Any other rule is a *simpagation*. In case the guard is the built-in *true*, it and the vertical bar may be omitted. A guard (or single built-in atom) is *logical* if it contains only logical predicates. Guards are removed from application instances as they are *a priori* satisfied. The following definition will become useful later on when we consider confluence.

Definition 5. *Consider two application instances* $r_i = (A_i \setminus B_i <=> C_i)$, $i = 1, 2$. *We say that* r_1 *is* blocking r_2 *whenever* $B_1 \cap (A_2 \cup B_2) \neq \emptyset$.

For this to be the case, r_1 must be a simplification or simpagation. Intuitively, it means that if r_1 has been applied to a state, it is not possible subsequently to apply r_2. In the following definition of execution states for CHR, irrelevant details of the state representation are abstracted away using principles of [14]. To keep notation consistent with Sect. 2, we use letters such as x, y, etc. for states.

Definition 6. *A* state representation *is a pair* $\langle S, T \rangle$, *where*

- S *is a finite, indexed set of atoms called the* constraint store,
- T *is a set of application records called the* propagation history.

Two state representations S_1 *and* S_2 *are* isomorphic, *denoted* $S_1 \equiv S_2$ *whenever one can be derived from the other by a renaming of variables and a consistent replacement of indices (i.e., by a 1-1 mapping). When* Σ *is the set of all state representations, a* state *is an element of* $\Sigma/_{\equiv} \cup \{failure, error\}$, *i.e., an equivalence class in* Σ *induced by* $\equiv$ *or one of two special states; applying the failure (error) substitution to a state yields the failure (error) state. To indicate a given state, we may for simplicity mention one of its representations.*

A query q *is a conjunction of constraints, which is also identified with an initial state* $\langle q', \emptyset \rangle$ *where* q' *is an indexed version of* q.

To make statements about, say, two states x, y and an instance of a rule r, we may do so mentioning state representatives x', y' and application instance r' having recurring indices. The following notions becomes useful in Sect. 5, when we go into more detail on how to prove confluence modulo equivalence,

Definition 7. *An* extension *of a state* $\langle S, R \rangle$ *is a state of the form* $\langle S\sigma \cup S^+, R \cup R^+ \rangle$ *for suitable* σ, S^+ *and* R^+; *an* I-extension *is one that satisfies* I; *and a state is said to be* I-extendible *if it has one or more extensions that are* I-states.

In contrast to [2,5,6], we have excluded global variables, which refer to those of the original query, as they are easy to simulate: A query $q(X)$ is extended to $global('X', X), q(X)$, where $global/2$ is a new constraint predicate; $'X'$ is a constant that serves as a name of the variable. The value *val* for X is found in the final state in the unique constraint $global('X', val)$. References [2,5,6] use a state component for constraints waiting to be processed, plus a separate derivation step to introduce them into the constraint store. We avoid this as the derivations made under either premises are basically the same. Our derivation relation is defined as follows; here and in the rest of this paper, $\uplus$ denotes union of disjoint sets.

Definition 8. *A derivation step* $\mapsto$ *from one state to another can be of two types: by rule* $\overset{r}{\mapsto}$ *or by built-in* $\overset{b}{\mapsto}$, *defined as follows.*

Apply: $\langle S \uplus H_1 \uplus H_2, T \rangle \overset{r}{\mapsto} \langle S \uplus H_1 \uplus C, T' \rangle$
   whenever there is an application instance r *of the form* $H_1 \setminus H_2 \mathrel{<=>} C$ *with* applied(r) $\notin T$, *and* T' *is derived from* T *by (1) removing any application record having an index in* H_2 *and (2) adding* applied(r) *in case* r *is a propagation.*
Built-in: $\langle \{b\} \uplus S, T \rangle \overset{b}{\mapsto} \langle S, T \rangle \cdot Exe(b)$.

A state z *is final for query* q, *whenever* $q \overset{*}{\mapsto} z$ *and no step is possible from* z.

The removal of certain application records in Apply steps means to keep only those records that are essential for preventing repeated application of the same rule to the same constraints (identified by their indices).

As noticed by [6], introducing an invariant makes more programs confluent, as one can ignore unusual states that never appear in practice. An invariant may also make it easier to characterize an equivalence relation for states.

Definition 9. *An invariant is a property* $I(\cdot)$ *which may or may not hold for a state, such that for all states* x, y, $I(x) \wedge (x \mapsto y) \Rightarrow I(y)$. *A state* x *for which* $I(x)$ *holds is called an* I-state, *and an* I-derivation *is one starting from an* I-state. *A program is* I-terminating *whenever all* I-derivations are terminating. *A set of allowed queries* Q *may be specified, giving rise to an invariant* reachable$_Q(x)$ $\Leftrightarrow$ $\exists q \in Q: q \overset{*}{\mapsto} x$.
   A (state) equivalence is an equivalence relation $\approx$ *on the set of* I-states.

The central Theorem 1 applies specifically for CHR programs equipped with invariant I and equivalence relation $\approx$. When $\approx$ is identity, it coincides with a theorem of [6] for observable confluence. If, furthermore, $I \Leftrightarrow true$, we obtain the classical confluence results for CHR [1].

The following definition is useful when considering confluence for programs that use Prolog built-ins such as "is/2".

Definition 10. *A logical predicate* p *is complete with respect to invariant* I *(or, for short, is* I-complete) *whenever, for any atom* b *with predicate symbol* p *in some* I-state, *that* $Exe(b) \neq error$.

A logical guard (or a built-in atom) is also called I-complete, whenever all its predicates are I-complete. We use the term I-incomplete for any such notion that is not I-complete.

As promised earlier, "is/2" is complete with respect to an invariant that guarantees groundness of the second argument of any call to "is/2".

Example 2. Our semantics permits CHR programs that define constraints such as Prolog's `dif/2` constraint and a safer version of `is/2`.

```
dif(X,Y) <=> X==Y | fail.
dif(X,Y) <=> X\=Y | true.
X safer_is Y <=> ground(Y) | X is Y.
```

5 Proving Confluence Modulo Equivalence for CHR

We consider here ways to prove the local confluence properties α and β from which confluence modulo equivalence may follow, cf. Theorem 1. The corners in the following definition generalize the critical pairs of [2]. For ease of usage, we combine the common ancestor states with the pairs, thus the notion of corners corresponding to the "given parts" of diagrams for the α and β properties, cf. Fig. 1a. The definitions below assume a given I-terminating program with invariant I and state equivalence $\approx$. Two states x and x' are *joinable modulo* $\approx$ whenever there exist states z and z' such that $x \mapsto^* z \approx z' {}^*\!\!\leftharpoondown x'$.

Definition 11. *An α-corner consists of I-states x, y and y' with $y \neq y'$ and two derivation steps such that $y \overset{\gamma}{\leftharpoondown} x \overset{\delta}{\mapsto} y'$. An α-corner is* joinable modulo $\approx$ *whenever y and y' are joinable modulo $\approx$.*

A β-corner consists of I-states x, x' and y with $x \neq x'$ and a derivation step such that $x' \approx x \overset{\gamma}{\mapsto} y$. A β-corner is joinable modulo $\approx$ whenever x' and y are joinable modulo $\approx$.

Joinability of α_1-corners holds trivially in a number of cases:

- when γ and δ are application instances, none blocking the other,
- when γ and δ are built-ins, both logical and I-complete, or having no common variables, or
- when, say, γ is an application instance whose guard is logical and I-complete, and δ is any built-in that has no common variable with the guard of γ.

These cases are easily identified syntactically. All remaining corners are recognized as "critical", which is defined as follows.

Definition 12. *An α-corner $y \overset{\gamma}{\leftharpoondown} x \overset{\delta}{\mapsto} y'$ is* critical *whenever one of the following properties holds.*

  α_1: *γ and δ are application instances where γ blocks δ (Definition 5).*
  α_2: *γ is an application instance whose guard is extra-logical or I-incomplete, and δ is a built-in with $vars(g) \cap vars(\delta) \neq \emptyset$.*
  α_3: *γ and δ are built-ins with γ extra-logical or I-incomplete, and $vars(\gamma) \cap vars(\delta) \neq \emptyset$.*

A β-corner $x' \approx x \overset{\gamma}{\mapsto} y$ is critical *whenever the following property holds.*

– $x \neq x'$ and there exists no state y' and single derivation step δ such that $x' \overset{\delta}{\mapsto} y' \approx y$.

Our definition of critical β-corners are motivated by the experience that often the δ step can be formed trivially by applying the same rule or built-in of γ in an analogous way to the state x'. By inspection and Theorem 1, we get the following.

Lemma 1. *Any non-critical corner is joinable modulo* $\approx$.

Theorem 2. *A terminating program is confluent modulo* $\approx$ *if and only if all its critical corners are joinable modulo* $\approx$.

5.1 Joinability of α_1-Critical Corners

Without invariant, equivalence and extra-logicals, the only critical corners are of type α_1; here [2] has shown that joinability of a finite set of minimal critical pairs is sufficient to ensure local confluence. In the general case, it is not sufficient to check such minimal states, but the construction is still useful as a way to group the cases that need to be considered. We adapt the definition of [2] as follows.

Definition 13. *An* α_1-critical pattern (with evaluated guards) *is of the form*

$$\langle S_1\sigma_1, \emptyset \rangle \overset{r_1}{\dashv} \langle S, \emptyset \rangle \overset{r_2}{\mapsto} \langle S_2\sigma_2, R \rangle$$

whenever there exist, for $k = 1, 2$, *indexed rules* $r_k = (A_k \setminus B_k \mathrel{<=>} g_k \mid C_k)$, *and*

$$R = \begin{cases} \{a\} & \text{whenever } r_2 \text{ is a propagation with application record } a, \\ \emptyset & \text{otherwise.} \end{cases}$$

The remaining entities are given as follows.

– *Let* $H_k = A_k \cup B_k$, $k = 1, 2$, *and split* B_1 *and* H_2 *into disjoint subsets by* $B_1 = B_1' \uplus B_1''$ *and* $H_2 = H_2' \uplus H_2''$, *where* B_1' *and* H_2' *must have the same number of elements* ≥ 1.
– *The set of indices used in* B_1' *and* H_2' *are assumed to be identical, and any other index in* r_1, r_2 *unique, and* σ *is a most general unifier of* B_1' *and a permutation of* H_2'.
– $S = A_1\sigma \cup B_1\sigma \cup A_2\sigma \cup B_2\sigma$, *with* S *being* I-extendible,
– $S_k = S \setminus B_k\sigma \cup C_k\sigma$, $k = 1, 2$,
– g_k *is logical with* $\sigma_k = Exe(g_k\sigma) \notin \{error, failure\}$ *for* $k = 1, 2$.

An α_1-critical pattern (with delayed guards) *is of the form*

$$\langle S_1, \emptyset \rangle \overset{r_1}{\dashv} \langle S, \emptyset \rangle \overset{r_2}{\mapsto} \langle S_2, R \rangle,$$

where all parts are defined as above, except in the last step, that one of g_k *is extra-logical or its evaluation by* Exe *results in error; the guards* $g_k\sigma$ *are recognized as the* unevaluated guards.

Definition 14. *An α_1-critical corner $y \overset{r_1}{\leftharpoondown} x \overset{r_2}{\mapsto} y'$ is* covered by *an α_1-critical pattern*

$$\langle S_1, \emptyset \rangle \overset{r_1}{\leftharpoondown} \langle S, \emptyset \rangle \overset{r_2}{\mapsto} \langle S_2, R \rangle,$$

whenever x is an I-extension of $\langle S, \emptyset \rangle$.

Analogously to previous results on confluence of CHR [2], we can state the following.

Lemma 2. *For a given I-terminating program with invariant I and equivalence $\approx$, the set of critical α_1-patterns is finite, and any critical α_1-corner is covered by some critical α_1-pattern.*

The requirement of definition 13, that a critical α_1-corner needs to be I-extendible, means that there may be fewer patterns to check than if classical confluence is investigated. Examples of this is used for when showing confluence of the Union-Find program, Sect. 7 below. We can reuse the developments of [2] and joinability results derived by their methods, e.g., using automatic checkers for classical confluence [12].

Lemma 3. *If a critical α_1-pattern π (viewed as an α_1-corner) is joinable modulo the identity equivalence, then any α_1-corner covered by π is joinable under any I and $\approx$.*

This means that we may succeed in showing confluence modulo $\approx$ under I in the following way for a program without critical α_2, α_3 and β corners.

- Run a classical confluence checker (e.g., [12]) to identify which classical, critical pairs that are not joinable. Those such that do not correspond to I-extendible α_1 patterns can be disregarded.
- Those critical α_1-patterns that remain need separate proofs, which may succeed due to the stronger antecedent given by I and the weakening of the joinability consequent by an equivalence relation.

Example 3 (Example 1, continued). We consider again the one line program of Example 1 that collects a items into a set, represented as a list. Suitable invariant and equivalence are given as follows; the propagation history can be ignored as there are no propagations.

I: $I(x)$ holds if and only if $x = \{\texttt{set}(L)\} \cup Items$, where *Items* is a set of $\texttt{item/1}$ constraints whose argument is a constant and L a list of constants.

$\approx$: $x \approx x'$ if and only if $x = \{\texttt{set}(L)\} \cup Items$ and $x' = \{\texttt{set}(L')\} \cup Items$ where *Items* is a set of $\texttt{item/1}$ constraints and L is a permutation of L'.

There are no built-ins and thus no critical α_2- or α_3-patterns. There is only one critical α_1-pattern, namely

$$\{\texttt{set([B|L])}, \texttt{item(A)}\} \leftharpoondown \{\texttt{set(L)}, \texttt{item(A)}, \texttt{item(B)}\} \mapsto \{\texttt{set([A|L])}, \texttt{item(B)}\}.$$

The participating states are not I-states as A, B and L are variables; the set of all critical α_1-corners can be generated by different instantiations of the variables, discarding those that lead to non-I-states. We cannot use Lemma 3 to prove joinability as the equivalence is $\approx$ essential. Instead, we can apply a general argument that goes for any I-extension of this pattern. The common ancestor state in such an I-extension is of the form $\{\texttt{set}(L), \texttt{item}(A)\} \cup$ *Items*, and joinability is shown by applying the rule to the two "wing" states (not shown) to form the two states $\{\texttt{set}([B, A, |L])\} \cup$ *Items* $\approx \{\texttt{set}([A, B, |L])\} \cup$ *Items*. To show confluence modulo $\approx$, we still need to consider the β-corners which we return to in Example 5 below.

5.2 About Critical α_2-, α_3- and β-Corners

It is not possible to characterize the sets of all critical α_2-, α_3- and β-corners by finite sets of patterns of mini-states in the same way as for α_1.

The problem for α_2 and α_3 stems from the presence of extra-logical or incomplete built-ins. Here the existence of one derivation step from a given state S does not imply the existence of another, analogous derivation step from an extension $S\sigma \cup S^+$. This is demonstrated by the following example.

Example 4. Consider the following program that has extra-logical guards.

```
r1:  p(X) <=> var(X)    | q(X).
r2:  p(X) <=> nonvar(X) | r(X).
r3:  q(X) <=> r(X).
```

There are no propagation rules, so we can identify states with multisets of constraints. The invariant I is given as follows, and the state equivalence is trivial identity so there are no critical β-corners to consider.

> $I(S)$: S is a multiset of p, q and r constraints and built-ins formed by the "=" predicate. Any argument is either a constant or a variable.

The meaning of equality built-ins is as defined in Sect. 3 above.

It can be argued informally that this program is I-confluent as all userdefined constraints will eventually become r constraints unless a failure occurs due to the execution of equality built-ins; the latter can only be introduced in the initial query, so if one derivation leads to failure, all terminated derivations do. Termination follows from the inherent stratification of the constraints.

To prove this formally, we consider all critical corners and show them joinable. One group of critical α_2-corners are of the following form, (1)

$$S_1 = (\{q(x), x = a\} \uplus S) \overset{r_1}{\leftarrowtail} (\{p(x), x = a\} \uplus S) \overset{=}{\mapsto} (\{p(a)\} \uplus S) = S_2;$$

x is a variable, a a constant and S an arbitrary set of constraints such that I is maintained. Any such corner is joinable, which can be shown as follows, (2)

$$S_1 \overset{=}{\mapsto} S_1' \overset{r_2}{\mapsto} \{r(a)\} \uplus S \overset{r_2}{\leftarrowtail} S_2;$$

The remaining critical α_2-corners form a similar group.

$$\{q(x), x = y\} \uplus S \overset{r_1}{\underset{=}{\leftarrowtail}} \{p(x), x = y\} \uplus S \overset{=}{\mapsto} \{p(x)\} \uplus S;$$

x and y variables, r_1 and S and S an arbitrary set of constraints such that I is maintained. Joinability is shown by a similar argument that goes for this entire group. The only critical corners are those α_2 cases that have been considered, so the program is confluent.

We notice, however, that the derivation steps in (1) and (2) are possible only due to the assumptions about the permitted instances of x, a and S. The symbol a, for example, is not a variable in a formal sense, neither is it a constant, but a meta-variable or placeholder of the sort that mathematicians use all the time. This means that we cannot reduce the formulas (1) and (2) to refer to derivations over mini-states, with proper variables as placeholders, as then r_2 can never apply.

To see critical α_3-corners, we change I into I' by allowing also var constraints in a state. One group of such corners will have the following shape.

$$\{\text{var}(a)\} \uplus S \overset{=}{\underset{=}{\leftarrowtail}} \{\text{var}(x), x = a\} \uplus S \overset{\text{var}}{\mapsto} \{x = a\} \uplus S$$

x is a variable, a a constant and S an arbitrary set of constraints such that I' is maintained. For, e.g., $S = \emptyset$, this corner is obviously not joinable, so the program is not confluent (module equivalence) under I'. As above, we observe that the set of critical α_3 corners cannot be characterized by a finite set of mini-states.

The β property needs to be considered when the state equivalence is non-trivial, as in the following example

Example 5 (Examples 1 and 3, continued). To check the β property, we notice that any β-corner is of the form

$$\{\text{set}(L'), \text{item}(A)\} \uplus \textit{Items} \approx \{\text{set}(L), \text{item}(A)\} \uplus \textit{Items} \mapsto \{\text{set}([A|L])\} \uplus \textit{Items}$$

where L and L' are lists, one being a permutation of the other. Applying the rule to the "left wing" state leads to $\{\text{set}([A|L'])\} \cup \textit{Items}$ which is equivalent (wrt. $\approx$) to the "right wing" state; there are thus no critical β-corners. Together with results for critical α-corners above, we have now shown local confluence modulo $\approx$ for the sets-as-lists program, and as the program is clearly I-terminating, it follows that it is confluent modulo $\approx$.

6   Confluence of Viterbi Modulo Equivalence

Dynamic programming algorithms produce solutions to a problem by generating solutions to a subproblem and iteratively extending the subproblem and its solutions (until the original problem is solved). The Viterbi algorithm [20] finds a most probable path of state transitions in a Hidden Markov Model (HMM) that produces a given emission sequence Ls, also called the *decoding* of Ls; see [7] for a background on HMMs. There may be exponentially many paths but an early pruning strategy ensures linear time. The algorithm has been studied in CHR by [4], starting from the following program; the "@" operator is part of the implemented CHR syntax used for labelling rules.

```
:- chr_constraint path/4, trans/3, emit/3.

expand @ trans(Q,Q1,PT), emit(Q,L,PE), path([L|Ls],Q,P,PathRev) ==>
    P1 is P*PT*PE | path(Ls,Q1,P1,[Q1|PathRev]).

prune @ path(Ls,Q,P1,_) \ path(Ls,Q,P2,_) <=> P1 >= P2 | true.
```

The meaning of a constraint $\mathtt{path}(Ls,q,p,R)$ is that Ls is a remaining emission sequence to be processed, q the current state of the HMM, and p the probability of a path R found for the already processed prefix of the emission sequence. To simplify the program, a path is represented in reverse order. Constraint $\mathtt{trans}(q,q',pt)$ indicates a transition from state q to q' with probability pt, and $\mathtt{emit}(q,\ell,pe)$ a probability pe for emitting letter ℓ in state q.

Decoding of a sequence Ls is stated by the query "HMM, $\mathtt{path}(Ls,\mathtt{q0},1,[\,])$" where HMM is an encoding of a particular HMM in terms of $\mathtt{trans}$ and $\mathtt{emit}$ constraints. Assuming HMM and Ls be fixed, the state invariant I is given as reachability from the indicated query. The program is I-terminating, as any new $\mathtt{path}$ constraint introduced by the expand rule has a first argument shorter than that of its predecessor. Depending on the application order, it may run in between linear and exponential time, and [4] proceeds by semantics preserving program transformations that lead to an optimal execution order.

The program is not confluent in the classical sense, i.e., without an equivalence, as the $\mathtt{prune}$ rule may need to select one out of two different and equally probable paths. A suitable state equivalence may be defined as follows.

Definition 15. Let $\langle HMM \sqcup PATHS_1, T\rangle \approx \langle HMM \sqcup PATHS_2, T\rangle$ whenever: For any indexed constraint $(i\colon \mathtt{path}(Ls,q,P,R_1)) \in PATHS_1$ there is a corresponding $(i\colon \mathtt{path}(Ls,q,P,R_2)) \in PATHS_2$ and vice versa.

The built-ins used in guards, $\mathtt{is}/2$ and $\mathtt{>=}/2$, are logical and I-complete, so there are no α_2- or α_3-critical corners. For simplicity of notation, we ignore the propagation histories. There are three critical α_1 patterns to consider:

(i) $y \overset{\text{prune}}{\leftarrowtail} x \overset{\text{prune}}{\mapsto} y'$, where x contains two $\mathtt{path}$ constraints that may differ only in their last arguments, and y and y' differ only in which of these constraints that are preserved; thus $y \approx y'$.

(ii) $y \overset{\text{prune}}{\leftarrowtail} x \overset{\text{expand}}{\mapsto} y'$ where $x = \{\pi_1, \pi_2, \tau, \eta\}$, $\pi_i = \mathtt{path}(L,q,P_i,R_i)$ for $i = 1,2$, $P_1 \geq P_2$, and τ, η the $\mathtt{trans}$ and $\mathtt{emit}$ constraints used for the expansion step. Thus $y = \{\pi_1, \tau, \eta\}$ and $y' = \{\pi_1, \pi_2, \pi_2', \tau, \eta\}$ where π_2' is expanded from π_2. To show joinability, we show the stronger property of the existence of a state z with $y \overset{*}{\mapsto} z \overset{*}{\leftarrowtail} y'$. We select $z = \{\pi_1, \pi_1', \tau, \eta\}$, where π_1' is expanded from

π_1.[1] The probability in π_1' is greater or equal to that of π_2', which means that a pruning of π_2' is possible when both are present. Joinability is shown as follows.

$$y \overset{\text{expand}}{\mapsto} z \overset{\text{prune}}{\hookleftarrow} \{\pi_1, \pi_1', \pi_2, \tau, \eta\} \overset{\text{prune}}{\hookleftarrow} \{\pi_1, \pi_1', \pi_2, \pi_2', \tau, \eta\} \overset{\text{expand}}{\hookleftarrow} y'$$

(iii) As case *ii* but with $P_2 \geq P_1$ and $y = \{\pi_2, \tau, \eta\}$; proof similar and omitted.

Thus all α-critical corners are joinable. There are no critical β corners, as whenever $x' \approx x \overset{r}{\mapsto} y$, the rule r can apply to x' with an analogous result, i.e., there exists a state y' such that $x' \overset{r}{\mapsto} y' \approx y$. This finishes the proof of confluence modulo $\approx$.

7   Confluence of Union-Find Modulo Equivalence

The Union-Find algorithm [19] maintains a collection of disjoint sets under union, with each set represented as a tree. It has been implemented in CHR by [16] who proved it nonconfluent using critical pairs [2]. We have adapted a version from [6], extending it with a new token constraint to be explained; let UF_{token} refer to our program and UF_0 to the original without token constraints.

```
union     @ token, union(A,B) <=> find(A,X), find(B,Y), link(X,Y).
findNode @ A ~> B \ find(A,X) <=> find(B,X).
findRoot @ root(A) \ find(A,X) <=> A=X.
linkEq    @ link(A,A) <=> token.
link      @ root(A) \ link(A,B), root(B) <=> B ~> A, token.
```

The ~> and root constraints, called *tree constraints*, represent a set of trees. A finite set T of ground tree constraints is *consistent* whenever: for any constant a in T, there is either one and only one root(a) $\in T$, or a is connected via a unique chain of ~> constraints to some r with root(r) $\in T$. We define $sets(T)$ to be the set of sets represented by T, formally: the smallest equivalence relation over constants in T that contains the reflexive, the transitive closure of ~>; $set(a, T)$ refers to the set in $sets(T)$ containing constant a.

The *allowed queries* are ground and of the form $T \cup U \cup \{\text{token}\}$, where T is a consistent set of tree constraints, and U is a set of constraints union(a_i, b_i), where a_i, b_i appear in T. The token constraint is necessary for triggering the union rule, so it needs to be present in the query to get the process started; it is consumed when one union operation starts and reintroduced when it has finished (as marked by the linkEq or link rules), thus ensuring that no two union operations overlap in time. The invariant I is defined by reachability from these queries. By induction, we can show the following properties of any I-state S.

---

[1] It may be the case that π_1' was produced and pruned at an earlier stage, so the propagation history prevents the creation of π_1' anew. A detailed argument can show, that in this case, there will be another constraints π_1'' in the store similar to π_1' but with a $\geq$ probability, and π_1'' can be used for pruning π_2' and obtain the desired result in that way.

– Either $S = T \cup U \cup \{\text{token}\}$, where T is a consistent set of tree constraints
  and U a set of **union** constraints whose arguments are in T, or
– $S = T \cup U \cup \{\text{link}(A_1, A_2)\} \cup F_1 \cup F_2$ where T, U are as in the previous case,
  and for $i = 1, 2$,
  - if A_i is a constant, $F_i = \emptyset$, otherwise
  - $F_i = \{\text{find}(a_i, A_i)\}$ or $F_i = \{(a_i = A_i)\}$ for some constant a_i.

As shown by [16], UF_0 is not confluent in the classical sense, which can be related
to the following issues.

(i) When the detailed steps of two **union** operations are intertwined in an unfor-
   tunate way, the program may get stuck in a state where it cannot finish the
   operation as shown in the following derivation.
   ```
   root(a), root(b), root(c), union(a,b), union(b,c) ⟼*
   root(a), root(b), root(c), link(a,b), link(b,c) ⟼
   b ~> a, root(a), root(c), link(b,c)
   ```
(ii) Different execution orders of the **union** operations may lead to different data
    structures (representing the same sets). This is shown in the following deriva-
    tions from a query $q_0 = \{\text{root(a)}, \text{root(b)}, \text{root(c)}, \text{union(a,b)}, \text{union(b,c)}\}$.
    ```
    q₀ ⟼ root(a), root(c), b ~> a, union(b,c) ⟼* root(a), b ~> a, c ~> a
    q₀ ⟼* root(a), root(b), c ~> b, union(a,b) ⟼ root(b), b ~> a, c ~> b
    ```

We proceed, now, to show that UF_{token} is confluent modulo an equivalence $\approx$,
defined as follows; letters U and T refer to sets of **union** and of tree constraints.

– $T \cup U \cup \{\text{token}\} \approx T' \cup U \cup \{\text{token}\}$ whenever $sets(T) = sets(T')$.
– $T \cup U \cup \{\text{link}(A_1, A_2)\} \cup F_1 \cup F_2 \approx T' \cup U \cup \{\text{link}(A_1', A_2')\} \cup F_1' \cup F_2'$ whenever
  $sets(T) = sets(T')$ and for $i = 1, 2$, that
  - if A_i is a constant and (by I) $F_i = \emptyset$, then A_i' is a constant, $set(A_i, T) = set(A_i', T')$ and $F_i' = \emptyset$
  - if A_i is a variable and $F_i = \{\text{find}(a_i, A_i)\}$ for some constant a_i, then $F_i' = \{\text{find}(a_i', A_i')\}$ and $set(a_i, T) = set(a_i', T')$,
  - if A_i is a variable, $F_i = \{(a_i = A_i)\}$ for some constant a_i with $\text{root}(a_i) \in T$ then $F_i' = (a_i' = A_i')\}$, $\text{root}(a_i') \in T'$ and $set(a_i, T) = set(a_i', T')$.

There are no critical α_2- and α_3-patterns. The α_1-patterns (critical pairs) of
UF_{token} are those of UF_0 and a new one, formed by an overlap of the **union**
rule with itself as shown below. We reuse the analysis of [16] who identified all
critical pairs for UF_0; by Lemma 3, we consider only those pairs, they identified
as non-joinable.

In [16], eight non-joinable critical pairs are identified; the first one ("the unavoid-
able" pair) concerns issue (ii). Its ancestor state $\{\text{find}(B, A), \text{root}(B), \text{root}(C),$
$\text{link}(C, B)\}$, is excluded by I: any corner covered, B and C must be ground, thus
also the **link** constraint, which according to I excludes a **find** constraint. This can
be traced to the effect of our **token** constraint, that forces any **union** to complete
its detailed steps, before a next **union** may be entered. However, issue (ii) pops up
in the new α_1-pattern for UF_{token}, $y \hookleftarrow x \mapsto y'$ where:

$$x = \{\text{token}, \text{union}(A, B), \text{union}(A', B')\}$$
$$y = \{\text{find}(A, X), \text{find}(B, Y), \text{link}(X, Y), \text{union}(A', B')\}$$
$$y' = \{\text{find}(A', X'), \text{find}(B', Y'), \text{link}(X', Y'), \text{union}(A, B)\}$$

To show joinability of any corner covered by this pattern means to find z, z' such that $y \overset{*}{\mapsto} z \approx z' \overset{*}{\hookleftarrow} y'$. This can be done by, from y, first executing all remaining steps related to $\text{union}(A, B)$ and then the steps relating to $\text{union}(A', B')$ to reach a state $z = T \cup U \cup \{\text{token}\}$. In a similar way, we construct $z' = T' \cup U \cup \{\text{token}\}$, starting with the steps relating to $\text{union}(A', B')$ followed by those of $\text{union}(A, B)$. It can be proved by induction that $sets(T) = sets(T')$, thus $z \approx z'$.

Next, [16] identifies three critical pairs, that imply inconsistent tree constraints. The authors argue informally that these pairs will never occur for a query with consistent tree constraints. As noticed by [6], this can be formalized using an invariant. The last four pairs of [16] relate to issue (i) above; [16] argues these to be avoidable, referring to procedural properties of implemented CHR systems (which is a bit unusual in a context concerning confluence). In [6], those pairs are avoided by restricting allowed queries to include only a single union constraint; we can allow any number of those, but avoid the problem due to the control patterns imposed by the token constraints and formalized in our invariant I.

This finishes the argument that UF_{token} satisfies the α property, and by inspection of the possible derivation steps one by one (for each rule and for the "$=$" constraint), it can be seen that there are no critical β corners. Thus UF_{token} is locally confluent modulo $\approx$, and since tree consistency implies termination, it follows that UF_{token} is confluent modulo $\approx$.

8   Discussion and Detailed Comments on Related Work

We already commented on the foundational work on confluence for CHR by [2], who, with reference to Newman's lemma, devised a method to prove confluence by inspecting a finite number of critical pairs. This formed also the foundation of automatic confluence checkers [2,5,12] (with no invariant and no equivalence).

The addition of an invariant I in the specification of confluence problems for CHR was suggested by [6]. The authors considered a construction similar to our α_1-corners and critical α_1-patterns. They noted that critical α_1-patterns usually do not satisfy the invariant, so they based their approach on defining a collection of corners based on I-states as minimal extensions of such patterns. Local confluence, then, follows from joinability of this collection of minimally extended states. However, there are often infinitely many such minimally extended states; this happens even for a natural invariant such as groundness when infinitely many terms are possible, as is the case in Prolog based CHR versions. We can use this construction (in cases where it is finite!) to further cluster the space of our critical corners, but our examples worked quite well without this.

Of other work concerned with confluence for CHR, we may mention [10,15] which considered confluence for non-terminating CHR programs. We may also refer to [18] that gives an overview of CHR related research until 2010, including confluence.

9   Conclusion and Future Work

We have introduced confluence modulo equivalence for CHR, which allows a much larger class of programs to be characterized as confluent in a natural way, thus increasing the practical relevance of confluence for CHR.

We demonstrated the power of the framework by showing confluence modulo equivalence for programs that use a redundant data representation (the set-as-lists and Union-Find programs) and a dynamic programming algorithm (the Viterbi program); all these are out of scope of previous confluence notions for CHR. With the new operational semantics, we can also handle extra-logical and incomplete built-in predicates, and the notational improvements obtained by this semantics may also promote new applications of and research on confluence.

As a first steps towards semi- or fully automatic proof methods, it is important to notice that classical joinability of a critical pair – as can be decided by existing confluence checkers such as [12] – provide a sufficient condition for joinability modulo any equivalence. Thus only classically non-joinable pairs – in our terminology α_1 patterns – need to be examined in more details involving the relevant equivalence; however, in some cases there may also be critical α_2, α_3 and β patterns that need to be considered.

While the set of critical α_1-patterns can be characterized by a finite collection of patterns consisting of mini-states tied together by derivations, the same things is not possible for the other sorts of critical patterns. In our examples, we used semi-formal patterns, whose meta-variables or placeholders are covered by side-conditions such as "x is a variable" and "a is a constant". However, this must be formalized in order to approach automatic or semi-automatic methods. A formal and machine readable language for specifying invariants and equivalences will also be an advantage in this respect.

References

1. Abdennadher, S.: Operational semantics and confluence of constraint propagation rules. In: Smolka, G. (ed.) CP 1997. LNCS, vol. 1330, pp. 252–266. Springer, Heidelberg (1997)
2. Abdennadher, S., Frühwirth, T.W., Meuss, H.: On confluence ofconstraint handling rules. In: Freuder, E.C. (ed.) CP 1996. LNCS, vol. 1118, pp. 1–15. Springer, Heidelberg (1996)
3. Aho, A.V., Sethi, R., Ullman, J.D.: Code optimization and finite Church-Rosser systems. In: Rustin, R. (ed.) Design and Optimization of Compilers, pp. 89–106. Prentice-Hall, Englewood Cliffs (1972)
4. Christiansen, H., Have, C.T., Lassen, O.T., Petit, M.: The Viterbi algorithm expressed in Constraint Handling Rules. In: Van Weert, P., De Koninck, L. (eds.) Proceedings of the 7th International Workshop on Constraint Handling Rules. Report CW 588, pp. 17–24. Katholieke Universiteit Leuven, Belgium (2010)
5. Duck, G.J., Stuckey, P.J., García de la Banda, M., Holzbaur, C.: The refined operational semantics of constraint handling rules. In: Bart, D., Vladimir, L. (eds.) ICLP 2004. LNCS, vol. 3132, pp. 90–104. Springer, Heidelberg (2004)

6. Duck, G.J., Stuckey, P.J., Sulzmann, M.: Observable confluence for constraint handling rules. In: Dahl, V., Niemelä, I. (eds.) ICLP 2007. LNCS, vol. 4670, pp. 224–239. Springer, Heidelberg (2007)
7. Durbin, R., Eddy, S., Krogh, A., Mitchison, G.: Biological Sequence Analysis: Probabilistic Models of Proteins and Nucleic Acids. Cambridge University Press, Cambridge (1999)
8. Frühwirth, T.W.: Theory and practice of Constraint Handling Rules. J. Logic Progr. 37(1–3), 95–138 (1998)
9. Frühwirth, T.W.: Constraint Handling Rules. Cambridge University Press, Cambridge (2009)
10. Haemmerlé, R.: Diagrammatic confluence for Constraint Handling Rules. TPLP 12(4–5), 737–753 (2012)
11. Huet, G.P.: Confluent reductions: abstract properties and applications to term rewriting systems: abstract properties and applications to term rewriting systems. J. ACM 27(4), 797–821 (1980)
12. Langbein, J., Raiser, F., Frühwirth, T.W.: A state equivalence and confluence checker for CHRs. In: Weert, P.V., Koninck, L.D. (eds.) Proceedings of the 7th International Workshop on Constraint Handling Rules. Report CW 588, pp. 1–8. Katholieke Universiteit Leuven, Belgium (2010)
13. Newman, M.: On theories with a combinatorial definition of "equivalence". Ann. Math. 43(2), 223–243 (1942)
14. Raiser, F., Betz, H., Frühwirth, T.W.: Equivalence of CHR states revisited. In: Raiser, F., Sneyers, J. (eds.) Proceedings of the 6th International Workshop on Constraint Handling Rules, Report CW 555, pp. 33–48. Katholieke Universiteit Leuven, Belgium (2009)
15. Raiser, F., Tacchella, P.: On confluence of non-terminating CHR programs. In: Djelloul, K., Duck, G.J., Sulzmann, M. (eds.) CHR 2007, pp. 63–76. Porto, Portugal (2007)
16. Schrijvers, T., Frühwirth, T.W.: Analysing the CHR implementation of union-find. In: Wolf, A., Frühwirth, T.W., Meister, M. (eds.) W(C)LP. Ulmer Informatik-Berichte, vol. 2005-01, pp. 135–146. Universität Ulm, Ulm (2005)
17. Sethi, R.: Testing for the Church-Rosser property. J. ACM 21(4), 671–679 (1974)
18. Sneyers, J., Weert, P.V., Schrijvers, T., Koninck, L.D.: As time goes by: Constraint Handling Rules. TPLP 10(1), 1–47 (2010)
19. Tarjan, R.E., van Leeuwen, J.: Worst-case analysis of set union algorithms. J. ACM 31(2), 245–281 (1984)
20. Viterbi, A.J.: Error bounds for convolutional codes and an asymptotically optimum decoding algorithm. IEEE Trans. Inform. Theory 13, 260–269 (1967)

Exhaustive Execution of CHR Through Source-to-Source Transformation

Ahmed Elsawy[1]($\boxtimes$), Amira Zaki[1,2], and Slim Abdennadher[1]

[1] German University in Cairo, Cairo, Egypt
{ahmed.el-sawy,amira.zaki,slim.abdennadher}@guc.edu.eg
[2] Ulm University, Ulm, Germany

Abstract. Constraint Handling Rules (CHR) is a committed-choice rule-based programming language. Rules rewrite a global multi-set of constraints to another. Overlapping sets of constraints within the rules and the order of constraints within rules and queries entail different derivation paths. In this work, a novel operational strategy is proposed which enables a high-level form of execution control that empowers a comprehensive and customizable execution strategy. It allows full space exploration for any CHR program, thus finding all possible results to a query which is interesting for many non-confluent programs. The proposed transformation is performed as a source-to-source transformation from any CHR program to one utilizing disjunction to force an exhaustive explorative execution strategy. The work is complemented by formal arguments to prove the correctness and completeness of the transformation.

Keywords: Constraint Handling Rules · Execution flow control · Exhaustive execution · Search space exploration · Source-to-source transformation

1 Introduction

Constraint Handling Rules (CHR) is a rule-based programming language based on a set of multi-headed guarded rewrite rules [9]. The rules operate on a global multi-set of constraints to rewrite them from one multi-set of constraints to another. The rules are applied exhaustively until a final state is attained; where no more rules are applicable. The execution of a program on an initial query is known as a derivation. Derivations follow a committed-choice manner, where a chosen rule cannot be retracted. Rules could operate on overlapping sets of constraints. The order of constraints within the rules, the order of constraints within the input query, and the actual order of program rules chosen for execution are factors defined by the operational semantics that determine the derivation path.

For a given CHR program and for any initial state, if all derivations from that state result in the same final state then the program is confluent. On the other hand, non-confluent programs produce different final states for a given query depending on the chosen derivation path. Non-confluent programs are common

© Springer International Publishing Switzerland 2015
M. Proietti and H. Seki (Eds.): LOPSTR 2014, LNCS 8981, pp. 59–73, 2015.
DOI: 10.1007/978-3-319-17822-6_4

specially with agent-oriented programming. The Blocks World example [8,12] can be modelled in CHR to describe the behaviour of an agent in a world of objects.

Example 1. **Blocks World**

```
rule-1 @ get(X), empty <=> hold(X).
rule-2 @ get(X), hold(Y) <=> hold(X), clear(Y).
```

The constraint `hold(X)` represents that the agent holds an object `X`. The constraint `empty` represents that the agent holds nothing, the constraint `clear(Y)` denotes that the object `Y` is not held, and the constraint `get(X)` represents the agent's action to get an object `X`. The program consists of two simplification rules having overlapping heads, both rules have a `get(X)` constraint on the left-hand side. The simplification rules rewrite matched left-hand side constraints to the right-hand side constraints. With the abstract operational semantics and an initial query '`empty, get(box), get(cup)`', two disjoint final states can be reached:

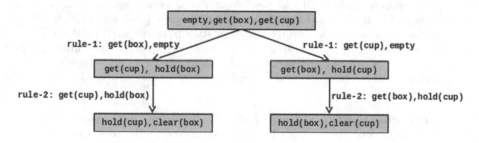

Fig. 1. Derivation tree for Example 1 with query: `empty,get(box),get(cup)`.

The CHR implementation of the K.U. Leuven system [10], which follows the refined operational semantics but with additional fixed orders, entails that only one final state is reached '`hold(cup), clear(box)`'. However, changing the order of program rules, the order of constraints within an input query or within the rules, yields different final states for a non-confluent program. For example, executing the program with the reordered query: '`empty, get(cup), get(box)`' entails the final state '`hold(box), clear(cup)`'.

In order to change the program execution flow, source-to-source transformations are used to facilitate a straightforward implementation on top of existing CHR implementations whilst exploiting the optimizations of current CHR compilers [14]. For example, such transformations can be used to transform CHR with user-defined priorities into regular CHR [4,6] and to a tree-based semantics with a breadth-first search [5].

In this work, we capture the essence of the tree-based semantics [5] by introducing branching derivations through identifying overlapping rule-heads.

The tree-based semantics makes use of the disjunctive branching points within a program to generate CHR constraints to encode nodes and edges.

The presented approach is based on transforming the input program to a program that utilizes Constraint Handling Rules with Disjunction ($CHR^\vee$); which is an extension of CHR [3]. $CHR^\vee$ facilitates relaxing the committed-choice behaviour of CHR by allowing disjunctive bodies and hence backtracking over alternatives. Application of a transition rule generates a branching derivation which can be utilized to overcome the committed-choice execution of CHR.

This work introduces an exhaustive execution strategy as a source-to-source transformation of the CHR program rules. It finds all possible results to a query for non-confluent programs. This execution platform is extensible to adopt different search strategies on a program's search space. For implementing constraint solvers, this could definitely enable optimized search techniques [5]. The angelic semantics of CHR [13] is a preliminary similar work aimed to explore all possible execution choices using derivation nets. However, no implementation nor definition of an operational semantics was provided. The transformation proposed here aims to reach an angelic semantics however by generating search trees similar to the tree-based semantics without explicitly encoding edges and nodes.

In this work, we establish a strong correspondence of the proposed transformation to the operational semantics of CHR to prove the soundness and completeness of the proposed transformation.

Moreover, as a practical application of this work, the exhaustive execution by transformation is used as a top-level execution strategy required for the inverse execution of CHR [15]. The inverse execution of CHR presents a simple inversion transformation, that changes a CHR program to its inverse one by interchanging rule sides. For the blocks world example mentioned earlier, that would mean a program which given an output state 'hold(cup), clear(box)' generates all possible input states that lead to it. With a normal CHR execution implementation only one input is uncovered; thus a different exhaustive execution strategy is required. A primitive unformalized execution transformation was proposed in [15]. In this work, we extend the work of [15] by using the proposed transformation for executing inverse CHR programs.

The paper proceeds by providing some background information in Sect. 2, followed by the proposed transformation in Sect. 3. The proof of the soundness and completeness of the transformation is given in Sect. 4. Then an application of the proposed transformation is given in Sect. 5, followed by some concluding remarks in Sect. 6.

2 Constraint Handling Rules

2.1 Syntax

Constraint Handling Rules (CHR) [9,10] consist of guarded rewrite rules that perform conditional transformation of multi-sets of constraints, known as a constraint store. There are three types of rules: simplification, propagation and simpagation rules. All rules have an optional unique identifier *rule-id* separated from the rule body by @. For all rules, H_k and/or H_r form the rule head. H_k and H_r are sequences of user-defined CHR constraints that are kept and removed from the constraint store respectively. *Guard* is the rule guard consisting of a sequence of built-in constraints, and B is the rule body comprising of CHR and built-in constraints. The generalized simpagation rules are of the form:

$$rule\text{-}id \text{ @ } H_k \setminus H_r \Leftrightarrow Guard \mid B$$

The two other types of rules are special cases of the generalized simpagation rule. Simplification rules have no kept head constraints H_k while propagation rules do not remove any head constraints. These rules are of the form:

Simplification rule: $rule\text{-}id \text{ @ } H_r \Leftrightarrow Guard \mid B$
Propagation rule: $rule\text{-}id \text{ @ } H_k \Rightarrow Guard \mid B$

Propagation and simpagation rules can be represented as simplification rules by adding kept constraints to the rule bodies; in its abstract sense this introduces the notion of non-termination. However this issue can be avoided by using a refined operational semantics which uses a token store as a history for propagated constraints [1]. For simplicity, in this paper the focus will be only with simplification rules, however the work can be extended in a straightforward manner by adding the notion of a propagation history.

Constraint Handling Rules with Disjunction ($CHR^\vee$) [3] is an extension which allows disjunctive rule bodies and enables a backtrack search over alternatives. A simplification rule with two possible bodies ($B_1 \vee B_2$) is of the form:

$$rule V\text{-}id \text{ @ } H_r \Leftrightarrow Guard \mid B_1 \vee B_2$$

2.2   Operational Semantics

An operational semantics of CHR describes the behaviour of a program in terms of a state transition system which models the execution of a program. An initial state is an arbitrary one and a final state is a terminal one where no further transitions are possible.

The abstract operational semantics ω_{va} of CHR is formulated as a state transition system, where states are goals comprising a conjunction of built-in and CHR constraints, and a transition defines an execution state and its subsequent execution state. An initial state is an arbitrary one and a final state is a terminal one where no further transitions are possible. The abstract operational semantics includes one transition rule shown below, where P is a CHR program with rules r and CT is the constraint theory for the built-in constraints.

Apply
$$(H_k \wedge H_r \wedge C) \mapsto^r_{apply} (H_k \wedge Guard \wedge B \wedge C)$$
if there is an instance of a rule r in P with new local variables $\bar{x}$ such that:
$$r \text{ @ } H_k \setminus H_r \Leftrightarrow Guard \mid B \text{ and } CT \models \forall(C \rightarrow \exists \bar{x} \; Guard)$$

The extended operational semantics for $CHR^\vee$ operates on a disjunction of CHR states known as a configuration: $s_1 \vee s_2 \vee \cdots \vee s_n$. The semantics includes two transitions; one containing the original apply transition which is applicable to a single state. The second transition is a split transition which is applicable to any goal containing a disjunction. It leads to a branching derivation entailing two states, where each state can be processed independently. The second split transition required by $CHR^{\vee}$ is shown below.

Split

$$((H_1 \vee H_2) \wedge C) \vee S \mapsto_{split} (H_1 \wedge C) \vee (H_2 \wedge C) \vee S$$

Such a derivation can be visualized as a tree; consisting of a set of nodes and a set of directed edges connecting these nodes [11]. The root node corresponds to the initial state or goal. An intermediate node represents a midway non-final state within a derivation. A leaf node represents a successful or failed final state. Edges between nodes denote the applied transitions. Due to the clarity offered by search trees for such derivations, in this work all examples will be depicted using trees.

The refined operational semantics ω_r is a theoretical operational semantics that makes execution considerably more deterministic by establishing an order for goal constraints [7]. It is more closely related to actual implementations of CHR. Due to the limited space of this paper, we refer to the ω_r as defined in [7].

3   Transformation

In this section, we describe how to transform a CHR program P to a CHR program P^T, such that execution of program P^T under the refined operational semantics ω_r produces all possible final results. The transformation is divided into three sets of rules:

Definition 1. *Modified Source Rules. For every rule* $r @ H \Leftrightarrow Guard \mid B$ *in program P with $H = c_1(X_{11}, ..., X_{1n_1})...c_l(X_{l1}, ..., X_{ln_l})$, a modified source rule r_t is added to the transformed program, defined as follows:*

Modified Source Rules

$r_t \quad @ \quad depth(Z), c_1{}^T(X_{11}, ..., X_{1n_1}, y_1, Z), ..., c_l{}^T(X_{l1}, ..., X_{ln_l}, y_l, Z)$
$$\Leftrightarrow Guard \mid B, depth(Z+1)$$

As a result, every modified source rule r_t in P^T corresponds to rule r in P. Every constraint $c(X_1, ..., X_n)$ in the head of rule r is transformed to constraint $c^T(X_1, ..., X_n, y, Z)$ in rule r_t. The argument y represents the yth occurrence of constraint c/n. Let m be the number of occurrences of c/n in the source program, in the transformed program the first occurrence of $c(X_1, ..., X_n)$ is transformed to $c^T(X_1, ..., X_n, 1, Z)$, the second occurrence of $c(X_1, ..., X_n)$ is transformed to $c^T(X_1, ..., X_n, 2, Z)$, and the last occurrence of $c(X_1, ..., X_n)$ is transformed to $c^T(X_1, ..., X_n, m, Z)$. The argument Z and constraint $depth/1$ are explained in the **Assignment Rules**.

Definition 2. *Assignment Rules. For every constraint $c(X_1, ..., X_n)$ that appears in a rule head in program P, an assignment rule is added to the transformed program, defined as follows:*

Assignment Rules

$$assign_c \quad @ \quad depth(Z) \setminus c(X_1, ..., X_n)$$
$$\Leftrightarrow \; c^T(X_1, ..., X_n, 0, Z) \vee ... \vee c^T(X_1, ..., X_n, m, Z)$$

An assignment rule simplifies constraint $c(X_1, ..., X_n)$ to constraint $c^T(X_1, ..., X_n, Y, Z)$. The argument Y is used to have at most one possible matching between the transformed constraint and the constraints that appear in the heads of the **Modified Source Rules**. The domain of Y is from 0 to m, in which m is the number of occurrences of constraint c/n in the rule heads of program P. The argument 0 is used to prevent an active constraint $c^T(X_1, ..., X_n, 0, Z)$ from being matched during its current state in the tree, thus allowing it to be matched later in the derivation tree. Constraint $depth/1$ represents the depth of a state in the derivation tree. The depth is increased when a modified source rule is fired.

Definition 3. Reset Rules. *For every constraint $c(X_1, ..., X_n)$ that appears in a rule head in program P, a reset rule is added to the transformed program, defined as follows:*

Reset Rules

$$reset_c \quad @ \quad depth(Z) \setminus c^T(X_1, ..., X_n, 0, Z')$$
$$\Leftrightarrow \; Z' < Z \mid c(X_1, ..., X_n)$$

The purpose of reset rules is to reset the unmatched constraints $c^T(X_1, ..., X_n, 0, Z')$ if a newly state in the tree is derived. The constraint $c^T(X_1, ..., X_n, 0, Z')$ will be reset to constraint $c(X_1, ..., X_n)$, thus allowing it to be re-assigned by the **Assignment Rules**.

Finally, the rule *start* $\Leftrightarrow$ *depth*(0) is added to the transformed program to trigger the constraints needed for execution. Moreover, a constraint *start* is added at the end of the initial query.

Example 1 (continued). The transformation of program P:

```
start <=> depth(0).
%%%%%%%%%%%%%%%%%%% Reset Rules %%%%%%%%%%%%%%%%%%%%%%
reset_empty @ depth(Z) \ empty_t(0,Z1) <=> Z1 < Z | empty.
reset_get @ depth(Z) \ get_t(X,0,Z1) <=> Z1 < Z | get(X).
reset_hold @ depth(Z) \ hold_t(X,0,Z1) <=> Z1 < Z | hold(X).
%%%%%%%%%%%%%%%%%%%% Assignment Rules %%%%%%%%%%%%%%%%%%%%%%
assign_empty @ depth(Z) \ empty
                  <=> empty_t(0,Z) ; empty_t(1,Z).
assign_get @ depth(Z) \ get(X)
                  <=> get_t(X,0,Z) ; get_t(X,1,Z) ; get_t(X,2,Z).
assign_hold @ depth(Z) \ hold(X)
                  <=> hold_t(X,0,Z) ; hold_t(X,1,Z).
%%%%%%%%%%%%%%%%%%%% Modified Source Rules %%%%%%%%%%%%%%%%%%%%%%
rule-1_t @ depth(Z), get_t(X,1,Z), empty_t(1,Z)
                  <=> hold(X), Z1 is Z+1, depth(Z1).
rule-2_t @ depth(Z), get_t(X,2,Z), hold_t(Y,1,Z)
                  <=> hold(X), clear(Y), Z1 is Z + 1, depth(Z1).
```

Figure 2 shows the derivation tree of the transformed program when executed by the query 'empty,get(box),get(cup),start'. The redundant states in the tree are represented by grey arrows. States are duplicated in the tree by the Assignment Rules, because these rules create variants to the same state. For example, the two states 'depth(0),empty_t(1,0),get_t(box,0,0),get_t (cup,0,0)' and 'depth(0),empty_t(0,0),get_t(box,0,0),get_t(cup,0,0)' are variants of the same state.

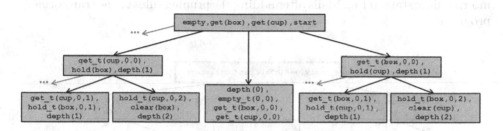

Fig. 2. Derivation tree for Example 1 after transformation.

Every state in the derivation tree of the source program in Fig. 1, has a corresponding result (final state) in the transformed program. Thus, we extend the transformation with pruning rules to remove every result in the transformed program that corresponds to an intermediate (non-final) state in the source program. To map a state in the transformed program to a state in the source program, every constraint $c^T(X_1, ..., X_n, \_, \_)$ is mapped to constraint $c(X_1, ..., X_n)$ and constraint $depth/1$ is removed. This mapping can be done by adding bridge rules [3] to the transformed program to relate constraints between the source and transformed programs. These rules are not described here due to their clarity.

Definition 4. Intermediate States in P. *Let S be a state in program P, S is intermediate if there exists a derivation $S \mapsto S'$.*

Definition 5. Pruning Rules. *A result S^T in program P^T has to be removed if it corresponds to an intermediate state in P. To remove S^T, for every rule r @ H $\Leftrightarrow$ Guard | B in the source program P, a corresponding pruning-r rule is added to the transformed program P^T, defined as follows:*

Pruning Rules
$$pruning\text{-}r \; @ \; end, H^T \; \Leftrightarrow \; Guard \; | \; fail$$

Every constraint $c(X_1, ..., X_n)$ in H is transformed to constraint $c^T(X_1, ..., X_n, \_, \_)$ in H^T. Moreover, the rule $start \; \Leftrightarrow \; depth(0)$ is transformed to $start \; \Leftrightarrow \; depth(0), end$, therefore constraint $end/0$ will become active only when all Reset Rules, Assignment Rules, and Modified Source Rules cannot be triggered anymore. If a Pruning Rule can be fired on state S^T and the *Guard* holds, this implies that rule r can be applied on state S that corresponds to state S^T; state S^T is removed by having a derivation $S^T \mapsto fail$.

Example 1 (transformation with pruning rules)
`start <=> depth(0).` will be modified to `start <=> depth(0), end`.
Moreover the following rules will be added to the transformed program:

`pruning-rule-1 @ end, get_t(X,_,_), empty_t(_,_) <=> fail.`
`pruning-rule-2 @ end, get_t(X,_,_), hold_t(Y,_,_) <=> fail.`

The derivation tree for the query 'empty, get(box), get(cup), start' is depicted in Fig. 3. As shown in the figure, every result that corresponds to an intermediate state in Fig. 1 fails after adding the pruning rules to the transformed program.

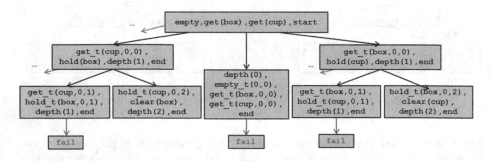

Fig. 3. Derivation tree for Example after pruning.

4   Soundness and Completeness of the Transformation

In this section, soundness and completeness of the transformation are proved. The abstract operational semantics will be used for the derivations throughout the proofs. However for the completeness proof, derivations of the transformed program are based on the refined operational semantics ω_r, because the current CHR compiler is based on this semantics. Therefore it proves that the transformed program produces all possible results when executed under the semantics of the CHR compiler.

Definition 6. *Equivalence of two CHR states.* *Let S be a CHR state derived from the initial query $G \mapsto {}^*S$ in program P, and $W = S_1 \vee ... \vee S_n$ be a configuration derived from the initial query $(G \wedge depth(0)) \mapsto {}^*W$ in program P^T. CHR state S is equivalent to CHR state S' in $\{S_1, ..., S_n\}$ according to the following inductive definition. $S \equiv S'$ iff*
base case:
  – *$(S = G_{builtin}) \wedge (S' = (G_{builtin} \wedge depth(Z)))$, in which $G_{builtin}$ is a conjunction of built-in constraints.*
inductive step:
  – *$(S = (c(X_1, ..., X_n) \wedge S_r)) \wedge (S' = (c^T(X_1, ..., X_n, Y, Z) \wedge S'_r)) \wedge (S_r \equiv S'_r)$*
  – *$(S = (c(X_1, ..., X_n) \wedge S_r)) \wedge (S' = (c(X_1, ..., X_n) \wedge S'_r)) \wedge (S_r \equiv S'_r)$*

Definition 7. Soundness of a $CHR^\vee$ configuration. *Given an initial query G and a configuration $W = S_1 \vee ... \vee S_n$ derived from $(G \wedge depth(0)) \mapsto {}^*W$ in program P^T. W is sound if and only if for every CHR state S_i in $\{S_1, ..., S_n\}$ there exists a derivation in program P, where $G \mapsto {}^*S$ and $S \equiv S_i$.*

Theorem 1. *Given a CHR Program P, its corresponding transformed program P^T, and an initial query G. P^T is sound with respect to P, if and only if every derived configuration W in program P^T from the query $(G \wedge depth(0))$ is sound.*

Proof.
Base Case. $S = G$, $W = (G \wedge depth(0))$, in which G is the initial query. Since $S \equiv W$, then the initial configuration W is sound.

Induction Hypothesis. Let $W = S_1 \vee ... \vee S_k$ be a configuration derived from $(G \wedge depth(0)) \mapsto {}^*W$, assume that W is sound.

Induction Step. We prove that if $W \mapsto W'$, then W' is sound.
Without loss of generality, we assume that the fired rule is applied on the first CHR state S_1 in W.

Case 1. A $reset_c$ rule is applicable on W and $CT \models G_{builtin} \rightarrow (Z' < Z)$

$W = ((depth(Z) \wedge c^T(X_1, ..., X_n, 0, Z') \wedge C) \vee (S_2 \vee ... \vee S_k))$	Because a $reset_c$ rule is applicable on W
$W' = (depth(Z) \wedge c(X_1, ..., X_n) \wedge C) \vee (S_2 \vee ... \vee S_k)$	Firing a $reset_c$ rule on W, $W \mapsto {}^{reset_c}_{apply} W'$
$S \equiv (depth(Z) \wedge c^T(X_1, ..., X_n, 0, Z') \wedge C)$	According to the hypothesis there exists a derivation $G \mapsto {}^*S$
$S = (c(X_1, ..., X_n) \wedge C'')$ and $C'' \equiv (C \wedge depth(Z))$	According to the definition of equivalence
$(depth(Z) \wedge c(X_1, ..., X_n) \wedge C)$ is sound	Because $S \equiv (depth(Z) \wedge c(X_1, ..., X_n) \wedge C)$
Therefore W' is sound	

Case 2. An $assign_c$ rule is applicable on W

$W = ((c(X_1, ..., X_n) \wedge depth(Z) \wedge C) \vee (S_2 \vee ... \vee S_k))$	Because an $assign_c$ rule is applicable on W
$W' = (((c^T(X_1, ..., X_n, 0, Z) \vee ... \vee c^T(X_1, ..., X_n, l, Z)) \wedge depth(Z) \wedge C) \vee (S_2 \vee ... \vee S_k))$	Firing an $assign_c$ rule on W, $W \mapsto {}^{assign_c}_{apply} W'$
$W'' = ((c^T(X_1, ..., X_n, 0, Z) \wedge depth(Z) \wedge C) \vee ... \vee (c^T(X_1, ..., X_n, l, Z) \wedge depth(Z) \wedge C) \vee (S_2 \vee ... \vee S_k))$	A split is applied on $W' \mapsto {}_{split} W''$
$S \equiv (depth(Z) \wedge c(X_1, ..., X_n) \wedge C)$	According to the hypothesis there exists a derivation $G \mapsto {}^*S$
$S = (c(X_1, ..., X_n) \wedge C')$ and $C' \equiv (C \wedge depth(Z))$	According to the definition of equivalence
Therefore W'' is sound	Because $\forall_{0 \leq i \leq l}((c^T(X_1, ..., X_n, i, Z) \wedge depth(Z) \wedge C) \equiv S)$

Case 3. A Modified rule r_t is applicable on W and $CT \models G_{builtin} \rightarrow Guard$

$W = ((depth(Z) \wedge c_1^T(X_{11}, ..., X_{1n_1}, Y_1, Z)$ $\wedge ... \wedge c_l^T(X_{l1}, ..., X_{ln_l}, Y_l, Z) \wedge C) \vee (S_2 \vee$ $... \vee S_k))$	Because a modified rule r_t is applicable on W
$S \equiv (depth(Z) \wedge c_1^T(X_{11}, ..., X_{1n_1}, Y_1, Z)$ $\wedge ... \wedge c_l^T(X_{l1}, ..., X_{ln_l}, Y_l, Z) \wedge C)$	According to the hypothesis there exists a derivation $G \mapsto {}^*S$
$S = (c_1(X_{11}, ..., X_{1n_1}) \wedge ... \wedge$ $c_l(X_{l1}, ..., X_{ln_l}) \wedge C')$ and $C' \equiv (C \wedge$ $depth(Z))$	According to the definition of equivalence
$W' = ((B \wedge depth(Z + 1) \wedge Guard \wedge C) \vee$ $(S_2 \vee ... \vee S_k))$	Firing rule r_t on W, $W \mapsto {}^{r_t}_{apply} W'$
$S' = (Guard \wedge C' \wedge B)$	Firing rule r on S, $S \mapsto {}^r_{apply} S'$
Therefore W' is sound	Because $(B \wedge depth(Z+1) \wedge Guard \wedge$ $C) \equiv S'$

$\square$

For simplicity, we define a state in ω_r as the tuple $\langle A, S, B \rangle$ [7]. The stack A is a sequence of built-in constraints and CHR constraints. The store S is a set of CHR constraints. The built-in store B is a conjunction of built-in constraints.

Definition 8. *Let S be a CHR state derived from the initial query $G \mapsto {}^*S$ in program P and $S^T = \langle A, C, B \rangle$ be a state derived from the initial state $\langle G{+}{+}[depth(0)], \phi, true \rangle$ in program P^T, in which $++$ is the append operator defined for sequences. Let E be the property defined on S and S^T, $E(S, S^T)$ holds iff $A = [depth(Z)]$, $depth(Z) \in C$, $(C \wedge B) \equiv S$, and $Z' < Z$ holds for all transformed constraints $c^T(X_1, ..., X_n, Y, Z')$ in C.*

Theorem 2. *Given a CHR Program P, its corresponding transformed program P^T, and an initial query G. P^T is complete if and only if for every CHR state S derived from the initial query $G \mapsto {}^*S$ in program P, there exists a state S^T derived from the initial state $\langle G{+}{+}[depth(0)], \phi, true \rangle$ in program P^T such that $\langle G{+}{+}[depth(0)], \phi, true \rangle \mapsto {}^*S^T$ and $E(S, S^T)$ holds.*

Proof.

Base Case. There exists a derivation to state S^T in P^T and $E(G, S^T)$ holds.

$\langle G{+}{+}[depth(0)], \phi, true \rangle$	G is the original query, $G = G_c \cup G_b$
$\mapsto {}^* \langle [depth(0)], G_c, G_b \rangle$	All CHR constraints in G will be activated, added to the CHR store, then dropped from the stack, and all built-in constraints will be solved and removed from the stack
$\mapsto {}_{activate}$ $S^T = \langle [depth(0)], \{depth(0)\} \cup G_c, G_b \rangle$	$depth(0)$ becomes the active constraint

Since $E(G, \langle [depth(0)], \{depth(0)\} \cup G_c, G_b \rangle)$ holds, then the base case holds.

Induction Hypothesis. Let S be a state in P derived from the initial query G and S^T be a state in P^T derived from the initial state $\langle G{+}{+}[depth(0)], \phi, true \rangle$, assume that $E(S, S^T)$ holds.

Induction Step. We prove that for a derivation $S \mapsto S'$, there exists a derivation $S^T \mapsto {}^* S_1{}^T$ and $E(S', S_1{}^T)$ holds. For simplicity of the proof, we define the following:

- Let $r @ H \Leftrightarrow Guard \mid B$ be a rule in program P that applies on $S = H \wedge H_{rest}$ such that $S = (H \wedge H_{rest}) \mapsto {}^r_{apply} S' = (H_{rest} \wedge B \wedge Guard)$.
- Let map be a function that maps every constraint in H to a partner constraint in S.
- Let $r_t @ depth(Z), H^T \Leftrightarrow Guard \mid B, depth(Z+1)$ be a modified source rule in program P^T, that corresponds to rule r in program P.
- Let $occurrence$ be a function that is applied on a transformed constraint c^T, such that $occurrence(c^T(X_1, ..., X_n, Y, Z)) = Y$.
- Let dep be a function that is applied on a transformed constraint c^T, such that $dep(c^T(X_1, ..., X_n, Y, Z)) = Z$.
- Let $assign$ be a function that maps a constraint c/n in S to an integer value according to the following:
  - if there exists a constraint d in H such that $map(d) = c$, then $assign(c) = occurrence(d^T(X_1, ..., X_m, Y, Z))$, in which d^T is the transformed constraint in H^T that corresponds to constraint d
  - otherwise, $assign(c) = 0$

$S^T = \langle [depth(Z)], \{depth(Z)\} \cup H_c \cup H_{c^T}, H_b \rangle$	According to the hypothesis $dep(c^T) < Z$ for all constraints in H_{c^T} and $(depth(Z) \wedge H_c \wedge H_{c^T} \wedge H_b) \equiv S$ holds.
$\mapsto^{reset_c}_{apply}$ $\langle [c]{+}{+}[depth(Z)], \{depth(Z)\} \cup H_c \cup \{c\} \cup H_{c^T} \backslash \{c^T\}, H_b \rangle$ $\mapsto^{assign_c}_{apply} \langle [c_1^T \vee ... \vee c_m^T]{+}{+}[depth(Z)], \{depth(Z)\} \cup H_c \cup H_{c^T} \backslash \{c^T\}, H_b \rangle$ $\mapsto_{split} \langle [c^T]{+}{+}[depth(Z)], \{depth(Z)\} \cup H_c \cup H_{c^T} \backslash \{c^T\}, H_b \rangle$	A $reset_c$ rule will be applied on constraint c^T in H_{c^T}, constraint c^T will be removed from the CHR store and its corresponding constraint c will be added to the top of the stack, constraint c will be active and will fire rule $assign_c$. Due to the existence of disjunction in the body of the $assign_c$ rule, a split will be applied. The path that will lead to a correct state derivation is the one with $occurrence(c^T) = assign(c)$. For simplicity the reset of disjuncts will be removed from the stack.
$\mapsto_{activate} \langle [c^T]{+}{+}[depth(Z)], \{depth(Z)\} \cup H_c \cup H_{c^T}, H_b \rangle$ $\mapsto_{drop} \langle [depth(Z)], \{depth(Z)\} \cup H_c \cup H_{c^T}, H_b \rangle$	c^T will become active, thus added to the CHR store. Then if it can not fire a rule it will be dropped from the stack
	The two rules $reset_c$ and $assign_c$ will be applied repeatedly till all constrains c^T with $dep(c^T) < Z$ in H_{c^T} are reset and then assigned, or till rule r_t can be fired. The second case is covered by the first one, thus only the first case is considered.

$\mapsto^{assign_c}_{apply} \langle [c_1^T \vee ... \vee c_m^T]++$ $[depth(Z)], depth(Z) \cup H_c \backslash \{c\} \cup$ $H_{c^T}, H_b \rangle$	Since r_t cannot be fired and all constraints c^T in H_{c^T} with $dep(c^T) < Z$ are reset, rule $assign_c$ will be applied on constraint c in H_c. The state derivation that will be chosen is the one with $occurrence(c^T) = assign(c)$ and the reset of disjuncts are removed.
$\mapsto_{split} \langle [c^T]++[depth(Z)],$ $depth(Z) \cup H_c \backslash \{c\} \cup H_{c^T}, H_b \rangle$	
$\mapsto_{drop} \langle [depth(Z)], \{depth(Z)\} \cup$ $H_c \backslash \{c\} \cup H_{c^T} \cup \{c^T\}, H_b \rangle$	c^T will be activated and added to the CHR store, if c^T can not fire rule r_t then it will be dropped from the stack. Constraint c is removed from H_c and its corresponding constraint c^T is added to H_{c^T}, thus $S \equiv depth(Z) \wedge (H_c \backslash \{c\}) \wedge (H_{c^T} \cup \{c^T\}) \wedge H_b$ holds.
$\mapsto^* \langle [c^T]++[depth(Z)],$ $\{depth(Z)\} \cup H'_c \cup H'_{c^T} \cup H^T, H_b \rangle$	The rule $assign_c$ will be fired repeatedly till constraint c^T at the top of the stack can fire rule r_t and $S = (H \wedge H_{rest}) \equiv depth(Z) \wedge H'_c \wedge H'_{c^T} \wedge H^T \wedge H_b$ and since H corresponds to H^T then $H_{rest} \equiv depth(Z) \wedge H'_c \wedge H'_{c^T} \wedge H_b$
$\mapsto^{r_t}_{apply} \langle B++[depth(Z+1)], H'_c \cup$ $H'_{c^T}, H_b \wedge Guard \rangle$	Applying rule $depth(Z), H^T \Leftrightarrow Guard \mid B, depth(Z+1)$
$\mapsto^* S_1^T = \langle [depth(Z+1)],$ $\{depth(Z+1)\} \cup H'_c \cup H'_{c^T} \cup B_c, H_b \wedge$ $Guard \wedge B_b \rangle$	All CHR constraints in B will be moved to the CHR store and all built-in constraints will be solved and added to the built-in store and $B = B_c \cup B_b$

Since $H_{rest} \equiv depth(Z+1) \wedge H'_c \wedge H'_{c^T} \wedge H_b$ and $B \equiv depth(Z+1) \wedge B_c \wedge B_b$, then $H_{rest} \wedge B \wedge Guard \equiv depth(Z+1) \wedge H'_c \wedge H'_{c^T} \wedge H_b \wedge B_c \wedge B_c \wedge Guard$, and since $dep(c^T) < Z+1$ holds for all constraints c^T in H'_{c^T}, then $E(S', S_1^T)$ holds.

□

There are two sources of non-determinism in the refined operational semantics in contrast with the actual implementation of CHR. The first source of non-determinism is the order of the constraints added to the top of the stack when the transition **Solve+Wake** is fired, the completeness proof assumes only that the constraints are added to the top of the stack, therefore, the order of the constraints does not matter. The second source of non-determinism is choosing the partner constraints from the CHR store when the transition **Apply** is fired, in the completeness proof **Assignment Rules** and **Reset Rules** are applied on arbitrary constraints c and c^T respectively. Moreover, the completeness proof shows that if a rule r is applicable on CHR state S, then there exists a path in the derivation tree of the transformed program, such that every constraint in the head of rule r_t has only one matching in the CHR store and the rest of the constraints in state S are assigned the occurrence 0, which implies that the only applicable **Modified Source Rule** is rule r_t and that there is only one way to match the constraints in the CHR store with the constraints in the head of rule r_t. Therefore, it is implied that the transformed program produces all final results when executed under the semantics of the CHR compiler, although the proof is based on the refined operational semantics.

5   Inverse Execution Using Proposed Transformation

The inverse execution of CHR presented in [15] presents a simple inversion transformation, that changes a forward CHR program to an inverse one by interchanging rule sides. Thus for any forward simplification rule $(H_r \Leftrightarrow Guard \mid B_c, B_b)$, where B_b are the built-in body constraints and B_c are the CHR body constraints and $B_b \cup B_c = B$, an inverse rule would be of the form:

$$inv - simpf@B_c \Leftrightarrow B_b, Guard \mid H_r$$

The inverse programs are then transformed using the proposed approach in order to achieve an exhaustive traversal of the inverse program that reaches all possible input states. However, since all intermediate states are also possible input states, thus pruning rules are removed from the transformed program. In some applications (such as for fault analysis) this step may not be necessary, as retaining pruning rules results in only uncovering the root output states.

Example 2. **Blocks World in Reverse** - If one wishes to run the blocks world (Example 1) in reverse, to retrace an agent's steps, then we require the inverse transformation of the program code. The forward run for the query `empty`, `get(box)`, `get(cup)` was previously depicted in Fig. 1. Using the simple inversion transformation [15], the inverse program becomes:

```
inv-rule-1 @ hold(X) <=> get(X), empty.
inv-rule-2 @ hold(X), clear(Y) <=> get(X), hold(Y).
```

A run of the inverse program for the previously generated result of 'hold(cup), clear(box)', would produce a single result of 'get(cup), empty, clear(box)', which is not the actual input that was used during the forward run. The inverse program is now transformed using the proposed exhaustive execution transformation (but without the pruning rules) to become as follows:

```
start <=> depth(0).
%%%%%%%%%%%%%%%%%%%%% Reset Rules %%%%%%%%%%%%%%%%%%%%%
reset_hold  @ depth(Z) \ hold_t(X,0,Z1) <=> Z1 < Z | hold(X).
reset_clear @ depth(Z) \ clear_t(X,0,Z1) <=> Z1 < Z | clear(X).
%%%%%%%%%%%%%%%%%%%%% Assignment Rules %%%%%%%%%%%%%%%%%%%%%
assign_hold  @ depth(Z) \ hold(X)
                <=> hold_t(X,0,Z) ; hold_t(X,1,Z) ; hold_t(X,2,Z).
assign_clear @ depth(Z) \ clear(X) <=> clear_t(X,0,Z) ; clear_t(X,1,Z).
%%%%%%%%%%%%%%%%%%%%% Modified Source Rules %%%%%%%%%%%%%%%%%%%%%
inv-rule-1_t @ depth(Z), hold_t(X,1,Z)
                <=> get(X), empty, Z1 is Z + 1, depth(Z1).
inv-rule-2_t @ depth(Z), hold_t(X,2,Z), clear_t(Y,1,Z)
                <=> get(X), hold(Y), Z1 is Z + 1, depth(Z1).
```

Running the inverse and transformed program yields the search tree depicted in Fig. 4 (note that the `start` constraint is added to trigger the exhaustive execution). The resultant states in the figure represent all possible inputs to the forward program and amongst them is the particular input used in Example 1. Any final state reached by the transformed program is a valid input to a forward run that generates the goal 'hold(cup), clear(box)'.

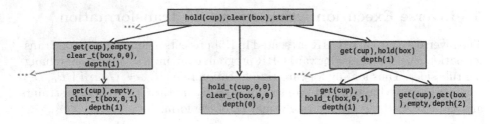

Fig. 4. Derivation tree for Example 2 with query: 'hold(cup), clear(box), start'

6 Conclusion

A source-to-source transformation was proposed, which expresses any CHR program as one utilizing disjunction, to force an exhaustive explorative execution strategy. It enables a high-level form of execution control that empowers a comprehensive execution while retaining expressive power. It is particularly useful for non-confluent programs with overlapping heads, as it enables finding all possible results to a query. The operational semantics of CHR features a "don't care non-determinism" where all choices will lead to a successful derivation, so we do not care which one is chosen. The proposed transformation changes it to a "don't know non-determinism", where some of the choices will lead to a successful search but we do not know which one beforehand. This change enables exploring the search space generated during any derivation.

The proposed transformation focuses on simplification rules, since propagation and simpagation rules can be represented as simplification rules with a token store history. Therefore, in the future the transformation can be extended to transform all CHR rule types.

The execution platform proposed makes extensive use of the disjunctive operator of $CHR^\vee$, and produces a comprehensive search tree for any query. In the future, this transformation can be easily extended with implementations of search strategies, such as the breadth-first transformation [5] due to the presence of disjunction in the rule bodies. The integration and customization of other search strategies can also be incorporated into the transformation.

Despite the pruning rules, the transformation still produces many redundant states (or nodes) which have already been visited in the search tree. Optimizations can be devised to eliminate nodes (or sub-trees) that have already been visited, this could be implemented by the use of a traversal history mechanism.

Furthermore, it would be interesting to add priorities to the branches of the search trees generated and hence enable a priority-based execution. For agent-based programs, this would allow introducing a kind of reasoning whilst performing the various actions.

References

1. Abdennadher, S.: Operational semantics and confluence of constraint propagation rules. In: Smolka, Gert (ed.) CP 1997. LNCS, vol. 1330, pp. 252–266. Springer, Heidelberg (1997)
2. Abdennadher, S., Frühwirth, T.: Integration and optimization of rule-based constraint solvers. In: Bruynooghe, M. (ed.) LOPSTR 2004. LNCS, vol. 3018, pp. 198–213. Springer, Heidelberg (2004)
3. Abdennadher, S., Schütz, H.: CHR$^\vee$: a flexible query language. In: Andreasen, T., Christiansen, H., Larsen, H.L. (eds.) FQAS 1998. LNCS (LNAI), vol. 1495, pp. 1–14. Springer, Heidelberg (1998)
4. Betz, H., Raiser, F., Frühwirth, T.: A complete and terminating execution model for constraint handling rules. In: Proceedings of 26th International Conference on Logic Programming, pp. 597–610 (2010)
5. Koninck, L. D., Schrijvers, T., Demoen, B.: Search strategies in CHR(Prolog). In: Leuven, K.U. (ed.) Proceedings of 3rd Workshop on Constraint Handling Rules, pp. 109–124. Technical report CW 452 (2006)
6. Koninck, L. D., Schrijvers, T., Demoen, B.: User-definable rule priorities for CHR. In: Proceedings of 9th International Conference on Principles and Practice of Declarative Programming, PPDP 2007, pp. 25–36. ACM (2007)
7. Duck, G.J., Stuckey, P.J., García de la Banda, M., Holzbaur, C.: The refined operational semantics of constraint handling rules. In: Demoen, B., Lifschitz, V. (eds.) ICLP 2004. LNCS, vol. 3132, pp. 90–104. Springer, Heidelberg (2004)
8. Duck, C.J., Stuckey, P.J., Sulzmann, M.: Observable confluence for constraint handling rules. In: Dahl, V., Niemelä, I. (eds.) ICLP 2007. LNCS, vol. 4670, pp. 224–239. Springer, Heidelberg (2007)
9. Frühwirth, T.: Constraint Handling Rules. Cambridge University Press, New York (2009)
10. Frühwirth, T., Raiser, F. (eds.): Constraint Handling Rules: Compilation, Execution, and Analysis. Books on Demand, Norderstedt (2011)
11. De Koninck, L., Schrijvers, T., Demoen, B.: A flexible search framework for CHR. In: Schrijvers, T., Frühwirth, T. (eds.) Constraint Handling Rules. LNCS, vol. 5388, pp. 16–47. Springer, Heidelberg (2008)
12. Lam, E.S.L., Sulzmann, M.: Towards agent programming in CHR. In: Proceedings of 3rd CHR Workshop on CHR 2006, pp. 17–31 (2006)
13. Martinez, T.: Angelic CHR. In: Proceedings of the 8th Workshop on Constraint Handling Rules, CHR 2011, pp. 19–31 (2011)
14. Sneyers, J., Weert, P.V., Schrijvers, Tom., Koninck, L.D.: As time goes by: constraint handling rules - a survey of CHR research between 1998 and 2007. In: Theory and Practice of Logic Programming, pp. 1–47 (2010)
15. Zaki, A., Frühwirth, T., Abdennadher, S.: Towards inverse execution of constraint handling rules. In: Technical Communications of 29th International Conference on Logic Programming, vol. 13 (2013)

A Formal Semantics for the Cognitive Architecture ACT-R

Daniel Gall$^{(\boxtimes)}$ and Thom Frühwirth

Institute of Software Engineering and Compiler Construction,
University of Ulm, 89069 Ulm, Germany
{daniel.gall,thom.fruehwirth}@uni-ulm.de

Abstract. The cognitive architecture ACT-R is very popular in cognitive sciences. It merges well-investigated results of psychology to a unified model of cognition. This enables researchers to implement and execute domain-specific cognitive models. ACT-R is implemented as a production rule system. Although its underlying psychological theory has been investigated in many psychological experiments, ACT-R lacks a formal definition from a mathematical-computational point of view.

In this paper, we present a formalization of ACT-R's fundamental concepts including an operational semantics of the core features of its production rule system. The semantics abstracts from technical artifacts of the implementation. Due to its abstract formulation, the semantics is eligible for analysis. To the best of our knowledge, this operational semantics is the first of its kind.

Furthermore, we show a formal translation of ACT-R production rules to Constraint Handling Rules (CHR) and prove soundness and completeness of the translation mechanism according to our operational semantics.

Keywords: Computational psychology · Cognitive systems · ACT-R · Production rule systems · Constraint handling rules · Operational semantics

1 Introduction

Computational psychology is a field at the interface of psychology and computer science. It explores human cognition by implementing detailed computational models. The models are executable and hence capable of simulating human behavior. This enables researchers to conduct the same experiments with humans and a computational model to verify the behavior of the model. By this procedure, cognitive models are gradually improved. Furthermore, due to their executability, computational models have to be defined precisely. Hence, ambiguities which often appear in verbal-conceptual models can be eliminated.

Cognitive Architectures support the modeling process by bundling well-investigated research results from several disciplines of psychology to a unified theory. Domain-specific models are built upon such cognitive architectures. Ideally, cognitive architectures constrain modeling to only plausible domain-specific models.

© Springer International Publishing Switzerland 2015
M. Proietti and H. Seki (Eds.): LOPSTR 2014, LNCS 8981, pp. 74–91, 2015.
DOI: 10.1007/978-3-319-17822-6_5

Adaptive Control of Thought – Rational (ACT-R) is one of the most popular cognitive architectures [4]. It is implemented as a production rule system. Although its underlying psychological theory is well-investigated and verified in many psychological experiments, ACT-R lacks a formal definition of its production rule system from a mathematical-computational point of view. I.e. the main data structures and the resulting operational semantics suggested by the psychological theory are not defined properly. This led to a reference implementation full of assumptions and technical artifacts beyond the theory making it difficult to overlook. Furthermore, the lack of a formal operational semantics inhibits analysis of the models like termination or confluence analysis.

In this paper, we present a formalization of the fundamental concepts of ACT-R leading to an operational semantics. The semantics abstracts from many details and artifacts of the implementation. Additionally, aspects like time or external modules are ignored to concentrate on the basic state transitions of the ACT-R production rule system. Those abstractions lead to a short and concise definition of the semantics making it suitable for theoretical analysis of the main aspects of ACT-R. Nevertheless, the semantics is still closely related to the general computation process of ACT-R implementations as we exemplify by a simple model.

The formalization of ACT-R relates to the reference manual of the ACT-R production rule system [6]. However, due to the power of logic programming and CHR, an executable CHR version of ACT-R has been developed, which is very close to the formal description of the system. In this paper, we define the translation from ACT-R production rules to CHR rules formally. The translation is closely related to the translation process described informally in [10], but respects the changes necessary to correspond to the abstract semantics. Additionally, it is the first formal description of our translation mechanism which is described formally. Finally, we prove soundness and completeness of our translation according to our abstract operational semantics.

The paper is structured as follows: Sect. 2, covers the preliminaries. In Sect. 3, we first recapitulate the formalization of the basic notions of ACT-R and then present its abstract operational semantics. The formal translation of ACT-R to Constraint Handling Rules is shown in Sect. 4. Then, the translation is proved to be sound and complete in relation to our abstract operational semantics in Sect. 5. We conclude in Sect. 6.

2   Preliminaries

First, we cover some notational aspects and introduce ACT-R and CHR.

2.1   Notation

We assume some basic notions of first-order logic and logic programming like syntactic equality or unification. Substitution is denoted by $t[x/y]$ where all occurrences of the variable x in the term t are replaced by the variable y. For the sake of brevity, we treat logical conjunctions as multi-sets and vice-versa at some points. I.e. we use (multi-)set operators on logical conjunctions or multi-sets of terms as conjunctions. We use the relational notation for some functions and for binary relations we use infix notation.

2.2   ACT-R

First of all, we describe ACT-R informally. For a detailed introduction to ACT-R we refer to [4, 12] or [3]. Then we introduce a subset of its syntax and describe its operational semantics informally. The formalization is presented in Sect. 3.

ACT-R is a production rule system which distinguishes two types of knowledge: *declarative knowledge* holding static facts and *procedural knowledge* representing processes which control human cognition. For example, in a model of the game *rock, paper, scissors*, a declarative fact could be "The opponent played scissors", whereas a procedural information could be that a round is won, if we played rock and the opponent played scissors.

Declarative knowledge is represented as *chunks*. Each chunk consists of a symbolic name and labeled slots which hold symbolic values. The values can refer to other chunk names, i.e. chunks can be connected. Hence, a network of chunks can build complex constructs. The names of chunks are just symbols (which are only important for the modeler) but they get a meaning through their connections. For example, there can be chunks that represent numbers and are named *one, two*, ... In a model, such chunks could be the internal representation of the concept of numbers 1, 2, ... However, the names do not give them a meaning but are just helpful for the modeler. The meaning in such a model could come from other chunks, that link two number chunks to represent a count fact, for example. Such a count fact has the slots *first* and *second* and e.g. connects the chunks with name *one* and *two* in the *first* and *second* slot. This represents the concept of an ordering between numbers. To compare the concept of chunks with logic programming or Constraint Handling Rules, the names of chunks can be seen as constants (since they are just symbolic values) and the connections between chunks can relate to complex terms. Chunks are typed, i.e. the number and names of the slots provided by a chunk are determined by a type.

As usual for production rule systems, procedural knowledge is represented as rules of the form IF *conditions* THEN *actions*. Conditions match values of chunks, actions modify them.

ACT-R has a modular architecture. For instance, there is a declarative module holding the declarative knowledge or a visual module perceiving the visual field and controlling visual attention. Each module has a set of affiliated buffers which can hold at most one chunk at a time. For example, there is a retrieval buffer which is associated to the declarative module and which holds the last retrieved chunk from the declarative memory.

The procedural system consists of a *procedural memory* with a set of production rules. The conditions of a production rule refer to the contents of the buffers, i.e. they match the values of the chunk's slots.

There are three types of actions whose arguments are encoded as chunks as well: First of all, *buffer modifications* change the content of a buffer, i.e. the values of some of the slots of a chunk in a buffer. Secondly, the procedural module can state *requests* to external modules which then change the contents of buffers associated with them. Eventually, *buffer clearings* remove the chunk from a buffer. For the sake of brevity, we only regard buffer modifications and requests in this work. Nevertheless, our formalization and translation can be easily extended by other actions [10]. Additionally, to keep our definitions extensible and as general as possible, we refer to an arbitrary set of buffers instead of a concrete instantiation of the theory with defined buffers and modules. In the example section (Sect. 3.4) we show a concrete instantiation of our theory.

Syntax. We define the syntax of an ACT-R production rule over a set of symbols $\mathfrak{S}$ as follows. The set of symbols is possibly infinite and contains all the symbols that are valid to name objects in ACT-R (e.g. chunks, slot names or values). ACT-R does not know complex terms like in first order logic or in CHR. Such terms can rather be constructed by chunks that link the primitive symbols to a more complex construct. Additionally, there is a set of variable symbols $\mathfrak{V}$ that is disjoint from $\mathfrak{S}$.

Definition 1 (Production Rule). *An ACT-R production rule r has the form $LHS \Rightarrow RHS$) where LHS is a set of terms of the form $test(b, SVP)$ where the b values are called* buffers *and SVP is a set of pairs (s, v) where s refers to a slot and v refers to a value. Such a pair is called slot-value pair. Note that b and s must be constants from $\mathfrak{S}$, whereas v can be a variable or constant, i.e. has the domain $\mathfrak{S} \cup \mathfrak{V}$.*

RHS is a set of terms of the form $action(a, b, SVP)$, where $a \in \{mod, req\}$ (denoting either a modification *or a* request*). b again refers to a buffer and SVP is a set of slot-value pairs.*

The function vars takes a set of tests or actions and returns their variables. Note that the following must hold: $vars(RHS) \subseteq vars(LHS)$, i.e. no new variables must be introduced on the right hand side of a rule. Buffers appearing in RHS must also appear in LHS. However, slots on RHS are not required to appear on LHS. The buffers and slots of the RHS are assumed to be pairwise distinct and must refer to slots which are available for the chunk in the modified buffer (i.e. which exist for the chunk)[6].

We can ensure the condition that a modified slot must exist for a chunk by a typing system. An implementation of such a typing system compliant with the ACT-R reference [6] can be found in [10,11]. However, for this abstract paper we assume the rules to be valid. Note that we use a representation of production rules as sets of first-order terms which differs from the original ACT-R syntax. This allows for the use of typical set operators in the rest of the paper. It is easy to derive our syntactic representation from original ACT-R rules and vice-versa.

Informal Operational Semantics. A production rule as defined in Definition 1 is read as follows: The *LHS* of the rule are conditions matching the contents of the buffers. I.e. for a condition $test(b, \{(s_1, v_1), (s_2, v_2)\})$ the buffer

b is checked for a chunk with the value v_1 in its s_1 slot and the value v_2 in its s_2 slot. If all conditions on the LHS match, the rule can be applied, i.e. the chunks in the buffers are modified according to the specification on the RHS. For an action $action(mod, b, \{(s_1, v_1')\})$ the value in the slot s_1 of the chunk in buffer b is overwritten by the value v_1'. This type of action is called a *modification*. Since the buffers and slots on the RHS are pairwise distinct, there are no conflicting modifications.

A *request* of the form $action(req, b, \{(arg_1, argv_1), (arg_2, argv_2), \dots\})$ states a request to the corresponding module of buffer b. The arguments are defined by slot-value pairs, where the first part of the pair is the name of the argument and the second part its value. The request returns a pair $(c, \{(res_1, resv_1), \dots\})$ which represents a chunk c with corresponding slot-value pairs. This chunk is put into buffer b after the request has finished. Since arguments and result are chunks, the domain of the argument names, values and results is $\mathfrak{S}$.

Running Example: Counting. We investigate the first example from the official ACT-R tutorial [1] using our semantics and translation procedure. The model implements the cognitive task of counting by retrieving counting facts from the declarative memory. This method models the way how little children usually learn counting: They know that after one follows two, after two follows three, etc.

Example 1 (Production Rule). In the following, we define the production rule which counts to the next number. This rule has been derived from the ACT-R tutorial as mentioned above:

$$\{test(goal, \{(count, Num_1)\}),$$
$$test(retrieval, \{(first, Num_1), (second, Num_2)\})\}$$
$$\Rightarrow$$
$$\{action(mod, goal, \{(count, Num_2)\}),$$
$$action(req, retrieval, \{(first, Num_2)\})\}$$

The goal buffer is tested for slot *count* and Num_1 is bound to the value in this slot. The second test checks if there is a chunk in the retrieval buffer with Num_1 in its *first* slot and some number Num_2 in its *second* slot. If the conditions hold, the *goal* buffer is modified such that the *count* slot is updated with Num_2. Then the declarative memory is requested for a chunk which has Num_2 in its first slot.

2.3 Constraint Handling Rules

We recap the syntax and semantics of Constraint Handling Rules (CHR) shortly. For a detailed introduction to the language, we refer to [8].

Syntax. We briefly introduce a subset of the abstract CHR syntax as defined in [8]. Constraints are first-order logic predicates of the form $c(t_1, \dots, t_n)$ where the t values are first-order terms, i.e. function terms or variables. There are two distinct types of constraints: *built-in* and *user-defined* constraints. We constrain the allowed built-in constraints to *true, false* and the syntactic equality $=$.

Definition 2 (CHR Syntax). *A CHR program P is a finite set of rules. Simpagation rules have the form*

$$r \, @ \, H_k \backslash H_r \Leftrightarrow G \mid B.$$

r is an optional name of the rule, H_k and H_r are conjunctions of user-defined constraints (at least one of them is non-empty) called head constraints. *G is a conjunction of built-in constraints and is called the* guard. *Eventually, B a conjunction of built-in and user-defined constraints and called the* body *of the rule.*

Operational Semantics. The operational semantics of CHR is defined as a state transition system. Hence, we first define the notion of a CHR state and then introduce the so-called very abstract operational semantics of CHR [8,9].

Definition 3 (CHR State). *A CHR state is a goal, i.e. either true, false, a built-in constraint, a user-defined constraint or a conjunction of goals.*

Definition 4 (Head Normal Form). *A CHR rule is in* head normal form *(HNF) if each argument of a head constraint is a unique variable.*

A CHR rule can be put into HNF by replacing its head arguments t_i with a new variable V_i and adding the equations $V_i = t_i$ to its guard.

The operational semantics of CHR is defined upon a constraint theory CT which is nonempty, consistent and complete and contains at least an axiomatization of the syntactic equality $=$ together with the built-in constraints *true* and *false*.

Definition 5 (CHR Operational Semantics). *For CHR constraints H_k and H_r, built-in constraints G and constraints of both types R the following transition relation is defined:*

$$(H_k \wedge H_r \wedge G \wedge R) \mapsto_r (H_k \wedge C \wedge B \wedge G \wedge R)$$

if there is an instance with new variables $\bar{x}$ of a rule r in HNF,

$$r \, @ \, H'_k \backslash H'_r \Leftrightarrow C \mid B.$$

and $CT \models \forall \, (G \rightarrow \exists \bar{x} \, (C \wedge (H_k = H'_k) \wedge (H_r = H'_r)))$.

I.e., there is a state transition using the rule r, if (a part of) the built-in constraints G of the state imply that the guard holds and the heads the match.

For the successor state, the constraints in H_k are kept, the constraints in H_r are removed and the body constraints are added. Additionally, the state contains the constraints C from the guard. Since the rule is in HNF, the state contains equality constraints from the variable bindings of the matching $H_k = H'_k$ and $H_r = H'_r$.

3   Formalization of the ACT-R Production System

In this section, we formalize the core data structures of ACT-R formally. We follow the definitions from [10].

3.1   Chunk Stores

Intuitively, a chunk store represents a network of chunks. I.e., it contains a set of chunks. Each chunk has a set of slots. In the slots, there are symbols referring either to a name of another chunk (denoting a connection between the two chunks) or primitive elements (i.e. symbols which do not refer to another chunk).

Definition 6 (Chunk Store). *A* chunk-store *over a set of symbols* $\mathfrak{S}$ *is a tuple* $(\mathbb{C}, \mathrm{HasSlot})$, *where* $\mathbb{C}$ *is a finite set of chunk identifiers.* $\mathrm{HasSlot} : \mathbb{C} \times \mathfrak{S} \to \mathfrak{S}$ *is a partial function which receives a chunk identifier and a symbol referring to a slot. It returns the value of a chunk's slot. If a slot does not have a value,* $\mathrm{HasSlot}$ *is undefined (or in relational notation, if chunk c does not have a value in its slot s, then there is no v such that* $(c, s, v) \in \mathrm{HasSlot})$.

3.2   Buffer Systems

Buffer systems extend the definition of chunk stores by buffers. Each buffer can hold at most one chunk from its chunk store. This is modeled by the relation Holds in the following definition:

Definition 7 (Buffer System). *A* buffer system *with buffers* $\mathbb{B}$ *is a tuple* $(\mathbb{C}; \mathrm{HasSlot}; \mathrm{Holds})$, *where* $\mathbb{B} \subseteq \mathfrak{S}$ *is a finite set of buffer names,* $(\mathbb{C}, \mathrm{HasSlot})$ *is a chunk-store and* $\mathrm{Holds} : \mathbb{B} \to \mathbb{C}$ *a partial function that assigns every buffer at most one chunk that it holds. Buffers that do not appear in the* Holds *relation are called* empty.

3.3   The Operational Semantics of ACT-R

A main contribution of this work is the formal definition of an abstract operational semantics of ACT-R which is suitable for analysis. The semantics abstracts from details like timings, latencies and conflict resolution but introduces non-determinism to cover those aspects. This has the advantage that analysis is simplified since the details like timings are difficult to analyze and secondly to let those details exchangeable. For instance, there are different conflict resolution mechanisms for ACT-R which are interchangeable at least in our implementation of ACT-R as we have shown in [10]. However, for confluence analysis for example, the used conflict resolution mechanism does not matter since conflicts are resolved by some method. At some points though, we do not want to introduce rule conflicts and they are regarded as a serious error. An operational semantics making analysis possible to detect such conflicts in advance is capable of improving and simplifying the modeling process which is one of the goals of a cognitive architecture like ACT-R.

We define the operational semantics of ACT-R as a state transition system $(\mathbb{S}, \rightarrowtail)$. The state space $\mathbb{S}$ consists of states defined as follows:

Definition 8 (ACT-R States). $S := \langle \mathbb{C}; \text{HasSlot}; \text{Holds}; \mathbb{R} \rangle^{\mathcal{V}}$ *is called an ACT-R State. Thereby,* $(\mathbb{C}, \text{HasSlot}, \text{Holds})$ *form a buffer system of buffers* $\mathbb{B}$, $\mathcal{V}$ *is a set of variable bindings and* $\mathbb{R}$ *(the set of* pending requests*) is a subset of tuples* $\mathbb{B} \times 2^{\mathfrak{S} \times \mathfrak{S}}$, *i.e. tuples of the form* (b, SVP) *where* $b \in \mathbb{B}$ *and* SVP *is a set of slot-value pairs.* Initial states *are states where* $\mathbb{R} = \emptyset$.

Before we define the transitions $\rightarrowtail$, we introduce the notion of a holding buffer test and consequently a matching l.h.s. of a production rule in a state.

Definition 9 (Buffer Test). *A buffer test* t *of the form* $test(b, SVP)$ *holds in state* $S := \langle \mathbb{C}; \text{HasSlot}; \text{Holds}; \mathbb{R} \rangle^{\mathcal{V}}$, *written* $t \mathrel{\hat{\cong}} S$, *if* $\exists b^S \in \mathbb{B}, c^S \in \mathbb{C}$ *such that the variable bindings* $\mathcal{V}$ *of the state imply that* $b^S = b$, $\text{Holds}(b^S) = c^S$ *and* $\forall(s, v) \in SVP \; \exists s^S, v^S : (c^S, s^S, v^S) \in \text{HasSlot}$ *with* $s^S = s$ *and* $v^S = v$.

Definition 10 (Matching). *A set* T *of buffer tests matches a state* S, *written* $T \mathrel{\hat{\cong}} S$, *if all buffer tests in* T *hold in* S.

We define the following functions which simplify notations in the definition of the operational semantics. Since the behavior of a rule depends on the fact if a certain slot is modified or requested on r.h.s. of the rule, we introduce two functions to test this:

Definition 11 (Modified and Requested Slots). *For an ACT-R rule* r *the following functions are defined as follows:*

$$\overline{modified}_r(b, s) = \begin{cases} true & \text{if } \exists action(mod, b, SVP) \in RHS(r) \\ & \wedge \exists v : (s, v) \in SVP \\ false & otherwise \end{cases}$$

$$\overline{requested}_r(b) = \begin{cases} true & \text{if } \exists action(req, b, SVP) \in RHS(r) \\ false & otherwise \end{cases}$$

With the two functions from Definition 11, it can be tested, if a certain buffer is modified (in a certain slot) or requested. As a next step, we regard the actions of a production rule. An action adds or deletes information from the state. The following definition covers these aspects:

Definition 12 (Add and Delete Lists). *For an ACT-R rule* r *and a state* S, *we define the following sets:*

$$C_S(r) = \{(c, s, v) \in \text{HasSlot} \mid (b, c) \in \text{Holds}$$
$$\wedge \; action(mod, b, SVP) \in RHS(r) \wedge (s, v) \in SVP\}$$
$$mod\_del_S(r) = \{(c, s, v) \in \text{HasSlot} \mid (b, c) \in \text{Holds} \wedge \overline{modified}_r(b)\}$$
$$req\_del_S(r) = \{(b, c) \in \text{Holds} \mid \overline{requested}_r(b), c \in S\}$$

The functions *mod_add* and *mod_del* will overwrite modified slots by new values in the operational semantics, whereas the function *req_del* simply clears a buffer. As mentioned before, this happens when a request is stated and the buffer waits for its answer. The result of a request is module-dependent and is defined by a buffer-specific function $request_b : 2^{\mathfrak{S} \times \mathfrak{S}} \to \mathfrak{S} \times 2^{\mathfrak{S} \times \mathfrak{S}}$ which receives a finite set of slot-value pairs as input and produces a tuple with a symbol denoting a chunk name and a set of slot-value pairs. Hence, a request is stated by specifying a (partial) chunk derived from the slot-value pairs in the request action of a rule. Its result is again a chunk description (but also containing a name).

We now can define the transition relation $\rightarrowtail$ of our state transition system:

Definition 13 (Operational Semantics of ACT-R). *For a production rule* $r = (LHS \Rightarrow RHS)$ *the transition* $\rightarrowtail_r$ *is defined as follows.*

Rule Application: *If there is a fresh variant* $r' := r[\bar{x}/\bar{y}]$ *of rule* r *with variables* $\bar{x}$ *substituted by fresh variables* $\bar{y}$ *and* $\forall(\mathcal{V} \to \exists \bar{y}(LHS \hat{=} S))$ *then*

$$S := \langle \mathbb{C}; \text{HasSlot}; \text{Holds}; \mathbb{R} \rangle^{\mathcal{V}} \rightarrowtail_r \langle \mathbb{C}; \text{HasSlot}'; \text{Holds}'; \mathbb{R}' \rangle^{\mathcal{V} \cup (LHS(r') \hat{=} S)}$$

where
- $\text{Holds}' := \text{Holds} - req\_del_S(r')$
- $\text{HasSlot}' := \text{HasSlot} - mod\_del_S(r') \cup mod\_add_S(r')$
- $\mathbb{R}' = \mathbb{R} \cup \{(b, SVP) \mid action(req, b, SVP) \in RHS(r')\}$

We write $S \rightarrowtail_r S'$ *if the rule application transition is used by application of rule* r.

request *If the result of the request* $request_b(SVP^{in}) = (c, SVP)$ *then*
$$\langle \mathbb{C}; \text{HasSlot}; \text{Holds}; \mathbb{R} \cup (b, SVP^{in}) \rangle^{\mathcal{V}}$$
$$\rightarrowtail_r \langle \mathbb{C} \cup \{c\}; \text{HasSlot} \cup \bigcup_{(s,v) \in SVP}(c, s, v); \text{Holds} \cup \{(b, c)\}; \mathbb{R} \rangle^{\mathcal{V}}.$$
We write $S \rightarrowtail_{request} S'$ *if the request transition is used.*

Informally spoken: If the buffer tests on the l.h.s. match a state, the actions are applied. This means that chunks are modified by replacing parts of the HasSlot relation or chunks are requested by extending $\mathbb{R}$. If a request occurs as action of a rule, the requested buffer is cleared (i.e. the Holds relation is adapted) and a pending request is added to $\mathbb{R}$ memorizing the requested buffer and the arguments of the request in form of slot-value pairs. The variable bindings of the matching are added to the state, i.e. that the fresh variables from the rules are bound to the values from the state. The set of variable binding contains the equality predicates from the matching.

The request transition is possible as soon as a request has been stated, i.e. the last argument of the state is not empty. Then the arguments are passed to the corresponding $request_b$ function and the output chunk is put into the requested buffer.

Note that after a request has been stated the rule application transition might be used since other rules (testing other buffers) might be applicable. This non-deterministic formulation simulates the background execution of requests in the ACT-R reference implementation where a request can take some time until its results are present. During this time, other rules might fire.

3.4   Running Example: Operational Semantics of ACT-R

We exemplify the operational semantics of ACT-R by continuing with our running example – the counting model (Sect. 2.2). The actual instantiation of ACT-R is kept open in the formal semantics (Sect. 3.3): For instance, the semantics talks about an arbitrary set of buffers and corresponding request handling functions. In the following section, we describe the actual instantiation of ACT-R used in our running example. It is the default instantiation of ACT-R models which only use the declarative memory and the goal buffer without interaction with the environment.

ACT-R Instantiation. For our cognitive model, we need two buffers:

- the *goal buffer* taking track of the current goal and serving as memory for intermediate steps, and
- the *retrieval buffer* giving access to the declarative memory which holds all declarative knowledge of our model, i.e. all known numbers and number sequences.

This means, that the set of buffers is defined as follows: $\mathbb{B} = \{goal, retrieval\}$.

In ACT-R, declarative memory can be seen as an (independent) chunk store $DM = (\mathbb{C}_{DM}, \text{HasSlot}_{DM})$. In the following example, we show the initial content of the declarative memory for our counting model:

Example 2 (Ontent of Declarative Memory).

$$\mathbb{C}_{DM} = \{a, b, c, d, \dots\}$$
$$\text{HasSlot}_{DM} = \{(a, first, 1), (a, second, 2),$$
$$(b, first, 2), (b, second, 3),$$
$$\dots\}$$

A request to the retrieval buffer (and hence to the declarative memory) is defined as follows:

$$request_{retrieval}(SVP) = (c, SVP^{out}) \text{ if } c \in \mathbb{C}_{DM}$$
$$\forall (s, v) \in SVP : \exists (c, s', v') \in \text{HasSlot}_{DM}$$
$$\text{such that } s' = s \text{ and } v' = v.$$
$$SVP^{out} := \{(c', s', v') \in \text{HasSlot}_{DM} | c' = c\}$$

This means that a chunk from declarative memory is returned which has all slots and values (matches all conditions) in SVP.

Example Derivation. For our counting model, we use the previously defined ACT-R instantiation with a goal and a retrieval buffer. As described before, requests to the declarative memory return a matching chunk based on the arguments given in the request. The next step to an example derivation of the counting model is to define an initial state.

Example 3 (Initial State). The initial state is $S_1 := \langle \mathbb{C}, \text{HasSlot}, \text{Holds}, \emptyset \rangle^\emptyset$ with the following values:

$$\mathbb{C} = \{a, goalch\}$$
$$\text{HasSlot} = \{(a, first, 1), (a, second, 2),$$
$$(goalch, count, 1)\}$$
$$\text{Holds} = \{(goal, goalch), (retrieval, a)\}$$

This state has two chunks in its store: The chunk a which encodes the fact that 2 is successor of 1 and a *goalch* which has one slot *count* which is set to 1. This denotes that the current subgoal is to count from 1 to the next number.

We start the derivation from our initial state S_1. For better readability, we apply variable bindings directly in the state representation:

Example 4 (Derivation).

$$\langle\{a, goalch\}, \{(a, first, 1), (a, second, 2), (goalch, count, 1)\},$$
$$\{(goal, goalch), (retrieval, a)\}, \emptyset\rangle$$
$$\rightarrowtail_{count} \{a, goalch\}, \{(a, first, 1), (a, second, 2), (goalch, count, 2)\},$$
$$\{(goal, goalch))\}, \{(retrieval, \{(first, 2)\})\}\rangle$$
$$\rightarrowtail_{request} \{a, b, goalch\},$$
$$\{(a, first, 1), (a, second, 2), (b, first, 2), (b, second, 3), (goalch, count, 2)\},$$
$$\{(goal, goalch))\}, \emptyset\}\rangle$$
$$\cdots$$

It can be seen that as a first derivation step only the application of rule *count* is possible. After the application, only a request derivation step is possible, since the retrieval buffer is empty and hence the condition of rule *count* does not hold.

4   Translation of ACT-R Rules to CHR

In this section, we define a translation function $chr(\cdot)$ which translates ACT-R production rules and states to corresponding CHR rules and states. We show later on that the transition is sound and complete w.r.t. the abstract operational semantics of ACT-R. This enables the use of CHR analysis tools like the confluence test to analyze ACT-R models. The translation procedure is very close to the technical implementation given in [10]. Nevertheless, it is the first formal description of the translation process.

Definition 14 (Translation of Production Rules). *An ACT-R production rule r can be translated to a CHR rule $H_k \backslash H_r \Leftrightarrow G | B$ as follows. The translation is denoted as $chr(r)$.*

We introduce a set Θ which takes track of buffer-chunk mappings. We define H_k, H_r, B and Θ as follows:

– *For each* $test(b, SVP) \in LHS(r)$ *introduce a fresh variable* c *and set* $(b, c) \in \Theta$.
  There are two cases:
      case 1: *If* $requested_r(b)$, *then constraint* $buffer(b, c) \in H_r$.
      case 2: *If* $\neg requested_r(b)$, *then constraint* $buffer(b, c) \in H_k$.
  For each $(s, v) \in SVP$:
      case 1: *If* $modified_r(b, s)$, *then constraint* $chunk\_has\_slot(c, s, v) \in H_r$.
      case 2: *If* $\neg modified_r(b, s)$, *then constraint* $chunk\_has\_slot(c, s, v) \in H_k$.
– *For each* $action(a, b, SVP) \in RHS(r)$:
      case 1: *If* $a = mod$, *then for each* $(s, v) \in SVP$ *there is a constraint*
      $chunk\_has\_slot(c, s, v) \in B$ *where* $(b, c) \in \Theta$. *Additionally, if there is*
      no $test(b, SVP') \in LHS(r)$ *with* $(s, v) \in SVP'$, *then introduce fresh*
      variables c *and* v' *and set* $chunk\_has\_slot(c, s, v') \in H_r$ *and* $(b, c) \in \Theta$.
      case 2: *If* $a = req$, *then constraint* $request(b, SVP) \in B$

We assume a generic rule $request(b, SVP) \Leftrightarrow \ldots$ *in the program which imple-*
ments the request handling function $request_b$ *for every buffer* b. *The generation*
of such rules is given in Definition 16.

Note that the removed heads H_r are constructed by regarding the actions of the
rule. If slots are modified that are not tested on the left hand side as mentioned in
Definition 1, constraints with fresh, singleton variables as values are introduced.
Those and are not involved in the matching process of ACT-R rules (see Definition 13). Nevertheless, the corresponding constraints must be removed from
the store whis is why they appear in H_r. When writing an ACT-R rule it must
be ensured that only slots are modified which are part of the modified chunk as
required by Definition 1. In the CHR translation, such rules would never be able
to fire, since the respective constraint appearing in H_k can never be in the store.

    Informally, H_k contains all *buffer* and *chunk* constraints as well as all
chunk_has_slot constraints of the slots which are not modified on the r.h.s. In
contrast, H_r contains all *chunk_has_slot* constraints of the slots which appear
on the r.h.s., i.e. which are modified.

    We now have defined how our subset of ACT-R production rules can be
translated to CHR. In the following definition, we present the translation of
ACT-R states to CHR states.

Definition 15 (Translation of States). *An ACT-R state*

$$S := \langle \mathbb{C}; \text{HasSlot}; \text{Holds} \rangle^{\mathcal{V}}$$

can be translated to the corresponding CHR state (denoted by $chr(S)$*):*

$$\bigwedge\nolimits_{(b,c) \in \text{Holds}} buffer(b, c) \wedge$$
$$\bigwedge\nolimits_{(c,s,v) \in \text{HasSlot}} chunk\_has\_slot(c, s, v) \wedge$$
$$\bigwedge\nolimits_{(b,SVP) \in \mathbb{R}} request(b, SVP) \wedge \mathcal{V}$$

The Holds and the HasSlot relations are translated to *buffer* and *chunk_has_slot* constraints respectively. Pending requests appear as *request* constraints in the CHR state. The variable bindings $\mathcal{V}$ are represented by built-in equality constraints. The next definition shows how request functions are represented in the CHR program.

Definition 16 (Request Functions). *A request function request$_b$ can be translated to a CHR rule as follows:*

$$request(b, SVP^{in}) \Leftrightarrow$$
$$(c, SVP^{out}) = request_b(SVP^{in}) \; \wedge$$
$$buffer(b, c) \; \wedge$$
$$\forall (s, v) \in SVP^{out} : chunk\_has\_slot(c, s, v)$$

To continue our running example of the counting model, we show the translation of the production rule in Example 1 to CHR:

Example 5 (Translation of Rules). The rule *count* can be translated to the following CHR rule:

$$buffer(goal, C_1) \; \wedge$$
$$chunk\_has\_slot(C_2, first, Num_1) \; \wedge$$
$$chunk\_has\_slot(C_2, second, Num_2) \; \backslash$$
$$chunk\_has\_slot(C_1, count, Num_1) \; \wedge$$
$$buffer(retrieval, C_2)$$
$$\Rightarrow$$
$$chunk\_has\_slot(C_1, count, Num_2) \; \wedge$$
$$request(retrieval, \{(first, Num_2)\})$$

It can be seen that two new variables are introduced: C_1 which represents the chunk in the goal buffer and C_2 which represents the chunk in the retrieval buffer. The derivation of the program is equivalent to the ACT-R derivation in Sect. 3.4.

To analyze the program for confluence, the notion of observable confluence [7] is needed, since the definition of confluence is too strict: Intuitively, the program is (observably) confluent since there are no overlaps between the rule and the implicit *request* rule. However, there seems to be an overlap of the rule with itself. This overlap does not play a role, since both *buffer* and *chunk_has_slot* represent relations with functional dependency. Hence there is only one possibility to assign values to the variables and finding matching constraints if we only consider CHR representations of valid ACT-R states. However, the confluence test detects those states as non-joinable critical pairs, although they represent states that are not allowed in ACT-R. Hence, those states should not be considered in the confluence analysis, since they can never appear in a valid derivation. To formalize this intuitive observation, the invariants of the ACT-R formalization (like functional dependency of some of the relations) have to be formulated mathematically to allow for *observable confluence* analysis.

It can be seen that requests potentially produce non-determinism, since either another rule might fire or a request could be performed. Usually, in ACT-R programs, the goal buffer keeps track of the current state of the program and encodes if a request should be awaited or if another rule can fire. However, this leads to a more imperative thinking in the conditions of the rules, since the application sequence of rules is defined in advance.

5   Soundness and Completeness

In this section, we prove soundness and completeness of our translation scheme from Definition 14 and 15. I.e., we show that each transition of an ACT-R model in a certain state is also possible in the corresponding CHR program with the corresponding CHR state leading to the same results and vice versa. This is illustrated in Fig. 1. At first, we show that applicability is preserved by the translation and then extend this property to the soundness and completeness Theorem 1.

$$
\begin{array}{ccc}
S & \longmapsto & S' \\
chr(\cdot) \downarrow & & \downarrow chr(\cdot) \\
chr(S) & \mapsto & chr(S')
\end{array}
$$

Fig. 1. The proposition of Theorem 1. We show that applicability and actions are preserved by our translation.

Lemma 1 (Applicability). *If the production rule r is applicable in ACT-R state S, then the corresponding CHR rule $chr(r)$ is applicable in state $chr(S)$ and vice-versa.*

Proof. "$\Rightarrow$":

Let $S := \langle \mathbb{C}; \text{HasSlot}; \text{Holds}; \mathbb{R} \rangle^{\mathcal{V}}$. Since r is applicable in S, the following holds:

$$\forall (V \rightarrow \exists \bar{x}\, (LHS(r) \mathrel{\widehat{=}} s))$$

This implies that for every $test(b, SVP) \in LHS(r)$ $\exists b^S \in \mathbb{B}, c^S \in \mathbb{C} : b = b^S$ and $\forall (s, v) \in SVP$ $\exists s^S, v^S : (c^S, s^S, v^S) \in \text{HasSlot}$ with $s^S = s$ and $v^S = v$ according to Definitions 9 and 10.

By Definition 15, the state $chr(S)$ has the following constraints: For each $(b^S, c^S) \in \text{Holds}$ there is a constraint $buffer(b^S, c^S) \in chr(S)$ and for every $(c^S, s^S, v^S) \in \text{HasSlot}$ there is a constraint $chunk\_has\_slot(c^S, s^S, v^S) \in chr(S)$. Additionally, $\mathcal{V} \in chr(S)$.

This means that the following conditions hold. We refer to them by $(\star)$:

$\forall test(b, SVP) \in LHS(r)\ \exists buffer(b^S, c^S) \in chr(S)$ with $b^S = b$ and $c^S = c$

and

$\forall(s, v) \in SVP\ \exists chunk\_has\_slot(c^S, s^S, v^S) \in chr(S)$ with $s^S = s$ and $v^S = v$

Let $chr(r) = H_k \backslash H_r \Leftrightarrow B$ with $H := H_k \cup H_r$ be the translated CHR rule. For every $test(b, SVP)$ there is a constraint $buffer(b, c) \in H$ with a fresh variable c and for every $(s, v) \in SVP$ there is a constraint $chunk\_has\_slot(c, s, v) \in H$. Additionally, there are constraints $chunk\_has\_slot(c, s, v^*) \in H$ with a fresh variable $v*$ for slots which are modified on r.h.s but which do not appear on l.h.s. $chr(r)$ is applicable in $chr(S)$, if $\exists(G \to \bar{y}(H = H'))$ where H' are constraints in the state. Due to $(\star)$, this condition holds if we set $G = \mathcal{V}$ plus the bindings of the fresh v^* variables. Since for every test in the original ACT-R rule there are corresponding constraints in the state $chr(S)$ and in the rule $chr(r)$ the condition holds for all $chunk\_has\_slot$ constraints who have a correspondent test in $LHS(r)$. The other constraints have a matching partner in $chr(S)$ since a well-formed ACT-R rule only modifies slots which exist for the chunk according to Definition 1.

"$\Leftarrow$":

$chr(r)$ of form $H_k \backslash H_r \Leftrightarrow B$ with $H := H_k \cup H_r$ is applicable in state $chr(S) = \langle H' \wedge G \wedge R \rangle$. I.e. that $\forall(G \to (\exists \bar{x}(H = H')))$. Since $chr(r)$ is a translated ACT-R rule, it only consists of $buffer$ and $chunk\_has\_slot$ constraints. Since $H = H'$ there are matching constraints $H' \in chr(S)$, i.e. there is a matching M of the constraints in the state with the constraints in the rule. Set $unifier(LHS(r), S) = M$ and it follows that r is applicable in S.

Lemma 2 (Request Transitions). *For two ACT-R states S and S' and a CHR state S'', the two transitions $S \rightarrowtail_{request} S'$ and $chr(S) \mapsto_{request} S''$.*

Proof. "$\Rightarrow$":

$S \rightarrowtail_{request} S'$, i.e. $\mathbb{R} \neq \emptyset$ and there is some $(b^*, SVP^{in}) \in \mathbb{R}$. This means that in $chr(S)$ there is a constraint $request(b^*, SVP^{in})$ due to Definition 15.

There is a rule with head $request(SVP)$ for every function $request_b(b, SVP) = (c, SVP^{out})$ which implements this function (i.e. which adds $chunk\_has\_slot(c, s, v)$ constraints according to $(SVP)^{out}$ and a $buffer(b, c)$ constraint for a new chunk c). The $request(b^*, SVP^{in})$ constraint is removed from the store like (b^*, SVP^{in}) is removed from $\mathbb{R}$ according to Definition 13.

Hence, if the request transition is possible in S, the corresponding request rule is possible in $chr(S)$ and the resulting states $chr(S')$ and S'' are equivalent.

"$\Leftarrow$":

The argument is analogous to the other direction.

Lemma 3 (Soundness and Completeness of Rule Application). *For an ACT-R production rule r and two ACT-R states S and S' the transitions $S \rightarrowtail_r S'$ and $chr(S) \mapsto_r S''$ correspond to each other, i.e. $chr(S') = S''$.*

Proof. Let $chr(r) = r@H'_k \backslash H'_r \Leftrightarrow G|B$.

"$\Rightarrow$": According to Lemma 1, r is applicable in S iff $chr(r)$ is applicable in $chr(S)$. Let $chr(S) \mapsto_r S'' = (H_k \wedge C \wedge H_k = H'_k \wedge H_r = H'_r \wedge B \wedge G)$ (Definition 5). It remains to show that the resulting state $S'' = chr(S')$. Let $S = \langle \mathbb{C}; \text{HasSlot}, \text{Holds}, \mathbb{R}\rangle^{\mathcal{V}}$ and $S' = \langle \mathbb{C}; \text{HasSlot}', \text{Holds}', \mathbb{R}'\rangle^{\mathcal{V}'}$ be ACT-R states. Then

$$\text{Holds}' = \text{Holds} - req\_del_s(r)$$
$$\text{HasSlot}' = \text{HasSlot} - mod\_del_S(r) \cup mod\_add_S(r)$$
$$\mathbb{R}' = \mathbb{R} \cup \{(b, SVP)|action(req, b, SVP) \in RHS(r')\}$$

The corresponding CHR state $chr(S')$ contains the following constraints according to Definition 15:

$$\bigwedge\nolimits_{(b,c)\in \text{Holds}'} buffer(b, c) \wedge \bigwedge\nolimits_{(c,s,v)\in \text{HasSlot}'} chunk\_has\_slot(c, s, v) \wedge \mathcal{V}'$$

Since Holds', HasSlot' and $\mathbb{R}'$ are derived from Holds, HasSlot and $\mathbb{R}$, we have to check whether the corresponding *buffer* and *chunk_has_slot* constraints are removed and added to $chr(S)$ by $chr(r)$. For the CHR rule, the body B contains for every $action(mod, b, SVP) \in RHS(r)$, there is a constraint $chunk\_has\_slot(c, s, v) \in B$ according to Definition 14 which is therefore also added to s'' according to Definition 5. This corresponds to $mod\_add_s(r)$. According to Definition 14, a constraint $chunk\_has\_slot(c, s, v)$ appears in H_r if it is modified on $RHS(r)$ (independent of appearing in a test or not, see case 1.a). This corresponds to $mod\_del_S(r)$. A constraint $buffer(b, c)$ is in H_r, if $requested_r(b)$ is true. This corresponds to $req\_del_s(r)$. For each $action(req, b, SVP) \in RHS(r)$ there appears a constraint $request(b, SVP) \in B$ of the rule. This corresponds to the adaptation of $\mathbb{R}$ in S.

Hence, the state S'' is equivalent to $chr(S')$.

"$\Leftarrow$": Let $chr(S) \mapsto_r chr(S')$ and $S \rightarrowtail_r S''$. According to Lemma 1, $chr(r)$ is applicable in $chr(S)$ iff r is applicable in S. It remains to show that the resulting state $S'' = chr(S')$.

The removed constraints H_r in the CHR rule $chr(r)$ are either

(a) *chunk_has_slot* or
(b) *buffer* constraints.

In case (a) the constraints correspond to a modification action in $RHS(r)$. I.e., $modified_r(b, s)$ is true for a constraint $chunk\_has\_slot(c, s, v) \in chr(S)$ with $buffer(b, c) \in chr(S)$ iff it appears in H_r. This corresponds to $mod\_del_s(r)$. In case (b), $requested(b)$ is true for a constraint $buffer(b, c)$ if it appears in H_r according to Definition 14. This corresponds to $req\_del_s(r)$.

The added *chunk_has_slot* constraints of B in the CHR rule correspond directly to $mod\_add_S(r)$ by Definitions 14 and 12. The *request* constraints in B correspond directly to the adaptation in $\mathbb{R}'$.

Hence, $S'' = chr(S')$.

Theorem 1 (Soundness and Completeness). *Every ACT-R transition $s \longmapsto s'$ corresponds to a CHR transition $chr(S) \longmapsto_r chr(S')$ and vice versa. I.e., every transition (not only rule applications) possible in S is also possible in $chr(S)$ and leads to equivalent states.*

Proof. By Lemmas 3 and 2 the theorem follows directly.

6   Conclusion

In this paper, we have presented a formalization of the core of the production rule system ACT-R including an abstract operational semantics. Furthermore, we have shown a formal translation of ACT-R production rules to CHR. The translation is sound and complete.

The formalization of ACT-R is based on prior work. In [10] we have presented an informal description of the translation of ACT-R production rules to CHR rules. This informal translation has been implemented in a compiler transforming ACT-R models to CHR programs. Our implementation is modular and exchangeable in its core features as we have shown in [11] by exchanging the central part of the conflict resolution with four different methods. Although the implementation is very practical and covers a lot of practical details of the ACT-R implementations, it is not directly usable for analysis.

Our formalization of the translation process in this paper is very near to the practical implementation as it uses the same translation schemes for chunk stores, buffer systems and consequently states. Even the rules are a simplified version of our practical translation from [11]. However, it abstracts from practical aspects like time or conflict resolution. This is justifiable, since for confluence analysis, this kind of non-determinism in the operational semantics is useful. Additionally, as shown in our running example, the general computation process is reproduced closely by our semantics. Furthermore, due to the soundness and completeness of our translation, confluence analysis tools from CHR can be used on our models.

Hence, the contributions of this paper are

- an abstract operational semantics of ACT-R which is – to the best of our knowledge – the first formal representation of ACT-R's behavior,
- a formal description of our translation process (since in [10] a more technical description has been chosen),
- a soundness and completeness result of the abstract translation.

For the future, we want to extend our semantics such that it covers the more technical aspects of the ACT-R production rule system like time and conflict resolution. We then want to investigate how this refined semantics is related to our abstract operational semantics from this paper.

To overcome non-determinism, ACT-R uses a conflict resolution strategy. In [11] we have analyzed several conflict resolution strategies. A confluence test might be useful to reveal rules where the use of conflict resolution is undesired.

For the future, we want to investigate how the CHR analysis tools perform for our ACT-R semantics and how they might support modelers in testing their models for undesired behavior, since the informal application of the confluence test on our example is promising. We plan to lift the results for observable confluence of CHR to ACT-R models. Additionally, it could be interesting to use the CHR completion algorithm [2] to repair ACT-R models that are not confluent. We also want to investigate if the activation levels of ACT-R fit the soft constraints framework [5].

References

1. The ACT-R 6.0 tutorial. http://act-r.psy.cmu.edu/actr6/units.zip, http://act-r. psy.cmu.edu/actr6/units.zip (2012)
2. Abdennadher, S., Frühwirth, T.: On Completion of Constraint Handling Rules. In: Maher, Michael J., Puget, Jean-François (eds.) CP 1998. LNCS, vol. 1520, p. 25. Springer, Heidelberg (1998)
3. Anderson, J.R.: How can the human mind occur in the physical universe?. Oxford University Press, Oxford (2007)
4. Anderson, J.R., Bothell, D., Byrne, M.D., Douglass, S., Lebiere, C., Qin, Y.: An integrated theory of the mind. Psychol. Rev. **111**(4), 1036–1060 (2004)
5. Bistarelli, S., Frühwirth, T., Marte, M.: Soft constraint propagation and solving in CHRs. In: Proceedings of the 2002 ACM symposium on Applied computing. pp. 1–5. ACM (2002)
6. Bothell, D.: ACT-R 6.0 Reference Manual - Working Draft. Department of Psychology, Carnegie Mellon University, Pittsburgh, Pennsylvania 15213
7. Duck, G.J., Stuckey, P.J., Sulzmann, M.: Observable Confluence for Constraint Handling Rules. In: Dahl, V., Niemelä, I. (eds.) ICLP 2007. LNCS, vol. 4670, pp. 224–239. Springer, Heidelberg (2007)
8. Frühwirth, T.: Constraint Handling Rules. Cambridge University Press, Cambridge (2009)
9. Frühwirth, T., Abdennadher, S.: Essentials of Constraint Programming. Springer, Berlin (2003)
10. Gall, D.: A rule-based implementation of ACT-R using Constraint Handling Rules. Master Thesis, Ulm University (2013)
11. Gall, D., Frühwirth, T.: Exchanging conflict resolution in an adaptable implementation of ACT-R. Theor. Pract. Logic Program. **14**(4–5), 525–538 (2014)
12. Taatgen, N.A., Lebiere, C., Anderson, J.: Modeling paradigms in ACT-R. Cognition and Multi-Agent Interaction: From Cognitive Modeling to Social Simulation. Cambridge University Press, Cambridge (2006)

CHRAnimation: An Animation Tool for Constraint Handling Rules

Nada Sharaf[1][✉], Slim Abdennadher[1], and Thom Frühwirth[2]

[1] The German University in Cairo, Cairo, Egypt
[2] Ulm University, Ulm, Germany
{nada.hamed,slim.abdennadher}@guc.edu.eg,
thom.fruehwirth@uni-ulm.de

Abstract. Visualization tools of different languages offer its users with a needed set of features allowing them to animate how programs of such languages work. Constraint Handling Rules (CHR) is currently used as a general purpose language. This results in having complex programs with CHR. Nevertheless, CHR is still lacking on visualization tools. With Constraint Handling Rules (CHR) being a high-level rule-based language, animating CHR programs through animation tools demonstrates the power of the language. Such tools are useful for beginners to the language as well as programmers of sophisticated algorithms. This paper continues upon the efforts made to have a generic visualization platform for CHR using source-to-source transformation. It also provides a new visualization feature that enables viewing all the possible solutions of a CHR program instead of the don't care nondeterminism used in most CHR implementations.

Keywords: Constraint Handling Rules · Algorithm visualization · Algorithm animation · Source-to-source transformation

1 Introduction

Constraint Handling Rules (CHR) [1] is a committed-choice rule-based language with multi-headed rules. It rewrites constraints until they are solved. CHR has developed from a language for writing constraint solvers into a general purpose language. Different types of algorithms are currently implemented using CHR.

So far, visually tracing the different algorithms implemented in CHR was not possible. Such visual tools are important for any programming language. The lack of such tools makes it harder for programmers to trace complex algorithms that could be implemented with CHR. Although the tool provided through [2] was able to add some visualization features to CHR, it lacked generality. It was only able to visualize the execution of the different rules in a step-by-step manner. In addition to that, it was able to visualize CHR constraints as objects. However, the choice of the objects was limited and the specification of the parameters of the different objects was very rigid.

Thus the tool presented through this paper aims at providing a more general CHR visualization platform. In order to have a flexible tracer, it was decided

© Springer International Publishing Switzerland 2015
M. Proietti and H. Seki (Eds.): LOPSTR 2014, LNCS 8981, pp. 92–110, 2015.
DOI: 10.1007/978-3-319-17822-6_6

to use an already existing visualization tool. Such tools usually provide a wide range of objects and sometimes actions as well. As a proof of concept, we used Jawaa [3] throughout the paper. The annotation tool is available through: http://sourceforge.net/projects/chrvisualizationtool. A web version is also under development and should be available through http://met.guc.edu.eg/chranimation.

In addition to introducing a generic CHR algorithm visualization system, the tool has a module that allows the user to visualize the exhaustive execution of any CHR program forcing the program to produce all the possible solutions. This allows the user to trace the flow of a CHR program using some different semantics than the refined operational semantics [4] embedded in SWI-Prolog. The output of the visualization is a tree showing the different paths of the algorithm's solutions. The tree is the search tree for a specific goal. It is also linked to the visualization tool as shown in Sect. 8.

The paper is organized as follows: Sect. 2 introduces CHR. Section 3 shows some of the related work and why the tool the paper presents is different and needed. Section 4 shows the general architecture of the system. Section 5 introduces the details of the annotation module. The details of the transformation approach are presented in Sect. 6. Section 7 shows an example of the visualization of algorithms implemented through CHR. Section 8 shows how it was possible to transform CHR programs to produce all the possible solutions instead of only one. Finally, we conclude with a summary and directions for future work.

2   Constraint Handling Rules

A CHR program distinguishes between two types of constraints: CHR constraints introduced by the user and built-in constraints [5]. Any CHR program consists of a set of simpagation rules. Each rule has a head, a body and an optional guard. The head of any CHR rule consists of a conjunction of CHR constraints. The guard of a rule is used to set a condition for applying the rule. The guard can thus only contain built-in constraints. The body, on the other hand, can contain both CHR and built-in constraints [5]. A simpagation rule has the form:

$$optional\_rule\_name @ H_K \setminus H_R \Leftrightarrow G \mid B.$$

There are two types of head constraints. H_K is the conjunction of CHR constraint(s) that are kept on executing the rule. On the other hand, H_R are the CHR constraint(s) that are removed once the rule is executed. G is the optional guard that has to be satisfied to execute the rule. B is the body of the rule. The constraints in B are added to the constraint store once the rule is executed.

Using simpagation rules, we can distinguish between two types of rules. A *simplification rule* is a simpagation rule with empty H_K. Consequently, the head constraint(s) are removed on executing the rule. It has the following form:

$$optional\_rule\_name @ H_R \Leftrightarrow G \mid B.$$

On the other hand, a *propagation rule* is a simpagation rule with empty H_R. Thus, on executing a propagation rule, its body constraints are added to the constraint store without removing any constraint from the store. Its format is:

$$optional\_rule\_name @ H_K \Rightarrow G \mid B.$$

The following program extracts the minimum number out of a set of numbers. It consists of one rule: `extract_min @ min(X) \ min(Y) <=> Y>=X | true.`

As seen from the rule, the numbers are inserted through the constraint `min/1`. The rule `extract_min` is executed on two numbers X and Y if Y has a value that is greater than or equal to X. `extract_min` removes from the store the constraint `min(Y)` and keeps `min(X)` because it is a simpagation rule. Thus on consecutive executions of the rule, the only number remaining in the constraint store is the minimum one. For example, for the query `min(9)`, `min(7)`, `min(3)`, the rule is applied on `min(9)` and `min(7)` removing `min(9)`. It is then applied on `min(7)` and `min(3)` removing `min(7)` and reaching a fixed point where the rule is no longer applicable. At that point, the only constraint in the store is `min(3)` which is the minimum number.

3   Why "CHRAnimation"?

This section shows the need for the tool and its contribution. As introduced previously, despite of the fact that CHR has developed into a general purpose language, it lacked algorithm visualization and animation tools. Programmers of CHR used SWI-Prolog's "trace" option which produces a textual trace of the executed rules. Attempts focused on visualizing the execution of the rules. The tool provided through [2] is able to visualize the execution of the rules showing which constraints are being added and removed from the store. However, the algorithm the program implements did not affect the visualization in any means. Visual CHR [6] is another tool that is also able to visualize the execution of CHR programs. However, it was directed towards the Java implementation of CHR; JCHR [7]. To use the tool, the compiler of JCHR had to be modified to add the required features. Although [2] could be extended to animate the execution of different algorithms, the need of having static inputs remained due to the inflexibility of the provided tracer. The attempts provided through [8] and [9] also suffered from the problem of being tailored to some specific algorithms.

Thus compared to existing tools for CHR, the strength of the tool the paper presents comes from its ability to adapt to different algorithm classes. It is able to provide a generic algorithm animation platform for CHR programs. The tool eliminates the need to use any driver or compiler directives as opposed to [6,10] since it uses source-to-source transformation. Although the system adopts the concept of *interesting events* used in Balsa [11] and Zeus [12], the new system is much simpler to use. With the previous systems, algorithm animators had to spend a lot of time writing the views and specifying how the animation should take place. With *CHRAnimation*, it is easy for a user to add or change the animation. In addition, the animator could be the developer of the program or any CHR programmer. Thus this eliminates the need of having an animator with whom the developer should develop an animation plan ahead. Consequently, the tool could be easily used by instructors to animate existing algorithms to teach to students. The system provides an interactive tool. In other words, every time a new query is entered, the animation automatically changes. The animations thus do not have to be prepared in advance to show in a class room for example and are not just movie-based animations that are not influenced by the inputs of users similar to [13].

Unlike the available systems, the user does not need to know about the syntax and details of the visualization system in use. Using source-to-source transformation eliminates this since the programs are automatically modified without the need of manually instructing the code to produce visualizations. The only need is to specify, through the provided user interface, how the constraints should be mapped to visual objects. In Constraint Logic Programming (CLP), the available visualization tools (such as the tools provided through [14] and [15]) focused on the search space and how domains are pruned. Thus to the best of our knowledge, this is the first tool that provides algorithm animation and not algorithm execution visualization for logic programming.

4   System Architecture

The aim of *CHRAnimation* is to have a generic algorithm animation system. The system however should be able to achieve this goal without the need to manually instrument the program to produce the needed visualizations. *CHRAnimation* consists of modules separating the steps needed to produce the animations and keeping the original programs unchanged.

As seen from Fig. 1, the system has two inputs: the original CHR program P in addition to $Annot_{Cons}$, the output of the so-called "Annotation Module".

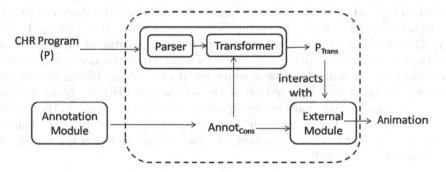

Fig. 1. Interactions between the modules in the system.

As a first step, the CHR program is parsed to extract the needed information. The transformation approach is similar to the one presented in [16] and [2]. Both approaches represent the CHR program using a set of constraints that encode the constituents of the CHR rule. For example `head(extract_min, min(Y), remove)` encodes the information that `min(Y)` is one of the head constraints of the rule named `extract_min` and that this constraint is removed on executing the rule. The CHR program is thus first parsed to automatically extract and represent the constituents of the rules in the needed "relational normal" form [16]. The *transformer* then uses this representation in addition to $Annot_{Cons}$ to convert the original CHR program (P) to another CHR program (P_{Tran}) with embedded visualization features as explained in more details in Sect. 6.

The *annotation module* is the component that allows the system to animate different algorithms while having a generic visual tracer. It allows users to define the visual states of the algorithm without having to go into any of the actual visualization details. The users are presented with a black-box module which allows them to define the needed visual output through the *interesting events* of the program. The module is explained in more details in Sect. 5. The output of the module ($Annot_{Cons}$) is used by different components of the system to be able to produce the corresponding animation.

P_{Trans} is a normal CHR program that users can run. Whenever the user enters any query to the system, P_{Trans} automatically communicates with an *external module* that uses $Annot_{Cons}$ to spontaneously produce an animation for the algorithm.

5 Annotation to Visualize CHR Algorithms

Algorithm animation represents the different states of the algorithm as pictures. The animation differs according to the interaction between such states [17]. As discussed before, the tool uses an existing tracer to overcome the problems faced in [2] in order to have a dynamic system that could be used with any algorithm type. The *annotation module* is built to achieve this goal while keeping a generic platform that is not tailored according to the algorithm type. Such module is needed to link between the different CHR constraints and the Jawaa objects/-commands. The idea is similar to the "interesting events" that Balsa [11] and Zeus [12] uses. This section introduces the basic functionalities of the annotation module which were first presented in [18] in addition to the new features that were added to accommodate for a wider set of algorithms. In the system, an interesting event is basically defined as the addition of CHR constraint(s) that leads to a change in the state of the algorithm and thus a change in the visualized data structure. For example, in sorting algorithms, every time an element in the list is moved to a new position, the list changes and thus the visualized list should change as well.

5.1 Basic Constraint Annotation

Constraint annotation is the basic building block of the annotation module. Users first identify the interesting events of a program. They could then determine the graphical objects that should be linked to them. For example, the program introduced in Sect. 2 represents a number through the constraint min/1 with its corresponding value. Adding or changing the min constraint is the interesting event in this algorithm. The annotation module provides its users with an interface through which they can choose to link constraint(s) with object(s) and/or action(s) as shown in Fig. 2.

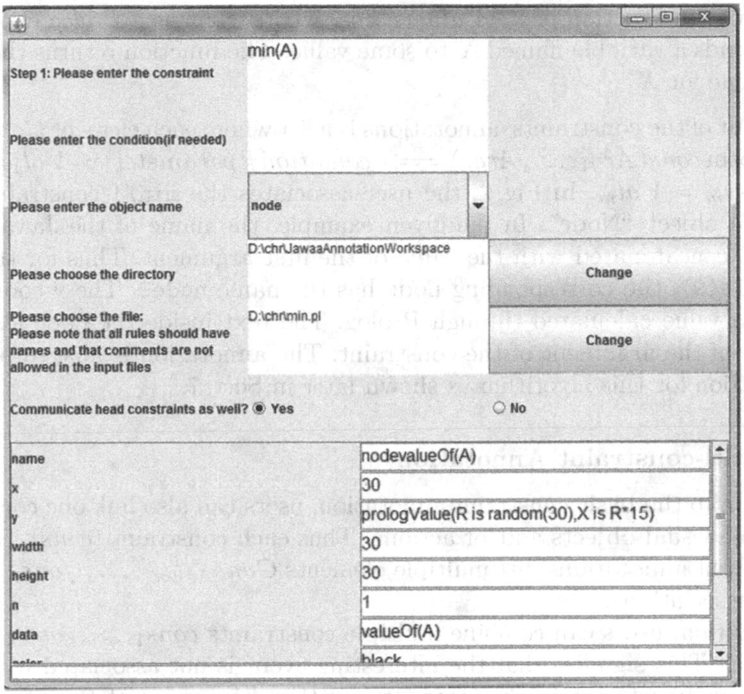

Fig. 2. Annotating the min constraint.

In order to have a dynamic system, the tool is automatically populated through a file that contains the available objects and actions and their corresponding parameters in the form $object\_name\#parameter_1\#\dots\#parameter_N$. For example, the line `circle#name#x#y#width#color#bkgrd`, adds the object `circle` as one of the available objects to the user. The `circle` object requires the parameters `name`, `x`, `y`, `width`, `color` and `bkgrd`. Users can then enter the name of the constraint and the corresponding annotation as shown in Fig. 2. The current system provides more annotation options than the prototype introduced in [18]. Users enter the constraint: $cons(Arg_1, \dots, Arg_n)$ representing the interesting event. With the current system, annotations can be activated according to defined conditions. Thus users provide some (Prolog) condition that should hold to trigger the annotation to produce the corresponding visualization. Users can then choose an object/action for annotation. This dynamically fills up the panel with the needed parameters so that users can enter their values. Parameter values can contain one or more of the following values (Val):

1. A constant c. The value can differ according to the type of the parameter. It could be a number, some text, …etc.
2. The built-in function $valueOf(Arg_i)$ to return the value of an argument (Arg_i).

3. The built-in function $prologValue(Expr)$ where $Expr$ is a $Prolog$ expression that binds a variable named X to some value. The function returns the binding value for X.

The output of the constraints' annotations is a list where each element $Cons_{Annot}$ has the form $cons(Arg_1, \ldots, Arg_n) ==> condition\#parameter_1 = Val_1\# \ldots \# parameter_m = Val_m$. In Fig. 2, the user associates the min/1 constraint with the Jawaa object "Node". In the given example, the name of the Jawaa node is "node" concatenated with the value of the first argument. Thus for the constraint min(9), the corresponding node has the name node9. The y-coordinate is random value calculated through Prolog. The text inside the node also uses the value of the argument of the constraint. The annotation is able to produce an animation for this algorithm as shown later in Sect. 7.

5.2   Multi-constraint Annotation

In addition to the basic constraint annotation, users can also link one constraint to multiple visual objects and/or actions. Thus each constraint ($cons$), can add to the output annotations' list multiple elements $Cons_{Annot_1}, \ldots, Cons_{Annot_n}$ if it has n associations.

In addition, users can combine multiple constraints $cons_1, \ldots, cons_N$ in one annotation. This signifies that the interesting event is not associated with having only one constraint in the store. It is rather having all of the constraints $cons_1, \ldots, cons_N$ simultaneously in the constraint store. Such annotations could thus produce and animate a color-mixing program for example. This kind of annotations adds to the annotations' list elements of the form:
$cons_1, \ldots, cons_n ==> annotation\_constraint_{cons_1, \ldots, cons_n}$.

5.3   Rule Annotations

In addition, users can choose to annotate CHR rules instead of only having annotations for constraints. In this case, the interesting event is the execution of the rule as opposed to adding a constraint to the store. This results in adding Jawaa objects and/or actions whenever a specific rule is executed. Thus, whenever a rule is annotated this way, a new step in the visual trace is added on executing the rule. Such annotation adds to the annotations' list an element of the form: $rule_i ==> annotation\_constraint_{rule_i}$. Rule annotations ignore the individual annotations for the constraints since the interesting event is associated with the rule instead. Therefore, it is assumed that the only annotation the user should visualize is the rule annotation since it accounts for all the constraints in the body. An example of this annotation is shown in Sect. 7.

6   Transformation Approach

The transformation mainly aims at interfacing the CHR programs with the entered annotations to produce the needed visual states. Thus the original program P is parsed and transformed into another program P_{Trans}. P_{Trans} performs the same functionality as P. However, it is able to produce an animation for the executed algorithm for any input query. As a first step, the transformation adds for every constraint constraint/N a rule of the form:

$$comm\_cons\_constraint @ constraint(X_1, X_2, ..., X_n) \Rightarrow check(status, false) \mid$$
$$communicate\_constraint(constraint(X_1, X_2, ..., X_n)).$$

This extra rule makes sure that every time a new constraint is added to the constraint store, it is communicated to the *external module*. Thus, in the case where the user had specified this constraint to be an interesting event (i.e. entered an annotation for it), the corresponding object(s)/action(s) is automatically produced. With such new rules, any new constraint added to the store is automatically communicated. Thus once the body constraints are added to the store, they are automatically communicated to the tracer.

The rules of the original program P can affect the visualization through the head constraints. To be more specific, head constraints removed from the store can affect the resulting visualization since their corresponding object(s) might need to be removed as well from the visual trace. Thus the transformer can instruct the new rules to communicate the head constraints[1].

The transformer also makes use of the output of the annotation module output ($Annot_{cons}$). Thus as a second step, the transformer adds for every compound constraint-annotation of the form:
$cons_1, ..., cons_n ==> annotation\_constraint_{cons_1,...,cons_n}$, a new rule of the form: $compound_{cons_1, ,cons_n} @$
$$cons_1(Arg_{cons_{1_1}}, ..., Arg_{cons_{1_x}}), ..., cons_n(Arg_{cons_{n_1}}, ..., Arg_{cons_{n_{ny}}})$$
$$\Rightarrow check(status, false) \mid annotation\_constraint_{cons_1,...,cons_n}(Arg1, ..., Arg_m).$$
The default case is to keep the constraints producing a propagation rule but the transformer can be instructed to produce a simplification rule instead. The annotation should be triggered whenever the constraints $cons_1, ..., cons_n$ exist in the store producing $annotation\_constraint_{cons_1,...,cons_n}$. This is exactly what the new rules ($compound_{cons_1,...,cons_n}$) do. They add to the store the annotation constraint whenever the store contains $cons_1, ..., cons_n$. The annotation constraint is automatically communicated to the tracer through the new $comm\_cons\_constraint\_name$ rules.

As a third step, the rules annotated by the user have to be transformed. The problem with rule annotations is that the CHR constraints in the body should be neglected since the whole rule is being annotated. Thus, even if the constraints were determined by the user to be interesting events, they have to be ignored since the execution of the rule includes them and the rule itself was annotated as an interesting event. Hence, to avoid having problems with this case, a generic *status* is used throughout P_{Trans}. In the transformed program, any rule annotated by the user changes the status to *true* at execution. All the new rules added by the transformer to P_{Trans} check that the status is set to *false* before communicating the corresponding constraint to the tracer. Consequently, such

---

[1] The tracer is able to handle the problem of having multiple Jawaa objects with the same name by removing the old object having the same name before adding the new one. This is possible even if the removed head constraint was not communicated.

rules are not triggered on executing an annotated rule since the guard check is always *false* in this case. Any rule $rule_i@H_K$, $H_K \Leftrightarrow G \mid B$ with the corresponding annotation $rule_i ==> annotation\_constraint_{rule_i}$ is transformed to: $rule_i@H_K$, $H_K \Leftrightarrow G \mid set(status, true)$, B, $annotation\_constraint_{rule_i}$, $set(status, false)$. In addition, the transformer adds the following rule to P_{Trans}:

$$comm\_cons_{annotation\_constraint_{rule_i}} @ \, annotation\_constraint_{rule_i} \Leftrightarrow$$
$$communicate\_constraint(annotation\_constraint_{rule_i}).$$

The new rule thus ensures that the events associated with the rule annotation are considered and that all annotations associated with the constraints in the body of the rule are ignored.

7   Examples

This section shows different examples of how the tool can be used to animate different types of algorithms.

Finding the Minimum Number in a Set is a CHR program consisting of one rule that is able to extract the smallest number out of a set of numbers as shown in Sect. 2. The interesting event in this program is adding and removing the constraint `min`. It was annotated using the basic constraint annotation producing the association: `min(A)==>node##name=nodevalueOf(A)#x=30#`
`y=prologValue(R is random(30), X is R*15)#width=30#height=30#n=1#`
`data=valueOf(A)#color=black#bkgrd=green#textcolor=black#type=CIRCLE`. The annotation links every `min` constraint to a Jawaa "Node" whose y-coordinate is randomly chosen through the function *prologValue*. The x-coordinate is fixed to a constant (30 in our case). As seen in Fig. 3, once a number is added to the store, the corresponding *node* is visualized. Once a number is removed from the store, its *node* object is removed. Thus by applying the rule, `extract_min`, the user gets to see in a step-by-step manner an animation for the program.

Bubble Sort is another algorithm that could be animated with the tool.
`start @ totalNum(T)<=> startBubbling, loop(1,1,T)`.

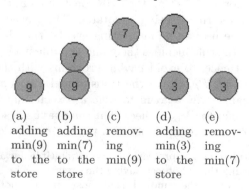

(a)	(b)	(c)	(d)	(e)
adding	adding	remov-	adding	remov-
min(9)	min(7)	ing	min(3)	ing
to the	to the	min(9)	to the	min(7)
store	store		store	

Fig. 3. Finding the minimum element of a set.

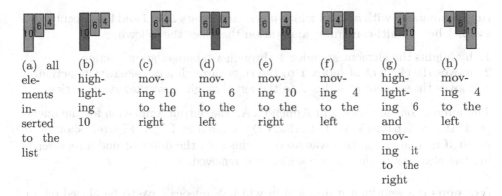

(a) all elements inserted to the list

(b) highlighting 10

(c) moving 10 to the right

(d) moving 6 to the left

(e) moving 10 to the right

(f) moving 4 to the left

(g) highlighting 6 and moving it to the right

(h) moving 4 to the left

Fig. 4. Sorting a list of numbers using rule annotations.

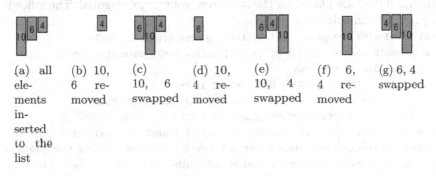

(a) all elements inserted to the list

(b) 10, 6 removed

(c) 10, 10, 6 swapped

(d) 10, 4 removed

(e) 10, 4 swapped

(f) 6, 4 removed

(g) 6, 4 swapped

Fig. 5. Sorting a list of numbers using constraint annotations only.

```
bubble @ startBubbling, loop(I,_,_) \ a(I,V), a(J,W) <=> I+1=:=J, V>W |
                                         a(I,W), a(J,V).
loop1 @ startBubbling\ loop(A,B,C) <=> A<C, B<C | A1 is A+1, loop(A1,B,C).
loop2 @ startBubbling \ loop(C,B,C) <=> B<C | B1 is B+1, loop(1,B1,C).
```

As seen from the program, the different elements of the list are entered using the constraint a/2. The rule `bubble` swaps two consecutive elements that are not sorted with respect to each other. Consequently, through multiple executions of this rule, the largest element is bubbled to the end. The constraint `loop/3` represents a pointer to the elements being compared. `loop1` advances the pointer one step through the list. `loop2` resets the pointer to the beginning of the list whenever one complete round of checks is done. The bubbling step is repeated T times where T is the number of elements in the list. There are thus two interesting events in this program. The first one is the insertion of an element to the list which is represented by the constraint a/2. The second interesting event is swapping two consecutive elements together through the rule `bubble`. The program has *three* annotations. The first one is a basic constraint annotation for a/2 constraint. The second annotation is a rule annotation for `bubble`. The rule is annotated with `swap/4` which has as arguments `I,V,J` and `W` consecutively. `swap/4` has a multi-constraint annotation that does the following:

1. highlights the element at index `I` through a "changeParam" action,
2. moves the element at index `I` to the right through a `moveRelative` action,
3. moves the element at index `J` to the left through a `moveRelative` action.

The annotations are shown in Appendix A. The output animation for the query `(a(1,10),a(2,6),a(3,4),totalNum(3))` is given in Fig. 4. Figure 5 shows the result if no rule annotation was used. In this case, the different nodes representing the elements in the list are added and removed.

*N*queens is a well-known problem in which N queens have to be placed on an N by N grid such that they do not attack each other. Two queens can attack each other if they are placed on the same row, column or diagonal. The following CHR program can solve this problem:

```
initial @ solve(N) <=> generate(1,N,List), queens(N,List), labelq.
add1 @ queens(N,Dom) <=> N>0 | N1 is N-1, in(N , Dom), queens(N1,Dom).
add2 @ queens(0,Dom) <=> true.
reduce @ in(N1 , [P]) \ in(N2 , Dom) <=> P1 is P-(N1-N2),
                P2 is P+(N1-N2), delete(Dom,P,D1), delete(D1,P1,D2),
                    delete(D2,P2,D3),Dom\==D3 | D3\==[], in(N2 , D3).
label @ labelq \ in(N , Dom)   <=> Dom=[\_,\_|\_] | member(P,Dom), in(N , [P]).
```

The model of the problem uses N variables each represented using the `queens/2` constraint. The value of every variable determines the row number. The index, on the other hand, determines the column number. For example if the value of the second queen is three, this means that the queen in the second column is placed in the third row. The domain of any queen is initialized to be from 1 to N using the predicate `generate/3`. As seen from the program the rule `initial` is used to initialize the solving process by adding the two constraints `queens/2` and `labelq` which enable finding a solution. As seen from the rule, the second argument of the `queens` constraint is set to be a list containing all the numbers from 1 till N. The two rules `add1` and `add2` are used to initialize the domains of all of the queens of the board using the previously computed list. The domain of every queen is represented using the `in/2` constraint. The rule `reduce` is used to prune the domains of the different queens. In order to execute the rule, the location of a specific queen has to be determined. This is represented by having a domain list with one element only. The rule removes from the domain of another queen any value that could lead to an attack. This ensures that whenever a location is chosen for this queen, it does not threaten the already labeled queen. Finally the rule `label` is used to search through the domains whenever domain pruning is not enough.

The visual board is initialized through specifying that the `solve` constraint is an interesting event. It should generate 16 rectangles in a board-like structure. To eliminate the need of entering 16 constraint annotations, users can now use the object *board* to annotate constraints and enter the number of squares it contains and their widths, heights, ...etc. Thus whenever the `solve` constraint is added

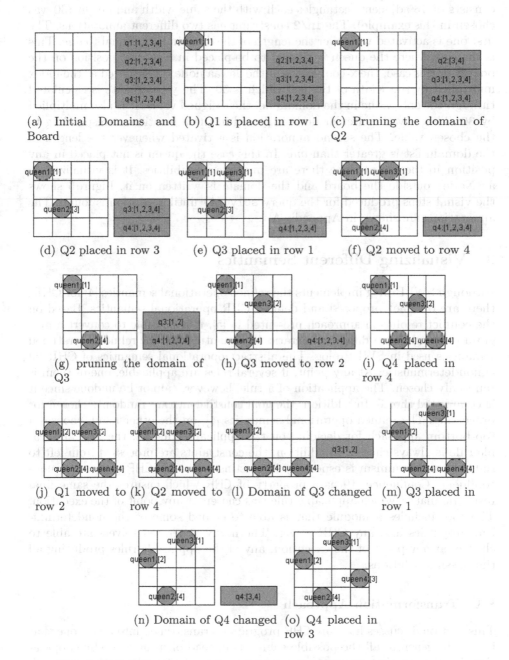

(a) Initial Domains and Board

(b) Q1 is placed in row 1

(c) Pruning the domain of Q2

(d) Q2 placed in row 3

(e) Q3 placed in row 1

(f) Q2 moved to row 4

(g) pruning the domain of Q3

(h) Q3 moved to row 2

(i) Q4 placed in row 4

(j) Q1 moved to row 2

(k) Q2 moved to row 4

(l) Domain of Q3 changed

(m) Q3 placed in row 1

(n) Domain of Q4 changed

(o) Q4 placed in row 3

Fig. 6. Visualizing the execution of the *n*queens algorithm for 4 queens.

to the store the 4-by-4 grid is visualized. The board, in the 4-queens problem, consists of 16 adjacent rectangles each with the same width and height (30 was chosen in this example). The in/2 constraint has two different annotations. The first one is activated whenever the length of the domain list is equal to *one*. This is the case where the queen is labeled to be placed in a specific position on the board. In this case, the x-coordinate of the Jawaa node is calculated as the index multiplied by the width of the cell which is 30. The y-coordinate is calculated through the only value in the domain i.e. the assigned value. It is also multiplied by 30. This way the circular node is placed in the a location corresponding to the chosen value. The second annotation is activated whenever the length of the domain list is greater than *one*. In this case the queen is not placed in any position in the board since there are multiple possibilities. It is visualized as a "Node" outside the board and the domain is written on it. Figure 6 shows the visual steps produced for the query solve(4) until a solution is found. The annotations are shown in Appendix A.

8   Visualizing Different Semantics

Although SWI-Prolog implements the refined operational semantics [4] for CHR, there are different proposed and defined CHR operational semantics. Based on the conflict resolution approach presented in [5], it is possible to convert a program running with a different operational semantics into the refined operational semantics used in SWI-Prolog. The abstract operational semantics of CHR [5] is nondeterministic. At any point, if several rules are applicable, one of them is randomly chosen. The application of a rule, however, cannot be undone since it is committed choice. In addition, the goal constituents are randomly chosen for processing. The refined operational semantics [4], on the other hand, chooses a top-bottom approach for deciding on the applicable rule i.e. the first applicable rule is always chosen. In addition, the constraints are processed from left to right. Nondeterminism is especially interesting when the CHR program is non-confluent. Confluence [19] is a property of CHR which ensures the same final result no matter which applicable rule was chosen at any point of the execution. The tool includes a module that is able to embed some of the nondeterminism properties into any CHR solver. The newly generated solvers are able to choose, at any point of the execution, any of the applicable rules producing all the possible solutions.

8.1   Transformation Approach

This section discusses how any CHR program is transformed into a new one that is able to generate all the possible solutions instead of using the refined operational semantics that generates only one solution. The transformation approach is based on the approaches presented in [5,20–22]. The main difference is that the new solver communicates some of the information to the visual tracer to be able to produce the needed visualization.

The transformed program starts each step by collecting the set of applicable rules with its corresponding head constraints. After the candidate list is built, the solver chooses one of the rules randomly using the built-in predicate `select/3`. The newly transformed program is thus a CHR [23] solver. For example a rule of the form:

`r1 @ Hk \ Hr <=> Guard | Body.`

generates two rules in the transformed program. The first generated rule is used to populate the candidate list. It is a propagation rule of the form:

`Hk, Hr ==> Guard | cand([(r1,[Hk,Hr])]).` The second rule is fired whenever this rule is chosen from the candidate list. It has the following form:

`Hk\fire((r1,[Hk,Hr])),Hr <=> Guard | communicate_heads_kept(Hk),`
`        communicate_heads_removed(Hr),communicate_body(Body),Body.`

In addition, the new program contains the following two rules:

`cand(L1),cand(L2) <=> append(L1,L2,L3) | cand(L3).`
`cand([H|T]),fire <=> select(Mem,[H|T],Nlist), fire(Mem),cand(NList),fire.`
The first rule ensures that the candidate list is correctly populated and incremented. The second rule, on the other hand, selects one of the elements of the candidate list at each step.
For example the program:

`:-chr_constraint sphere/2.`
`r1 @ sphere(X,red) <=> sphere(X,blue).`
`r2 @ sphere(X,red) <=> sphere(X,green).`
is transformed into

`:-chr_constraint sphere/2, fire/1, cand/1, fire/0.`
`r1_cand @ sphere(X,red) ==>cand([(r1,[sphere(X,red)])]).`
`r2_cand @ sphere(X,red) ==> cand([(r2,[sphere(X,red)])])`
`cand(L1),cand(L2) <=> append(L1,L2,L3), cand(L3).`
`cand([H|T]),fire <=>  select(Mem,[H|T] , NList), fire(Mem), cand(NList),fire.`
`r1 @ fire((r1,[sphere(X,red)])),sphere(X,red) <=>`
`                communicate_head_removed([sphere(X,red)]),`
`        communicate_body([sphere(X,blue)]), sphere(X,blue).`
`r2 @ fire((r2,[sphere(X,red)])),sphere(X,red) <=>`
`                communicate_head_removed([sphere(X,red)]),`
`        communicate_body([sphere(X,green)]), sphere(X,green).`

8.2   Visualization

With the refined operational semantics, the query `sphere(a,red)` results in executing `r1` adding to the store the new constraint `sphere(a,blue)`. Figure 7a shows the result of visualizing the execution of the solver with this query, using the tool presented in [2]. The CHR constraints remaining in the constraint store are shown in white and those removed are shown in red. The transformed program is able to generate the visual tree shown in Fig. 7b. Since there were two applicable rules, the output tree accounts for both cases by the different paths. Through SWI-Prolog the user can trigger this behavior using the ";" sign to search for more solutions.

Given the solver:
`rule1 @ sphere(X,red) <=> sphere(X,blue).`
`rule2 @ sphere(X,blue) <=> sphere(X,green).` The steps taken to execute the
query `sphere(b,blue), sphere(a,red)` with the solver are:

- First Solution
    1. `rule2` is fired replacing the constraint `sphere(b,blue)` by `sphere(b,green)`.
    2. `rule1` is then fired removing the constraint `sphere(a,red)` and adding the constraint `sphere(a,blue)`.
    3. Finally, `sphere(a,blue)` triggers `rule2` replacing it by `sphere(a,green)`.
- Second Solution
    1. Backtracking is triggered through the semicolon(;). We thus go back to the root and choose to apply `rule1` for `sphere(a,red)` producing the `sphere(a,blue)`.
    2. Afterwards, `rule2` is executed to replace `sphere(a,blue)` by `sphere(a,green)`.
    3. Finally, `rule2` is fired replacing `sphere(b,blue)` by `sphere(b,green)`.
- Third Solution
    1. This time when the user backtracks, execution goes back to the second level, applying `rule2` to replace `sphere(b,blue)` by `sphere(b,green)`.
    2. Afterwards, `rule2` replaces `sphere(a,blue)` by `sphere(a,green)`.

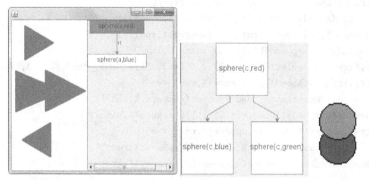

(a) Visualizing the execution of the solver.   (b) Showing all possible paths.   (c) Two random circles.

Fig. 7. Different options for visualizing the execution.

As seen from the tree in Fig. 8, the constraint store in the final states contains `sphere(a,green)`, `sphere(b,green)`. However, the paths taken are different. Once the user enters a query, the visual trees are automatically shown. In addition, whenever the user clicks on any node in the tree, the corresponding visual annotations are triggered. In this case the sphere can be mapped to a

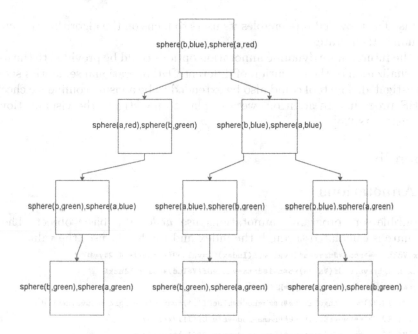

Fig. 8. Output tree.

Jawaa "circle" with a constant x-coordinate and a random y-coordinate and a background color that is equal to the value of the second argument of the constraint. If the user clicks on the node with the constraints (sphere(b,blue), sphere(a,green)), the system automatically connects the constraints to the previously introduced visual tracer that checks if any of the current constraints have annotations. This produces a visual state with two circles placed randomly as shown in Fig. 7c.

9 Conclusion

The paper introduced a new tool that is able to visualize different CHR programs by dynamically linking CHR constraints to visual objects. To have a generic tracing technique, the new system outsources the visualization process to existing tools. Intelligence is shifted to the transformation and annotation modules. Through the provided set of visual objects and actions, different algorithms could be animated. Such visualization features have proven to be useful in many situations including code debugging for programmers and educational purposes [24]. In addition, the paper explores the possibility of visualizing the execution of different operational semantics of CHR. It provides a module that is able to visualize the exhaustive execution of CHR and more importantly it links it to the annotated constraints. Thus, unlike the previously provided tools [15] for visualizing constraint programs, the focus is not just on the search space and the domains. The provided tool enables its users to focus on the algorithms executed to visualize their states.

In the future, more dynamic annotation options could be provided to the user. The visualization of the execution of different CHR operational semantics should be investigated. The tool could also be extended to be a visual confluence checker for CHR programs. In addition, we also plan to investigate the visualization of soft constraints [25].

Appendix

A   Annotations

The bubble sort program's annotations use *node* as a basic object. The x-coordinate is calculated through the index and the height uses the value.

```
a(Index,Value)==>node##name=nodevalueOf(Index)#x=valueOf(Index)*14+2#y=100#
width=12#height=valueOf(Value)*5#n=1#data=valueOf(Value)#color=black#
bkgrd=green#textcolor=black#type=RECT
swap(I1,V1,I2,V2)==>changeParam##name=nodevalueOf(I1)#paramter=bkgrd #newvalue=red
swap(I1,V1,I2,V2)==>moveRelative##name=nodevalueOf(I1)#x=14#y=0
swap(I1,V1,I2,V2)==>moveRelative##name=nodevalueOf(I2)#x=-14#y=0
swap(I1,V1,I2,V2)==>changeParam##name=nodevalueOf(I1)#paramter=bkgrd #newvalue=green
```

For the nqueens problem, the first set of annotations are the rectangles produced by the object "board" which users can choose through the interface. The number of vertical and horizontal squares (4 in our case), the initial x and y-coordinates (30 and 30 in our case), the squares' widths (30 in this case) in addition to the color chosen by the user automatically produces 16 associations for the `solve` constraint. The below constraints represent the rectangles that form the first two rows of the board and the two annotations for the `in` constraint.

```
solve(N)==>rectangle##name=rect1#x=30#y=30#width=30#height=30#color=black#bkgrd=white
solve(N)==>rectangle##name=rect2#x=60#y=30#width=30#height=30#color=black#bkgrd=white
solve(N)==>rectangle##name=rect3#x=90#y=30#width=30#height=30#color=black#bkgrd=white
solve(N)==>rectangle##name=rect4#x=120#y=30#width=30#height=30#color=black#bkgrd=white
in(N,List)==>node#length(valueOf(List),Len),Len is 1#name=nodevalueOf(N)#x=valueOf(N)*30
#y=prologValue(nth0(0,valueOf(List),El),X is El*30)#width=30#height=30#n=1#
data=queenvalueOf(N):valueOf(List)#color=black#bkgrd=green#textcolor=black#type=CIRCLE
in(N,List)==>node#length(valueOf(List),Len),Len > 1#name=nodevalueOf(arg0)#x=160#
y=valueOf(N)*30#width=90#height=30#n=1#data=qvalueOf(N):valueOf(List)#color=black#
bkgrd=green#textcolor=black#type=RECT
```

References

1. Frühwirth, T.: Theory and practice of constraint handling rules, special issue on constraint logic programming. J. Logic Program. **37**, 95–138 (1998)
2. Abdennadher, S., Sharaf, N.: Visualization of CHR through source-to-source transformation. In: Dovier, A., Costa, V.S. (eds.) ICLP (Technical Communications). LIPIcs, vol. 17, pp. 109–118. Schloss Dagstuhl - Leibniz-Zentrum fuer Informatik (2012)

3. Rodger, S.H.: Introducing computer science through animation and virtual worlds. In: Gersting, J.L., Walker, H.M., Grissom, S. (eds.) SIGCSE, pp. 186–190. ACM (2002)
4. Duck, G.J., Stuckey, P.J., García de la Banda, M., Holzbaur, C.: The refined operational semantics of constraint handling rules. In: Demoen, B., Lifschitz, V. (eds.) ICLP 2004. LNCS, vol. 3132, pp. 90–104. Springer, Heidelberg (2004)
5. Frühwirth, T.: Constraint Handling Rules. Cambridge University Press, Cambridge (2009)
6. Abdennadher, S., Saft, M.: A visualization tool for constraint handling rules. In: Kusalik, A.J. (ed.) WLPE (2001)
7. Schmauss, M.: An Implementation of CHR in Java, Master Thesis, Institute of Computer Science, LMU, Munich, Germany (1999)
8. Ismail, A.: Visualization of Grid-based and Fundamental CHR Algorithms, bachelor thesis, the Institute of Software Engineering and Compiler Construction, Ulm University, Germany (2012)
9. Said, M.A.: Animation of Mathematical and Graph-based Algorithms expressed in CHR, bachelor thesis, the Institute of Software Engineering and Compiler Construction, Ulm University, Germany (2012)
10. Stasko, J.: Animating algorithms with xtango. SIGACT News **23**, 67–71 (1992)
11. Brown, M.H., Sedgewick, R.: A system for algorithm animation. In: Proceedings of the 11th Annual Conference on Computer Graphics and Interactive Techniques, SIGGRAPH 1984, pp. 177–186. ACM, New York (1984)
12. Brown, M.: Zeus: a system for algorithm animation and multi-view editing. In: Proceedings of the 1991 IEEE Workshop on Visual Languages, pp. 4–9 (1991)
13. Baecker, R.M.: Sorting Out Sorting: A Case Study of Software Visualization for Teaching Computer Science, chap. 24, pp. 369–381. MIT Press, Cambridge (1998)
14. Smolka, G.: The definition of kernel oz. In: Podelski, A. (ed.) Constraint Programming: Basics and Trends. LNCS, vol. 910. Springer, Heidelberg (1995)
15. Meier, M.: Debugging constraint programs. In: Montanari, U., Rossi, F. (eds.) CP 1995. LNCS, vol. 976. Springer, Heidelberg (1995)
16. Frühwirth, T., Holzbaur, C.: Source-to-source transformation for a class of expressive rules. In: Buccafurri, F. (ed.) APPIA-GULP-PRODE, pp. 386–397 (2003)
17. Kerren, A., Stasko, J.T.: Algorithm animation. In: Diehl, S. (ed.) Dagstuhl Seminar 2001. LNCS, vol. 2269, pp. 1–17. Springer, Heidelberg (2002)
18. Sharaf, N., Abdennadher, S., Frühwirth, T. W.: Visualization of Constraint Handling Rules, CoRR, vol. abs/1405.3793 (2014)
19. Abdennadher, S., Frühwirth, T., Meuss, H.: On confluence of constraint handling rules. In: Freuder, E.C. (ed.) CP 1996. LNCS, vol. 1118. Springer, Heidelberg (1996)
20. Zaki, A., Frühwirth, T.W., Abdennadher, S.: Towards inverse execution of constraint handling rules. TPLP **13**(4-5) (2013) Online-Supplement
21. Abdennadher, S., Fakhry, G., Sharaf, N.: Implementation of the operational semantics for CHR with user-defined rule priorities. In: Christiansen, H., Sneyers, J. (eds.) Proceedings of the 10th Workshop on Constraint Handling Rules, pp. 1–12, Technical report CW 641, (2013)
22. Fakhry, G., Sharaf, N., Abdennadher, S.: Towards the implementation of a source-to-source transformation tool for CHR operational semantics. In: Gupta, G., Peña, R. (eds.) LOPSTR 2013, LNCS 8901. LNCS, vol. 8901, pp. 145–163. Springer, Heidelberg (2014)

23. Abdennadher, S., Schütz, H.: CHRv: a flexible query language. In: Andreasen, T., Christiansen, H., Larsen, H.L. (eds.) FQAS 1998. LNCS (LNAI), vol. 1495, pp. 1–14. Springer, Heidelberg (1998)
24. Hundhausen, C., Douglas, S., Stasko, J.: A meta-study of algorithm visualization effectiveness. J. Vis. Lang. Comput. **13**(3), 259–290 (2002)
25. Bistarelli, S., Frühwirth, T., Marte, M.: Soft constraint propagation and solving in chrs. In: Proceedings of the 2002 ACM Symposium on Applied Computing, pp. 1–5. ACM (2002)

Termination Analysis

Extending the 2D Dependency Pair Framework for Conditional Term Rewriting Systems

Salvador Lucas[1,2]($\boxtimes$), José Meseguer[1], and Raúl Gutiérrez[2]

[1] CS Department, University of Illinois at Urbana-Champaign,
Champaign, Illinois, USA
[2] DSIC, Universitat Politècnica de València, Valencia, Spain
slucas@dsic.upv.es

Abstract. Recently, a new *dependency pair framework* for proving operational termination of *Conditional Term Rewriting Systems* (CTRSs) has been introduced. We call it *2D* Dependency Pair (DP) Framework for CTRSs because it makes explicit and exploits the *bidimensional* nature of the termination behavior of conditional rewriting, where rewriting steps $s \to t$ and rewritings $s \to^* t$ (in zero or more steps) are defined for specific terms s and t by using an inference system where appropriate *proof trees* should be exhibited for such particular *goals*. In this setting, the *horizontal* component of the termination behavior concerns the existence of infinite sequences of rewriting steps, and the *vertical* component captures infinitely many *climbs* during the development of a proof tree for a *single* rewriting step. In this paper we extend the 2D DP Framework for CTRSs with several powerful *processors* for proving and *disproving* operational termination that specifically exploit the structure of conditional rules. We provide the first implementation of the 2D DP Framework as part of the termination tool MU-TERM. Our benchmarks suggest that, with our new processors, the 2D DP Framework is currently the most powerful technique for proving operational termination of CTRSs.

Keywords: Conditional rewriting · Dependency pairs · Operational termination · Program analysis

1 Introduction

In [7], the *Dependency Pair Framework for Term Rewriting Systems* [3] has been extended to prove *operational termination* [8] of *Conditional Term Rewriting Systems (CTRSs)*. We faithfully capture the *bidimensional nature* of infinite computations with CTRSs: there can be infinite sequences of rewriting steps (a *horizontal* dimension), but there can also be infinitely many failing attempts

Partially supported by NSF grant CNS 13-19109, EU (FEDER), Spanish MINECO projects TIN2010-21062-C02-02 and TIN 2013-45732-C4-1-P, and GV project PROMETEO/2011/052. Salvador Lucas' research was developed during a sabbatical year at the UIUC and was also supported by GV grant BEST/2014/026. Raúl Gutiérrez is also supported by Juan de la Cierva Fellowship JCI-2012-13528.

© Springer International Publishing Switzerland 2015
M. Proietti and H. Seki (Eds.): LOPSTR 2014, LNCS 8981, pp. 113–130, 2015.
DOI: 10.1007/978-3-319-17822-6_7

to satisfy the conditions of the rules when a *single* rewriting step is attempted (a *vertical* dimension). This *twofold* origin of infinite computations is captured by *two* sets of *2D Dependency Pairs* (2D DPs), see Sect. 3. Besides *characterizing* operational termination of CTRSs in terms of 2D DPs, a *2D DP Framework* for *mechanizing* proofs of operational termination of CTRSs is introduced. A central notion is that of *processor*, which *transforms* our *termination problems* into *sets* of *simpler* termination problems which can then be handled independently. This *divide and conquer* approach is paramount in the (2D) DP Framework.

In [7], only *four* processors were introduced, and *three* of them were close to well-known processors in the DP Framework: the *SCC processor* (which permits the use of graph techniques to *decompose* termination problems), the *Reduction Triple Processor* (which uses orderings to *simplify* termination problems) and a *shifting* processor which just *calls* the DP Framework for TRSs from the 2D DP Framework when TRSs, rather than CTRSs, are involved. As the benchmarks in this paper show, with those processors we are *unable* to outperform existing tools, like AProVE [5] and VMTL [11], that prove operational termination of CTRSs by using transformations into TRSs.

Example 1. The operational termination of the CTRS $\mathcal{R}$ [10, Example 17]:

$$a \rightarrow h(b) \quad (1) \qquad\qquad f(x) \rightarrow y \Leftarrow a \rightarrow h(y) \qquad\qquad (3)$$
$$a \rightarrow h(c) \quad (2) \qquad g(x,b) \rightarrow g(f(c),x) \Leftarrow f(b) \rightarrow x, x \rightarrow c \qquad (4)$$

cannot be proved with the processors in [7]. We define new processors to transform and analyze the (satisfaction of the) conditional part of the rules. They detect that x in (4) is bound only to $f(b)$ or c in any rewriting step with (4). This yields a more precise assessment of the role of the rule in the termination behavior and finally leads to a proof of operational termination of $\mathcal{R}$.

Our contributions in this paper are the following: (1) we refine the calculus of the graph that we use to represent termination problems (Sect. 4); (2) we adapt Hirokawa and Middeldorp's *subterm criterion* for TRSs [6] (Sect. 5); (3) we define new processors that exploit the simplification of the *conditional part* of the rules that participate in a given CTRS problem by either removing conditions that unify or by refining other conditions by narrowing (Sect. 6); (4) we extend and generalize the narrowing processor for TRSs in [3] (Sect. 7); (5) we introduce a new processor to specifically *disprove* operational termination of CTRSs (Sect. 8); then, (6) we describe the implementation of the 2D-DP framework as part of MU-TERM [2] and (7) provide some benchmarks showing that the 2D DP Framework with the new processors outperforms all existing tools for proving operational termination of CTRSs (Sect. 9).

2   Preliminaries

We use the standard notations in term rewriting (see, e.g., [9]). In this paper, $\mathcal{X}$ denotes a countable set of *variables* and $\mathcal{F}$ denotes a *signature*, i.e., a set of

$$(\text{Refl}) \quad \overline{t \to^* t} \qquad (\text{Cong}) \quad \frac{s_i \to t_i}{f(s_1, \ldots, s_i, \ldots, s_k) \to f(s_1, \ldots, t_i, \ldots, s_k)}$$
$$\text{for all } f \in \mathcal{F} \text{ and } 1 \leq i \leq k = ar(f)$$

$$(\text{Tran}) \quad \frac{s \to u \quad u \to^* t}{s \to^* t} \qquad (\text{Repl}) \quad \frac{\sigma(s_1) \to^* \sigma(t_1) \ \ \ldots \ \ \sigma(s_n) \to^* \sigma(t_n)}{\sigma(\ell) \to \sigma(r)}$$
$$\text{for } \ell \to r \Leftarrow s_1 \to t_1 \cdots s_n \to t_n \in \mathcal{R}$$
$$\text{and substitutions } \sigma.$$

Fig. 1. Inference rules for conditional rewriting

function symbols $\{f, g, \ldots\}$, each with a fixed *arity* given by a mapping $ar : \mathcal{F} \to$ $\mathbb{N}$. The set of terms built from $\mathcal{F}$ and $\mathcal{X}$ is $\mathcal{T}(\mathcal{F}, \mathcal{X})$. The symbol labeling the root of t is denoted as $root(t)$. The set of variables occurring in t is $Var(t)$. Terms are viewed as labelled trees in the usual way. *Positions* $p, q, \ldots$ are represented by chains of positive natural numbers used to address subterms $t|_p$ of t. The *set of positions* of a term t is $Pos(t)$. Given $\Delta \subseteq \mathcal{F}$, $Pos_\Delta(t)$ denotes the set of positions $p \in Pos(t)$ of subterms $t|_p$ of t that are rooted by a symbol in Δ (i.e., $root(t|_p) \in \Delta$). A substitution is a mapping from variables into terms which is homomorphically extended to a mapping from terms to terms. A conditional rule is written $\ell \to r \Leftarrow s_1 \to t_1, \cdots, s_n \to t_n$, where $\ell, r, s_1, t_1, \ldots, s_n, t_n \in \mathcal{T}(\mathcal{F}, \mathcal{X})$ and $\ell \notin \mathcal{X}$. As usual, ℓ and r are called the left- and right-hand sides of the rule, and the sequence $s_1 \to t_1, \cdots, s_n \to t_n$ (often abbreviated to c) is the *conditional part* of the rule. We often write $s_i \to t_i \in c$ to refer to the i-th atomic condition in c or $s \to t \in c$ if the position of the atomic condition in c does not matter. Rules $\ell \to r \Leftarrow c$ are classified according to the distribution of variables as follows: type 1 (or 1-rules), if $Var(r) \cup Var(c) \subseteq Var(\ell)$; type 2, if $Var(r) \subseteq Var(\ell)$; type 3, if $Var(r) \subseteq Var(\ell) \cup Var(c)$; and type 4, if no restriction is given. A CTRS $\mathcal{R}$ is a set of conditional rules; $\mathcal{R}$ is called an n-CTRS if it contains only n-rules; A 3-CTRS $\mathcal{R}$ is called *deterministic* if for each rule $\ell \to r \Leftarrow s_1 \to t_1, \ldots, s_n \to t_n$ in $\mathcal{R}$ and each $1 \leq i \leq n$, we have $Var(s_i) \subseteq Var(\ell) \cup \bigcup_{j=1}^{i-1} Var(t_j)$. Given $\mathcal{R} = (\mathcal{F}, R)$, we consider $\mathcal{F}$ as the disjoint union $\mathcal{F} = \mathcal{C} \uplus \mathcal{D}$ of symbols $c \in \mathcal{C}$ (called *constructors*) and symbols $f \in \mathcal{D}$ (called *defined functions*), where $\mathcal{D} = \{root(l) \mid (l \to r \Leftarrow c) \in R\}$ and $\mathcal{C} = \mathcal{F} - \mathcal{D}$. Terms $t \in \mathcal{T}(\mathcal{F}, \mathcal{X})$ such that $root(t) \in \mathcal{D}$ are called *defined* terms. We write $s \to_\mathcal{R} t$ (resp. $s \to_\mathcal{R}^* t$) iff there is a closed proof tree for $s \to t$ (resp. $s \to^* t$) using the inference system in Fig. 1. A proof tree T (for a CTRS $\mathcal{R}$, using the inference system in Fig. 1) is *closed* whenever it is finite and contains no open goals; it is *well-formed* if it is either an open goal, or a closed proof tree, or a derivation tree of the form $\frac{T_1 \ \ \cdots \ \ T_n}{G}$ where there is i, $1 \leq i \leq n$ such that $T_1, \ldots, T_{i-1}$ are closed, T_i is well-formed but not closed, and $T_{i+1}, \ldots, T_n$ are open goals. A proof tree T for $\mathcal{R}$ is a *proper prefix* of a proof tree T' if there are one or more open goals $G_1, \ldots, G_n$ in T such that T' is obtained from T by replacing each G_i by a derivation tree T_i with root G_i. We denote this as $T \subset T'$. An *infinite proof tree* for $\mathcal{R}$ is an infinite increasing chain of finite trees, that is, a sequence $\{T_i\}_{i \in \mathbb{N}}$ such that for all i, $T_i \subset T_{i+1}$. An infinite proof tree is *well-formed* if it is an ascending chain of well-formed finite proof trees. Intuitively, well-formed trees are the trees that an interpreter would incrementally build when trying to solve one condition at a time from left to right. A CTRS $\mathcal{R}$ is *operationally terminating* if no infinite well-formed tree for a goal $s \rightarrow^* t$ exists.

3   2D Dependency Pairs for CTRSs

The definition of 2D dependency pairs makes explicit the two dimensions of the (non)terminating behavior of CTRSs. We provide an intuition of the main ingredients in Definition 1 below. Complete technical details and further motivation can be found in [7]. In the following, given a signature $\mathcal{F}$ and $f \in \mathcal{F}$, we let $f^\sharp$ be a new fresh symbol (often called *tuple* symbol or *DP-symbol*) associated to a symbol f [1]. Let $\mathcal{F}^\sharp$ be the set of DP-symbols associated to symbols in $\mathcal{F}$. As usual, for $t = f(t_1, \ldots, t_k) \in \mathcal{T}(\mathcal{F}, \mathcal{X})$, we write $t^\sharp$ to denote the *marked* term $f^\sharp(t_1, \ldots, t_k)$.

1. The component $\mathsf{DP}_H(\mathcal{R})$ of our 2D DPs given below captures a *horizontal* dimension corresponding to the existence of infinite sequences of rewrite steps. Let $DRules(\mathcal{R}, t)$ be the set of (possibly conditional) rules in $\mathcal{R}$ defining $root(t)$ which *depend* on other defined symbols in $\mathcal{R}$:

$$DRules(\mathcal{R}, t) = \{\ell \rightarrow r \Leftarrow c \in \mathcal{R} \mid root(\ell) = root(t), r \notin \mathcal{T}(\mathcal{C}, \mathcal{X})\}$$

   This leads to our first component $\mathsf{DP}_H(\mathcal{R})$ of 2D DPs:

$$\mathsf{DP}_H(\mathcal{R}) = \{\ell^\sharp \rightarrow v^\sharp \Leftarrow c \mid \ell \rightarrow r \Leftarrow c \in R, r \trianglerighteq v, \ell \ntrianglerighteq v, DRules(\mathcal{R}, v) \neq \emptyset\}$$

2. The *vertical* dimension corresponds to (possibly empty) rewrite sequences infinitely often interrupted in steps with rules $\ell \rightarrow r \Leftarrow \bigwedge_{i=1}^{n} s_i \rightarrow t_i$ by failed attempts to evaluate some of the conditions. Given a term t, $Rules_C(\mathcal{R}, t)$ is the set of 'proper' conditional rules in $\mathcal{R}$ defining the root symbol of t:

$$Rules_C(\mathcal{R}, t) = \{\ell \rightarrow r \Leftarrow \bigwedge_{i=1}^{n} s_i \rightarrow t_i \in \mathcal{R} \mid root(\ell) = root(t), n > 0\}$$

   This set collects the rules that are primarily involved in such kind of transitions of the computations to *upper* levels. We let $URules(\mathcal{R}, t) = DRules(\mathcal{R}, t) \cup Rules_C(\mathcal{R}, t)$. This leads to our second group of 2D DPs:

$$\mathsf{DP}_V(\mathcal{R}) = \{\ell^\sharp \rightarrow v^\sharp \Leftarrow \bigwedge_{j=1}^{k-1} s_j \rightarrow t_j \mid \ell \rightarrow r \Leftarrow \bigwedge_{i=1}^{n} s_i \rightarrow t_i \in R,$$
$$\exists k, 1 \leq k \leq n, s_k \trianglerighteq v, \ell \ntrianglerighteq v, URules(\mathcal{R}, v) \neq \emptyset\}$$

   The defined subterms in the conditional part of the rules that originate the pairs in $\mathsf{DP}_V(\mathcal{R})$ are collected in the following set, which we use below:

$$V_C(\mathcal{R}) = \{v \mid \ell \rightarrow r \Leftarrow \bigwedge_{i=1}^{n} s_i \rightarrow t_i \in R,$$
$$\exists k, 1 \leq k \leq n, s_k \trianglerighteq v, \ell \ntrianglerighteq v, URules(\mathcal{R}, v) \neq \emptyset\}$$

The vertical transitions in computations can be separated by finite sequences of rewriting steps that correspond to the evaluation of the (instances of the) conditions $s_i \to t_i$ as reachability conditions $\sigma(s_i) \to^* \sigma(t_i)$. Since the (finite) rewrite sequences connecting vertical transitions must include proper conditional rules that promote a new vertical transition, we define:

$$MU(\mathcal{R}, t) = URules(\mathcal{R}, t) \cup \bigcup_{(l \to r \Leftarrow c) \in DRules(\mathcal{R}, t)} \;\; \bigcup_{v \in DSubterm(\mathcal{R}, r)} MU(\mathcal{R}, v)$$

and give our third (and last) group of dependency pairs, that are aimed to provide appropriate connections between pairs in $\mathsf{DP}_V(\mathcal{R})$ as follows:

$$\mathsf{DP}_{VH}(\mathcal{R}) = \bigcup_{w \in V_C(\mathcal{R})} \{\ell^\sharp \to v^\sharp \Leftarrow c \mid \ell \to r \Leftarrow c \in \overline{MU}(\mathcal{R}, w),$$
$$r \trianglerighteq v, \ell \ntrianglerighteq v, URules(\mathcal{R}, v) \neq \emptyset\}$$

where $\overline{MU}(\mathcal{R}, w) = \emptyset$ if $MU(\mathcal{R}, w)$ is a TRS (i.e., no rule has a conditional part) and $\overline{MU}(\mathcal{R}, w) = MU(\mathcal{R}, w)$ otherwise.

We can provide now the definition of 2D-Dependency pairs for CTRSs.

Definition 1 (2D-Dependency Pairs for CTRSs). *Let $\mathcal{R}$ be a CTRS. The 2D-dependency pairs (2D-DPs) of $\mathcal{R}$ are given by the triple $\mathsf{DP}_{2D}(\mathcal{R}) = (DP_H(\mathcal{R}), DP_V(\mathcal{R}), DP_{VH}(\mathcal{R}))$.*

Example 2. For $\mathcal{R}$ in Example 1,

$$\mathsf{DP}_H(\mathcal{R}): \qquad \mathsf{G}(x, \mathsf{b}) \to \mathsf{G}(\mathsf{f}(\mathsf{c}), x) \Leftarrow \mathsf{f}(\mathsf{b}) \to x, x \to \mathsf{c} \qquad (5)$$
$$\mathsf{DP}_V(\mathcal{R}): \qquad \mathsf{G}(x, \mathsf{b}) \to \mathsf{F}(\mathsf{b}) \qquad\qquad\qquad\qquad\qquad\qquad (6)$$

and $\mathsf{DP}_{VH}(\mathcal{R}) = \emptyset$. $\mathsf{DP}_H(\mathcal{R})$ is obtained by *marking* (here just *capitalizing* the root symbols of) the left- and right-hand sides $\mathsf{g}(x, \mathsf{b})$ and $\mathsf{g}(\mathsf{f}(\mathsf{c}), x)$ of (4) and keeping the conditional part. $\mathsf{DP}_V(\mathcal{R})$ is obtained from the left-hand side $\mathsf{g}(x, \mathsf{b})$ of (4) and the *first condition* $\mathsf{f}(\mathsf{b})$ of (4) which are *marked* and combined as a *new rule* whose conditional part is empty in this case.

Now, operational termination of CTRSs is characterized as the *absence* of infinite chains of the following kind [7, Definition 5].

Definition 2. *Let $\mathcal{P}, \mathcal{Q}, \mathcal{R}$ be CTRSs. A $(\mathcal{P}, \mathcal{Q}, \mathcal{R})$-chain is a sequence $(\alpha_i)_{i \geq 1}$ where $\alpha_i : u_i \to v_i \Leftarrow c_i \in \mathcal{P}$, together with a substitution σ satisfying that, for all $i \geq 1$, (1) $\sigma(s) \to^*_{\mathcal{R}} \sigma(t)$ for all $s \to t \in c_i$ and (2) $\sigma(v_i)(\to^*_{\mathcal{R}} \circ \xrightarrow{\Lambda}_{\overline{\mathcal{Q}}})^* \sigma(u_{i+1})$, where $s \xrightarrow{\Lambda}_{\overline{\mathcal{Q}}} t$ if $s = t$ or there is $\ell \to r \Leftarrow c \in \mathcal{Q}$ and a substitution θ such that $s = \theta(\ell)$, $t = \theta(r)$ and $\theta(u) \to^*_{\mathcal{R}} \theta(v)$ for all $u \to v \in c$. We assume $Var(\alpha_i) \cap Var(\alpha_j) = \emptyset$, if $i \neq j$ (rename the variables if necessary). A $(\mathcal{P}, \mathcal{Q}, \mathcal{R})$-chain is called* minimal *if for all $i \geq 1$, $\sigma(v_i)$ is $\mathcal{R}$-operationally terminating.*

Rules $\mathcal{Q}$ in $(\mathcal{P}, \mathcal{Q}, \mathcal{R})$ may contribute *connections* between $\sigma(v_i)$ and $\sigma(u_{i+1})$ using *root* steps only. We often speak of $(\mathcal{P}, \mathcal{Q}, \mathcal{R}, \mathsf{a})$-chains (or $(\mathcal{P}, \mathcal{Q}, \mathcal{R}, \mathsf{m})$-chains), if arbitrary (minimal) chains are considered.

3.1 2D Dependency Pair Framework for CTRSs

In the 2D DP Framework, a CTRS problem is a tuple $\tau = (\mathcal{P}, \mathcal{Q}, \mathcal{R}, e)$, where $\mathcal{P}$, $\mathcal{Q}$ and $\mathcal{R}$ are CTRSs, and $e \in \{a, m\}$ is a flag. We call τ *finite* if there is no infinite $(\mathcal{P}, \mathcal{Q}, \mathcal{R}, e)$-chain; τ is *infinite* if $\mathcal{R}$ is operationally nonterminating or there is an infinite $(\mathcal{P}, \mathcal{Q}, \mathcal{R}, e)$-chain. A *deterministic* 3-*CTRS* $\mathcal{R}$ is operationally terminating iff the *two* (initial) CTRS problems $\tau_H = (\mathsf{DP}_H(\mathcal{R}), \emptyset, \mathcal{R}, m)$ and $\tau_V = (\mathsf{DP}_V(\mathcal{R}), \mathsf{DP}_{VH}(\mathcal{R}), \mathcal{R}, m)$ are *finite* [7, Theorem 2]. A CTRS processor P maps CTRS problems into sets of CTRS problems. Alternatively, it can also return "no". P is *sound* if for all CTRS problems τ, if $\mathsf{P}(\tau) = \{\tau_1, \ldots, \tau_n\}$ for some $n \geq 0$ and for all i, $1 \leq i \leq n$, τ_i is finite, then τ is finite. P is *complete* if for all CTRS problems τ, if $\mathsf{P}(\tau) = \mathsf{no}$ or there is $\tau' \in \mathsf{P}(\tau)$ such that τ' is infinite, then τ is infinite.

Example 3. According to Example 2, for $\mathcal{R}$ in Example 1, $\tau_H = (\{(5)\}, \emptyset, \mathcal{R}, m)$ and $\tau_V = (\{(6)\}, \emptyset, \mathcal{R}, m)$. Some processors apply if $e = m$ only (e.g., Theorems 2 and 7). With m in τ_H and τ_V *more* processors can be potentially used.

In order to prove a deterministic 3-CTRS operationally terminating, we construct *two* trees whose inner nodes are labeled with CTRS problems, the leaves with "yes" or "no", and the roots with τ_H and τ_V, respectively. For every inner node n with label τ, there is a processor P satisfying one of the following conditions: (1) If $\mathsf{P}(\tau) = \mathsf{no}$, then n has just one child n' with label "no"; (2) If $\mathsf{P}(\tau) = \emptyset$ then n has just one child n' with label "yes"; (3) If $\mathsf{P}(\tau) = \{\tau_1, \ldots, \tau_k\}$ with $k > 0$, then n has exactly k children $n_1, \ldots n_k$ with labels $\tau_1, \ldots, \tau_k$, respectively. If *all leaves* of *both* trees are labeled with "yes" and all used processors are *sound*, then $\mathcal{R}$ is *operationally terminating*. If there is *a leaf* labeled with "no" in *one* of the trees, and all processors used on the path from the root to this leaf are *complete*, then $\mathcal{R}$ is *operationally nonterminating*.

Remark 1. In the following, when defining a processor P on a CTRS problem τ, we describe its specific action *under the specified conditions*. If such conditions do not hold, then we assume $\mathsf{P}(\tau) = \{\tau\}$ (it does nothing). Furthermore, we tacitly assume $\mathcal{P}$, $\mathcal{Q}$ and $\mathcal{R}$ to be *deterministic* 3-*CTRSs*.

4 Removing Useless Connection Pairs

Connections between rules in $\mathcal{P}$ become *arcs* in the *graph* $\mathsf{G}(\tau)$ associated to τ, whose set of *nodes* is $\mathcal{P}$; there is an arc from α to α' iff (α, α') is a $(\mathcal{P}, \mathcal{Q}, \mathcal{R}, e)$-chain [7, Definition 9]. We *estimate* them using *abstraction* and *unification* [7, Section 6.1], thus obtaining the *estimated* graph $\mathsf{EG}(\tau)$ [7, Definition 11]. Terms s and t *unify* (written $s =^? t$) if there is a substitution σ (called a *unifier*) such that $\sigma(s) = \sigma(t)$; when this happens, there is a *most general unifier* (*mgu*) θ of s and t which is unique for s and t up to *renaming* of variables. In the following, if terms s and t unify with *mgu* θ, we often write $s =^?_\theta t$ to make it explicit.

Example 4. Consider the following deterministic 3-CTRS $\mathcal{R}$ [7]:

$$g(a) \to c(b) \qquad b \to f(a) \qquad . \quad f(x) \to y \Leftarrow g(x) \to c(y)$$

where

$DP_H(\mathcal{R})$:	$G(a) \to B$	(7)	$DP_{VH}(\mathcal{R})$:	$G(a) \to B$		(9)
$DP_V(\mathcal{R})$:	$F(x) \to G(x)$	(8)		$B \to F(a)$		(10)

For $\tau_V = (\{(8)\},\{(9),(10)\},\ \mathcal{R},\mathsf{m})$, the *estimated graph* $EG(\tau_V)$ is

The 'component $\mathcal{Q}$' of τ_V is essential: the arc is settled because the right-hand side $G(a)$ of (8) 'root-rewrites' into $F(a)$ using $DP_{VH}(\mathcal{R})$: $G(a) \xrightarrow{\Lambda}_{(9)} B \xrightarrow{\Lambda}_{(10)} F(a)$. No rewriting with $\mathcal{R}$ is possible. Now, $F(a)$ *unifies* with the left-hand side $F(x)$.

In some cases, no pair in $\mathcal{Q}$ can be used for this purpose and we can safely *remove* $\mathcal{Q}$ from the CTRS problem. This is the focus of our first processor (Theorem 1). The following definitions are necessary. Following [4] (also [7], where we used it to *estimate* the graph), we let $\text{TCAP}_{\mathcal{R}}$ be $\text{TCAP}_{\mathcal{R}}(x) = y$, if x is a variable, and

$$\text{TCAP}_{\mathcal{R}}(f(t_1,\ldots,t_k)) = \begin{cases} f(\text{TCAP}_{\mathcal{R}}(t_1),\ldots,\text{TCAP}_{\mathcal{R}}(t_k)) & \text{if } \forall \ell \to r \Leftarrow c \in \mathcal{R}, \\ & \ell \text{ and } f(\text{TCAP}_{\mathcal{R}}(t_1),\ldots,\text{TCAP}_{\mathcal{R}}(t_k)) \text{ do not unify} \\ y & \text{otherwise} \end{cases}$$

where y is a fresh variable. We assume that ℓ shares no variable with $f(\text{TCAP}_{\mathcal{R}}(t_1)$, $\ldots,\text{TCAP}_{\mathcal{R}}(t_k))$. As discussed in [4,7], with $\text{TCAP}_{\mathcal{R}}$ we approximate reachability problems by means of *unification*.

Theorem 1 (Removing Connection Pairs). *Let $\mathcal{P}$, $\mathcal{Q}$ and $\mathcal{R}$ be CTRSs. Let $\mathcal{Q}_c = \{u' \to v' \Leftarrow c' \in \mathcal{Q} \mid \exists u \to v \Leftarrow c \in \mathcal{P}, \text{TCAP}_{\mathcal{R}}(v') =^? u\}$. If $\mathcal{Q}_c = \emptyset$, then*

$$P_{RQ}(\mathcal{P},\mathcal{Q},\mathcal{R},e) = \{(\mathcal{P},\emptyset,\mathcal{R},e)\}$$

is a sound and complete processor.

Example 5. For the CTRS $\mathcal{R}$ (from the *Termination Problem Data Base*, TPDB, file `TRS_Conditional/Mixed_CTRS/fib.xml`):

$$\begin{aligned} plus(x,y) &\to y' \Leftarrow x \to 0, y \to y' \\ plus(x,y) &\to s(plus(x',y')) \Leftarrow x \to s(x'), y \to y' \\ fib(0) &\to pair(0,s(0)) \\ fib(s(x)) &\to pair(z,plus(y,z)) \Leftarrow fib(x) \to pair(y,z) \end{aligned}$$

we have:

$DP_H(\mathcal{R})$:	$PLUS(x,y) \to PLUS(x',y') \Leftarrow x \to s(x'), y \to y'$	(11)
	$FIB(s(x)) \to PLUS(y,z) \Leftarrow fib(x) \to pair(y,z)$	(12)
$DP_V(\mathcal{R})$:	$FIB(s(x)) \to FIB(x)$	(13)

$\mathsf{DP}_{VH}(\mathcal{R}) = \mathsf{DP}_H(\mathcal{R})$, and $\tau_H = (\{(11),(12)\}, \emptyset, \mathcal{R}, \mathsf{m})$. For the rhs $\mathsf{PLUS}(x',y')$ of (11), we have that $\mathrm{TCAP}_{\mathcal{R}}(\mathsf{PLUS}(x',y')) = \mathsf{PLUS}(x'',y'')$ and $\mathsf{FIB}(\mathsf{s}(x))$ do *not* unify. A similar observation holds for the rhs $\mathsf{PLUS}(y,z)$ of (12). Thus, for $\tau_V = (\{(13)\}, \{(11),(12)\}, \mathcal{R}, \mathsf{m})$, $\mathsf{P}_{RQ}(\tau_V) = \{\tau_{V1}\}$, where $\tau_{V1} = (\{(13)\}, \emptyset, \mathcal{R}, \mathsf{m})$.

P_{RQ} can be seen as a refinement of P_{SCC} [7], which uses a similar approach.

Remark 2 (Notation). Given (possibly empty) sets of rules $\mathcal{R}, \mathcal{S}$ and a rule α : $\ell \to r \Leftarrow c$, we denote the (possible) *replacement* of α in $\mathcal{R}$ by the rules $\mathcal{S}$ as:

$$\mathcal{R}[\mathcal{S}]_\alpha = \begin{cases} (\mathcal{R} - \{\alpha\}) \cup \mathcal{S} & \text{if } \alpha \in \mathcal{R} \\ \mathcal{R} & \text{otherwise} \end{cases}$$

We let $c[c']_i$ be the condition obtained by replacing in c the i-th atomic condition $s_i \to t_i \in c$ by the sequence of conditions c'. If c' is empty, we write $c[\diamond]_i$.

5   Subterm Processor

In this section, we generalize the *subterm processor* for TRSs [6]. In the following, we write $s \unrhd t$ iff t is a subterm of s, and $s \rhd t$ iff $s \unrhd t$ and $s \neq t$.

Definition 3 (Root Symbols of a CTRS). *Let $\mathcal{R}$ be a CTRS. The set of root symbols in the left- and right-hand sides of the rules in $\mathcal{R}$ is:*

$$Root(\mathcal{R}) = \{root(\ell) \mid \ell \to r \Leftarrow c \in \mathcal{R}\} \cup \{root(r) \mid \ell \to r \Leftarrow c \in \mathcal{R}\}$$

Definition 4 (Simple Projection). *Let $\mathcal{R}$ be a CTRS. A simple projection for $\mathcal{R}$ is a (partial) function $\pi : Root(\mathcal{R}) \to \mathbb{N}$ such that $\pi(f) \in \{1, \ldots, ar(f)\}$ (π is undefined for constant symbols in $Root(\mathcal{R})$).*

We also use π to denote the mapping on terms given by $\pi(f(t_1, \ldots, t_k)) = t_{\pi(f)}$ if $f \in Root(\mathcal{R})$ and $k > 0$, and $\pi(t) = t$ for any other term t.

Given a simple projection π for a CTRS $\mathcal{R}$, we let $\pi(\mathcal{R}) = \{\pi(\ell) \to \pi(r) \mid \ell \to r \Leftarrow c \in \mathcal{R}\}$. Note that the conditions are just *dismissed*. Given a CTRS problem $(\mathcal{P}, \mathcal{Q}, \mathcal{R}, e)$, the subterm processor *removes* from $\mathcal{P}$ and $\mathcal{Q}$ those rules $u \to v \Leftarrow c$ whose left-hand side u contains an immediate subterm $\pi(u)$ which is a *strict superterm* of an immediate subterm $\pi(v)$ of v (i.e., $\pi(u) \rhd \pi(v)$). In the following result we say that a CTRS $\mathcal{R}$ is *collapsing* if there is a rule $\ell \to r \Leftarrow c \in \mathcal{R}$ such that r is a variable. We also write $\mathcal{D}_{\mathcal{R}}$ to denote the set of symbols which are *defined* by rules in $\mathcal{R}$, i.e., $\mathcal{D}_{\mathcal{R}} = \{root(\ell) \mid \ell \to r \Leftarrow c \in \mathcal{R}\}$.

Theorem 2 (Subterm Processor). *Let $\mathcal{P}$, $\mathcal{Q}$, and $\mathcal{R}$ be CTRSs such that $\mathcal{P}$ and $\mathcal{Q}$ are not collapsing, and $(Root(\mathcal{P}) \cup Root(\mathcal{Q})) \cap \mathcal{D}_{\mathcal{R}} = \emptyset$. Let π be a simple projection for $\mathcal{P} \cup \mathcal{Q}$, and $\alpha : u \to v \Leftarrow c \in \mathcal{P} \cup \mathcal{Q}$ be such that $\pi(\mathcal{P}) \cup \pi(\mathcal{Q}) \subseteq \unrhd$ and $\pi(u) \rhd \pi(v)$. Then,*

$$\mathsf{P}_\rhd(\mathcal{P}, \mathcal{Q}, \mathcal{R}, \mathsf{m}) = \{(\mathcal{P}[\emptyset]_\alpha, \mathcal{Q}[\emptyset]_\alpha, \mathcal{R}, \mathsf{m})\}$$

is a sound and complete processor.

Example 6. For $\tau_{V1}=(\{(13)\}, \emptyset, \mathcal{R}, \mathsf{m})$ in Example 5, with $\pi(\mathsf{FIB}) = 1, \mathsf{P}_{\triangleright}(\tau_{V1}) = \{(\emptyset, \emptyset, \mathcal{R}, \mathsf{m})\}$ because $\pi(\mathsf{FIB}(\mathsf{s}(x))) = \mathsf{s}(x) \triangleright x = \pi(\mathsf{FIB}(x))$. Since $(\emptyset, \emptyset, \mathcal{R}, \mathsf{m})$ is trivially finite, τ_V is also finite.

$\mathsf{P}_{\triangleright}$ does *not* apply to τ_H in Example 5 to remove (12). This is because no variable in the right-hand side $\mathsf{PLUS}(y, z)$ is in the corresponding left-hand side $\mathsf{FIB}(\mathsf{s}(x))$. Thus $\mathsf{s}(x)$ cannot be seen as a superterm of any subterm of $\mathsf{PLUS}(y, z)$.

6   Simplifying the Conditions of the Rules

The condition c of a rule $\ell \to r \Leftarrow c$ controls its *applicability* to specific redexes $\sigma(\ell)$ depending on the satisfaction of the instantiated condition $\sigma(c)$. However, dealing with the conditional part of the rules often requires specialized techniques (for instance, for solving *conditional constraints* as in the *removal triple processor* of [7, Theorem 10]) which can make proofs difficult. In other cases, the rules are just *useless*, because they cannot be applied at all, but establishing uselessness to take advantage of it in the analysis of termination is often difficult.

Example 7. Consider the CTRS $\mathcal{R}$ [9, Example 7.2.45]:

$$\mathsf{a} \to \mathsf{a} \Leftarrow \mathsf{b} \to x, \mathsf{c} \to x \quad (14) \qquad \mathsf{c} \to \mathsf{d} \Leftarrow \mathsf{d} \to x, \mathsf{e} \to x \quad (16)$$
$$\mathsf{b} \to \mathsf{d} \Leftarrow \mathsf{d} \to x, \mathsf{e} \to x \quad (15)$$

where $\mathsf{a}, \ldots, \mathsf{e}$ are *constants* and x is a *variable*. Powerful tools like AProVE do not find a proof of operational termination of $\mathcal{R}$ by using transformations. Our implementation of the processors in [7] cannot prove it either. Clearly, (15) and (16) *cannot* be used in any rewriting step (in any $(\mathcal{P}, \mathcal{Q}, \mathcal{R})$-chain) because d and e are irreducible and the only way to get the condition $\mathsf{d} \to x, \mathsf{e} \to x$ satisfied is the instantiation of x to both d and e, which is *not* possible. The only processor in [7] which specifically addresses the use of the information in the conditions of the rules is P_{UR} which removes *unsatisfiable rules* [7, Theorems 11 and 12], i.e., those that cannot be used in any chain because the conditional part is unsatisfiable. P_{UR} uses a well-founded *ordering* $\sqsupset$ which, under appropriate conditions, removes $\ell \to r \Leftarrow c$, if $\sigma(t) \sqsupset \sigma(s)$ holds for some $s \to t \in c$ and all substitutions σ. In our example, we should have $x \sqsupset \mathsf{d}$ or $x \sqsupset \mathsf{e}$ (for all instances of x). This is clearly impossible because, in particular, it would require $\mathsf{d} \sqsupset \mathsf{d}$ or $\mathsf{e} \sqsupset \mathsf{e}$, which contradicts well-foundedness of $\sqsupset$. Some *preprocessing* on the conditions is required before achieving a proof (see Examples 9 and 10 below).

The common feature of the new processors in this section is the *simplification* of the *conditional part* c of the rules in the CTRSs of a given CTRS problem.

6.1   Simplification by Unification

Some conditions $s \to t$ in the conditional part c of a rule $\ell \to r \Leftarrow c$ *cannot* start any (specific) rewriting computation before reaching the corresponding instance of t. They can then be viewed as *unification* problems $s =^? t$. Therefore, $s \to t$ can be *removed* from c if we instantiate the rule with the *most general unifier* θ of s and t. In the following, given a CTRS $\mathcal{R}$, we say that a non-variable term t is a *narrowing redex* (or a *narrex*, for short) if there is a rule $\ell \to r \Leftarrow c \in \mathcal{R}$ such that $Var(t) \cap Var(\ell) = \emptyset$ and t and ℓ unify. We say that $Narr_{\mathcal{R}}(t)$ holds if t contains a narrex, i.e., there is a non-variable position $p \in Pos_{\mathcal{F}}(t)$ such that $t|_p$ is an $\mathcal{R}$-narrex. In the following results, given a rule $\alpha : \ell \to r \Leftarrow c$ with n conditions, some i, $1 \leq i \leq n$, and a substitution θ, we let $\alpha_{\theta,i}$ be the rule $\alpha_{\theta,i} : \theta(\ell) \to \theta(r) \Leftarrow \theta(c[\diamond]_i)$.

Theorem 3 (Simplifying Unifiable Conditions). *Let $\mathcal{P}$, $\mathcal{Q}$, and $\mathcal{R}$ be CTRSs. Let $\alpha : \ell \to r \Leftarrow c \in \mathcal{P} \cup \mathcal{Q} \cup \mathcal{R}$ and $s_i \to t_i \in c$ such that: (1) s_i is linear, (2) $Narr_{\mathcal{R}}(s_i)$ does not hold, (3) $s_i =^?_{\theta} t_i$, (4) for all $s \to t \in c[\diamond]_i$, $Var(s_i) \cap Var(s) = \emptyset$, (5) for all $s \to t \in c$, $Var(s_i) \cap Var(t) = \emptyset$, and (6) $Var(s_i) \cap Var(r) = \emptyset$. Then,*

$$\mathsf{P}_{SUC}(\mathcal{P}, \mathcal{Q}, \mathcal{R}, e) = \{(\mathcal{P}[\{\alpha_{\theta,i}\}]_\alpha, \mathcal{Q}[\{\alpha_{\theta,i}\}]_\alpha, \mathcal{R}[\{\alpha_{\theta,i}\}]_\alpha, e)\}$$

is a sound and complete processor.

Remark 3. Requirement (5) in Theorem 3 (plus determinism of $\mathcal{P}$, $\mathcal{Q}$, and $\mathcal{R}$) implies that $Var(s_i) \subseteq Var(\ell)$. Indeed, every variable in s which is *not* in ℓ must occur in some of the (previous) t, which is forbidden by (5).

Example 8. For τ_H in Example 5, the *estimated graph* $\mathsf{EG}(\tau_H)$ is

With P_{SCC}, which *decomposes* a CTRS problem τ into as many CTRS problems as *Strongly Connected Components* (i.e., *maximal cycles*) are in $\mathsf{EG}(\tau)$ [7, Theorem 9], $\mathsf{P}_{SCC}(\tau_H) = \{\tau_{H1}\}$, where $\tau_{H1}=(\{(11)\}, \emptyset, \mathcal{R}, \mathsf{m})$. We use P_{SUC} to transform τ_{H1} into $\{\tau_{H2}\}$, where $\tau_{H2}=(\{(17)\}, \emptyset, \mathcal{R}, \mathsf{m})$ for the rule

$$\mathsf{PLUS}(\mathsf{s}(x'), y) \to \mathsf{PLUS}(x', y') \Leftarrow y \to y' \tag{17}$$

Now, $\mathsf{P}_{SUC}(\tau_{H2}) = \{\tau_{H3}\}$, for $\tau_{H3}=(\{(18)\}, \emptyset, \mathcal{R}, \mathsf{m})$ with

$$\mathsf{PLUS}(\mathsf{s}(x'), y') \to \mathsf{PLUS}(x', y') \tag{18}$$

$\mathsf{P}_\triangleright$ removes (18) from τ_{H3} (with $\pi(\mathsf{PLUS}) = 1$), to yield a singleton containing a trivially finite CTRS problem $\tau_{H4} = (\emptyset, \emptyset, \mathcal{R}, \mathsf{m})$ which (after an 'administrative' application of P_{SCC} to yield an empty set that becomes a leaf with label **yes** in the tree of the 2D DP Framework) proves τ_H in Example 5 finite. With the proof of finiteness of τ_V in Example 6, we conclude that $\mathcal{R}$ in Example 5 is *operationally terminating* (see Fig. 2).

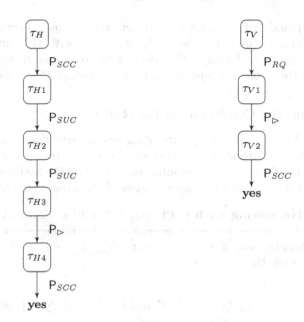

Fig. 2. Operational termination of $\mathcal{R}$ in example 5 in the 2D DP framework

Example 9. For $\mathcal{R}$ in Example 7, we have:

$$\mathsf{DP}_H(\mathcal{R}): \ A \to A \Leftarrow b \to x, c \to x \ (19) \qquad \mathsf{DP}_V(\mathcal{R}): \ A \to B \qquad (20)$$

$$A \to C \Leftarrow b \to x \qquad (21)$$

and $\mathsf{DP}_{VH}(\mathcal{R}) = \emptyset$. Therefore, $\tau_H = (\mathsf{DP}_H(\mathcal{R}), \emptyset, \mathcal{R}, \mathsf{m})$, and $\tau_V = (\mathsf{DP}_V(\mathcal{R}),$ $\mathsf{DP}_{VH}(\mathcal{R}), \mathcal{R}, \mathsf{m})$. We use P_{SUC} (twice) to simplify (15) and (16), thus transforming τ_H into $\{\tau_{H1}\}$, where $\tau_{H1}=(\{(19)\}, \emptyset, \{(14),(22),(23)\}, \mathsf{m})$ with

$$b \to d \Leftarrow e \to d \qquad (22) \qquad c \to d \Leftarrow e \to d \qquad (23)$$

Now, in contrast to P_{UR} in [7], our following processor uses a *syntactic* criterion to remove those rules that cannot be used due to the unsatisfiability of the conditional part of the rule.

Theorem 4 (Irreducible Conditions). *Let $\mathcal{P}$, $\mathcal{Q}$, and $\mathcal{R}$ be CTRSs. Let α : $\ell \to r \Leftarrow c \in \mathcal{P} \cup \mathcal{Q} \cup \mathcal{R}$ and $s \to t \in c$ be such that: (1) s is linear, (2) $Narr_{\mathcal{R}}(s)$ does not hold, and (3) s and t do not unify. Then,*

$$\mathsf{P}_{IC}(\mathcal{P}, \mathcal{Q}, \mathcal{R}, e) = \{(\mathcal{P}[\emptyset]_\alpha, \mathcal{Q}[\emptyset]_\alpha, \mathcal{R}[\emptyset]_\alpha, e)\}$$

is a sound and (if $\alpha \notin \mathcal{R}$ or $e = \mathsf{a}$) complete processor.

Example 10. For τ_{H1} in Ex. 9, $\mathsf{P}_{IC}(\tau_{H1}) = \{\tau_{H2}\}$, with $\tau_{H2}=(\{(19)\}, \emptyset, \{(16)\}, \mathsf{m})$. With P_{SUC}, we obtain $\tau_{H3}=(\{(19)\}, \emptyset, \{(16)\}, \mathsf{m})$, where

$$A \to A \Leftarrow c \to b \qquad (24)$$

Note that this could not be done directly on τ_H because $Narr_\mathcal{R}(\mathsf{b})$ is true. In contrast, $Narr_{\{(14)\}}(\mathsf{b})$ is *not*. Now, $\mathsf{P}_{IC}(\tau_{H3}) = \{(\emptyset, \emptyset, \{(14)\}, \mathsf{m})\}$, which is finite, thus proving τ_H in Example 9 finite as well. We prove τ_V finite by using P_{SCC} in [7] and then conclude operational termination of $\mathcal{R}$ in Example 7.

6.2   Narrowing the Conditions of the Rules

Reachability problems $\sigma(s) \rightarrow^* \sigma(t)$ are often investigated using *narrowing* and unification conditions directly over terms s and t, thus avoiding the 'generation' of the required substitution σ. In the following we define the notion of *narrowing* that we use for CTRSs as a suitable extension of the usual definition for TRSs.

Definition 5 (Narrowing with CTRSs). *Let $\mathcal{R}$ be a CTRS. A term s narrows to a term t (written $s \leadsto_{\mathcal{R},\theta} t$ or just $s \leadsto t$), iff there is a nonvariable position $p \in \mathcal{P}os_\mathcal{F}(s)$ and a rule $\ell \rightarrow r \Leftarrow \bigwedge_{i=1}^n s_i \rightarrow t_i$ in $\mathcal{R}$ (sharing no variable with s) such that:*

1. $s|_p =^?_{\theta_0} \ell$,
2. *for all i, $1 \le i \le n$, $\vartheta_{i-1}(s_i) \leadsto^*_{\mathcal{R},\theta_i} t'_i$ and $t'_i =^?_{\tau_i} \theta_i(\vartheta_{i-1}(t_i))$, where $\vartheta_0 = \theta_0$ and for all $i > 0$, $\vartheta_i = \tau_i \circ \theta_i \circ \vartheta_{i-1}$, and*
3. $t = \theta(s[r]_p)$, *where $\theta = \vartheta_n$.*

The reflexive and transitive closure $\leadsto^*$ of $\leadsto$ is $\leadsto^* = \bigcup_{i \ge 0} \leadsto^i$, where $s \leadsto^0_{\mathcal{R},\epsilon} s$, and $s \leadsto^n_{\mathcal{R},\theta_n} t$ if $s \leadsto_{\mathcal{R},\theta} u$, $u \leadsto^{n-1}_{\mathcal{R},\theta_{n-1}} t$, and $\theta_n = \theta_{n-1} \circ \theta$. In all narrowing steps we assume that a renamed rewrite rule $\ell \rightarrow r \Leftarrow c$ has been used in such a way that no variable in the rule occurs in any previous term in the sequence.

A CTRS $\mathcal{R}$ has *no strict overlaps* if for all $\alpha : \ell \rightarrow r \Leftarrow c \in \mathcal{R}$ and $p \in \mathcal{P}os_\mathcal{F}(\ell) - \{\Lambda\}$, there is no rule $\alpha' : \ell' \rightarrow r' \Leftarrow c'$ such that $Var(\ell) \cap Var(\ell') \ne \emptyset$ (rename the variables if necessary; α and α' can be the same rule) and $\ell|_p$ and ℓ' unify. In the following results, given a CTRS $\mathcal{R}$, we use the following notation:

- $Rules(\mathcal{R}, f) = \{l \rightarrow r \Leftarrow c \in \mathcal{R} \mid root(l) = f\}$ is the set of rules in $\mathcal{R}$ defining a symbol f and $Rules(\mathcal{R}, t) = \cup_{f \in \mathcal{F}(t)} Rules(\mathcal{R}, f)$ is the set of rules in $\mathcal{R}$ defining the symbols in term t.
- Given a term s, $N_1(\mathcal{R}, s)$ represents the set of one-step $\mathcal{R}$-narrowings issued from s: $N_1(\mathcal{R}, s) = \{(t, c, \theta) \mid s \leadsto_{\ell \rightarrow r \Leftarrow c, \theta} t, \ell \rightarrow r \Leftarrow c \in \mathcal{R}\}$.
- Given a rule $\alpha : \ell \rightarrow r \Leftarrow c$ with n conditions and i, $1 \le i \le n$, we let

$$\mathcal{N}(\mathcal{R}, \alpha, i) = \{\theta(\ell) \rightarrow \theta(r) \Leftarrow \theta(c)[\theta(c'), w \rightarrow \theta(t_i)]_i \mid s_i \rightarrow t_i \in c, (w, c', \theta) \in N_1(\mathcal{R}, s_i)\}$$

Theorem 5 (Narrowing the Conditions of Rules). *Let $\mathcal{P}$, $\mathcal{Q}$, and $\mathcal{R}$ be CTRSs. Let $\alpha : u \rightarrow v \Leftarrow c \in \mathcal{P} \cup \mathcal{Q}$ and $s_i \rightarrow t_i \in c$ be such that: (1) s_i is linear, (2) $Rules(\mathcal{R}, s_i)$ have no strict overlap, (3) $Var(s_i) \cap Var(t_i) = \emptyset$, and (4) s_i and t_i do not unify. Then,*

$$\mathsf{P}_{NC}(\mathcal{P}, \mathcal{Q}, \mathcal{R}, e) = \{(\mathcal{P}[\mathcal{N}(\mathcal{R}, \alpha, i)]_\alpha, \mathcal{Q}[\mathcal{N}(\mathcal{R}, \alpha, i)]_\alpha, \mathcal{R}, e)\}$$

is a sound and complete processor.

The unification requirement is essential for the correctness of P_{NC}.

Example 11. Consider the CTRS $\{a \to b, c \to d \Leftarrow a \to a\}$. The left-hand side a of the condition in the second rule narrows into b. But $c \to d \Leftarrow b \to a$ (as obtained by P_{NC}), now *forbids* the rewriting step $c \to d$.

In order to avoid the problem illustrated by the previous example, our next processor is able to remove the requirement of unification by combining the transformations in Theorems 3 and 5 into a single processor.

Theorem 6 (Simplification and Narrowing). *Let $\mathcal{P}$, $\mathcal{Q}$, and $\mathcal{R}$ be CTRSs. Let $\alpha : u \to v \Leftarrow c \in \mathcal{P} \cup \mathcal{Q}$ and $s_i \to t_i \in c$ be such that: (1) s_i is linear, (2) Rules($\mathcal{R}, s_i$) have no strict overlap, (3) $Var(s_i) \cap Var(t_i) = \emptyset$, (4) $s_i =_{\theta}^{?} t_i$, (5) for all $s_j \to t_j \in c$, $i \neq j$, $Var(s_i) \cap Var(s_j) = \emptyset$, (6) for all $s_j \to t_j \in c$, $Var(s_i) \cap Var(t_j) = \emptyset$, and (7) $Var(s_i) \cap Var(v) = \emptyset$. Then,*

$$P_{SUNC}(\mathcal{P}, \mathcal{Q}, \mathcal{R}, e) = \{(\mathcal{P}[\mathcal{N}(\mathcal{R}, \alpha, i) \cup \{\alpha_{\theta,i}\}]_{\alpha}, \mathcal{Q}[\mathcal{N}(\mathcal{R}, \alpha, i) \cup \{\alpha_{\theta,i}\}]_{\alpha}, \mathcal{R}, e)\}$$

is a sound and complete processor.

Example 12. For τ_H in Example 3, rule (5) is transformed by $P_{SUNC}(\tau_H) = \{\tau_{H1}\}$, with $\tau_{H1} = (\{(25),(26),(27)\}, \emptyset, \mathcal{R}, m)$ where

$$G(x, b) \to G(f(c), x) \Leftarrow a \to h(b), b \to x, x \to c \quad (25)$$
$$G(x, b) \to G(f(c), x) \Leftarrow a \to h(c), c \to x, x \to c \quad (26)$$
$$G(f(b), b) \to G(f(c), f(b)) \Leftarrow f(b) \to c \quad (27)$$

Using now P_{SUC} twice we obtain $\tau_{H2} = (\{(28),(29),(27)\}, \emptyset, \mathcal{R}, m)$, where

$$G(b, b) \to G(f(c), b) \Leftarrow a \to h(b), b \to c \quad (28)$$
$$G(c, b) \to G(f(c), c) \Leftarrow a \to h(c), c \to c \quad (29)$$

We use P_{IC} to *remove* (28) from τ_{H2} due to its condition's unsatisfiability. The graph for the obtained problem $\tau_{H3} = (\{(27),(29)\}, \emptyset, \mathcal{R}, m)$ is

With P_{SCC} we obtain $\tau_{H4} = P_{SCC}(\tau_{H3}) = (\{(27)\}, \emptyset, \mathcal{R}, m)$. The proof continues by proving τ_{H4} finite using the sequence P_{NC}, P_{IC}, P_{SUC}, and P_{NR} (see Theorem 7 below), with intermediate applications of P_{SCC}. The complete proof can be found in the benchmarks page (see below) under the label `jlap09-ex17.trs`.

7   Narrowing the Right-Hand Sides of Rules

As mentioned in Sect. 6.2, reachability problems are often approximated or *advanced* by narrowing. The connection between two rules $\alpha : u \to v \Leftarrow c$ and

$\alpha' : u' \to v' \Leftarrow c' \in \mathcal{P}$ within a $(\mathcal{P}, \mathcal{Q}, \mathcal{R})$-chain is a kind of reachability problem $\sigma(v)(\to_{\mathcal{R}}^* \circ \xrightarrow{\Lambda}_{\mathcal{Q}})^* \sigma(u')$, which can also be investigated using narrowing. If there is a (nontrivial, i.e., involving some rewriting with $\mathcal{R}$ or $\mathcal{Q}$) connection between α and α' as above, then after narrowing v into *all* its possible narrowings $v_1, \ldots, v_n$, the connection will be exhibited by some of the v_i. The good point is that connections between v and u that are obtained by the approximations, but which are unfeasible, may be *removed* by other processors (typically P_{SCC}), thus leading to a more precise analysis. In the following, given a CTRS $\mathcal{R}$, a rule $\alpha : u \to v \Leftarrow c$ and a narrowing step $v \rightsquigarrow_{\ell \to r \Leftarrow d, \theta} w$ on the right-hand side v of α, we say that $\theta(u) \to w \Leftarrow \theta(c), \theta(d)$ is a *narrowing* of α. Thus, we let

$$\mathcal{N}(\mathcal{R}, \alpha) = \{\theta(u) \to w \Leftarrow \theta(c), \theta(d) \mid v \rightsquigarrow_{\ell \to r \Leftarrow d, \theta} w, \alpha = u \to v \Leftarrow c, \ell \to r \Leftarrow d \in \mathcal{R}\}$$

Theorem 7 (Narrowing with $\mathcal{R}$). *Let $\mathcal{P}$, $\mathcal{Q}$, and $\mathcal{R}$ be CTRSs. Let $u \to v \Leftarrow c \in \mathcal{P}$ be such that v is linear, and for all $u' \to v' \Leftarrow c' \in \mathcal{P} \cup \mathcal{Q}$ (with possibly renamed variables), v and u' do not unify. Then,*

$$\mathsf{P}_{NR}(\mathcal{P}, \mathcal{Q}, \mathcal{R}, \mathsf{m}) = \{(\mathcal{P}[\mathcal{N}(\mathcal{R}, \alpha)]_\alpha, \mathcal{Q}, \mathcal{R}, \mathsf{m})\}$$

is a sound and complete processor.

In the following processor, we use the rules in $\mathcal{Q}$ to narrow the right-hand sides of pairs $u \to v \Leftarrow c \in \mathcal{P}$ at the root only. We now let

$$\mathcal{N}_\Lambda(\mathcal{R}, \alpha) = \{\theta(u) \to w \Leftarrow \theta(c), \theta(d) \mid v \xrightarrow{\Lambda}_{\ell \to r \Leftarrow d, \theta} w, \alpha = u \to v \Leftarrow c, \ell \to r \Leftarrow d \in \mathcal{R}\}$$

Theorem 8 (Narrowing with $\mathcal{Q}$). *Let $\mathcal{P}$, $\mathcal{Q}$, and $\mathcal{R}$ be CTRSs. Let $u \to v \Leftarrow c \in \mathcal{P}$ be such that v is linear, $Narr_{\mathcal{R}}(v)$ does not hold, and for all $u' \to v' \Leftarrow c' \in \mathcal{P}$ (with possibly renamed variables), v and u' do not unify. Then,*

$$\mathsf{P}_{NQ}(\mathcal{P}, \mathcal{Q}, \mathcal{R}, e) = \{(\mathcal{P}[\mathcal{N}_\Lambda(\mathcal{Q}, \alpha)]_\alpha, \mathcal{Q}, \mathcal{R}, \mathsf{a})\}$$

is a sound and complete processor.

Example 13. We apply P_{NQ} to τ_V in Example 4 to obtain $\mathsf{P}_{NQ}(\tau_V) = \{\tau_{V1}\}$ where $\tau_{V1} = (\{(30)\}, \{(9), (10)\}, \mathcal{R}, \mathsf{m})$ with

$$\mathsf{F}(\mathsf{a}) \to \mathsf{B} \tag{30}$$

And yet $\mathsf{P}_{NQ}(\tau_{V1}) = \{\tau_{V2}\}$ where $\tau_{V2} = (\{(31)\}, \{(9), (10)\}, \mathcal{R}, \mathsf{a})$ with

$$\mathsf{F}(\mathsf{a}) \to \mathsf{F}(\mathsf{a}) \tag{31}$$

This is an infinite CTRS problem, which we will handle with our last processor, introduced in the following section.

8 Detection of Infinite CTRS Problems

The following processor detects a simple kind of infinite CTRS problems.

Theorem 9 (Infinite Problem). *Let $\mathcal{P}$, $\mathcal{Q}$, and $\mathcal{R}$ be CTRSs. Let $u \to v \Leftarrow c \in \mathcal{P}$ and θ, ϑ be substitutions such that for all $s \to t \in c$, $\vartheta(s) \to_{\mathcal{R}}^* \vartheta(t)$ and $\vartheta(v) = \theta(\vartheta(u))$. Then,*

$$\mathsf{P}_{Inf}(\mathcal{P}, \mathcal{Q}, \mathcal{R}, e) = \mathsf{no}$$

is a sound and complete processor.

Note that ensuring the existence of a substitution ϑ that guarantees that the reachability conditions $\vartheta(s) \to_{\mathcal{R}}^* \vartheta(t)$ hold is essential.

Example 14. With τ_{V2} in Example 13 we have $\mathsf{P}_{Inf}(\tau_{V2}) = \mathsf{no}$, witnessing that $\mathcal{R}$ in Example 4 is not operationally terminating.

In Example 14 we easily proved τ_{V2} in Example 4 infinite because rule (31) has no conditional part. Thus, we do not need to check the condition $\vartheta(s) \to_{\mathcal{R}}^* \vartheta(t)$ prescribed in Theorem 9, and we actually let ϑ be the identity substitution. For pairs $u \to v \Leftarrow c \in \mathcal{P}$ where c is not empty, we can use the processor only if we find a substitution ϑ such that $\vartheta(s) \to_{\mathcal{R}}^* \vartheta(t)$ holds for all $s \to t \in c$. In general, this is not computable. In our current implementation, we apply P_{Inf} using a rule $u \to v \Leftarrow c$ only if there is a substitution ϑ such that $\vartheta(s) = \vartheta(t)$ for all $s \to t \in c$ and then we check whether $\vartheta(v) = \theta(\vartheta(u))$ for some substitution θ.

9 Experimental Evaluation

This is the first implementation of the 2D DP framework presented in [7]. It has been developed as part of the tool MU-TERM [2]. In order to assess the practical contributions of the results in this paper, we implemented two versions:

- MU-TERM 5.11 (WRLA'14), which includes the framework and processors defined in [7].
- MU-TERM 5.12 (LOPSTR'14), which extends the previous version with the processors defined in this paper.

Note that our processor P_{Inf} is the *first technique* for proving *operational nontermination* of CTRSs implemented in any termination tool[1]. We compared the MU-TERM implementations and the last version of the existing tools that handle conditional rewriting problems: VMTL 1.3 (http://www.logic.at/vmtl/) and APrOVE 2014 (http://aprove.informatik.rwth-aachen.de/). The experiments

---

[1] With regard to the negative proofs reported here for VMTL, the VMTL web site says that "non-termination means non-termination of the transformed TRS obtained by the conditional TRS through a transformation". Since the transformation used by VMTL is not complete (see [11]), such negative results do not imply operational nontermination of the original CTRS.

have been performed on an Intel Core 2 Duo at 2.4 GHz with 8 GB of RAM, running OS X 10.9.1 using a 60 s timeout. We considered examples from different sources: the CTRSs in the termination problem database[2], TPDB 8.0.7; the CTRSs in the VMTL webpage[3]; and the new CTRSs presented in [7] and in this paper. Results are summarized as follows[4]

Tool Version	Proved (YES/NO)	Av. YES	Av. NO
MU-TERM **5.11 (WRLA'14)**	15/33 (15/0)	0.35s	0s
MU-TERM **5.12 (LOPSTR'14)**	26/33 (21/5)	1.95s	1.02s
AProVE 2014	18/33 (18/0)	9.20s	0s
VMTL 1.3	18/33 (14/4)[3]	6.94s	1.05s[3]

The practical improvements revealed by the experimental evaluation are promising. We can prove (now) termination of 26 of the 33 examples, 11 more examples than our previous version, including 3 examples that cannot be proved by any other automatic tool. Furthermore, if we consider the 15 problems that can be proved by both MU-TERM versions, the new processors yield a faster tool, witnessed by a speedup of 1.26 with respect to the WRLA version. The new processors are very useful to simplify conditions on rules and are also very helpful to detect non-terminating chains. But there is still room for improvement, as we can see in the results obtained by AProVE and VMTL, where other processors based on transformations (Instantiation, Forward Instantiation, Rewriting, . . .) and polynomial interpretations using negative numbers can be useful to improve the results obtained in this paper.

10   Related Work and Conclusions

In [7], the generalization of the DP Framework to CTRSs is accomplished, including the definition of sets of *dependency pairs* which can be used to provide an independent description of infinite computations in the two (*horizontal* and *vertical*) dimensions. In particular, the 2D DP Framework was defined, but only a few processors were presented. Furthermore, such processors barely exploit the peculiarities of the 2D DP framework (e.g., the use of an additional set of pairs $\mathcal{Q}$ for connecting pairs in $\mathcal{P}$, or considering the *conditional part* of the rules to remove and transform them in $\mathcal{P}$, $\mathcal{Q}$, and $\mathcal{R}$). Finally, no implementation or experimental analysis of the 2D DP framework was available.

In this paper we have defined 8 new processors: P_{RQ}, which removes pairs from $\mathcal{Q}$ which are unable to establish any connection within a $(\mathcal{P}, \mathcal{Q}, \mathcal{R})$-chain,

---

[2] See http://www.termination-portal.org/wiki/TPDB/.

[3] See http://www.logic.at/vmtl/benchmarks-cond.html.

[4] detailed benchmarks can be found in http://zenon.dsic.upv.es/muterm/benchmarks/lopstr14/benchmarks.html.

$\mathsf{P}_{\rhd}$, which removes pairs from $\mathcal{P}$ and $\mathcal{Q}$ without paying attention to the structure of rules in $\mathcal{R}$, P_{SUC}, which faithfully removes unifiable conditions, P_{IC}, which removes rules containing conditions that cannot be used in any computation, P_{NC}, which transforms the conditional part of the rules by narrowing, P_{NR}, which transforms the right-hand sides of the rules in $\mathcal{P}$ by using *narrowing* with $\mathcal{R}$, P_{NQ}, which *narrows* with $\mathcal{Q}$ instead, and P_{Inf}, which provides a simple way to detect infinite CTRS problems. We have implemented all our processors (including the ones introduced in [7]). Among these processors, only $\mathsf{P}_{\rhd}$, P_{NR} and P_{Inf} have some analogue processor in the DP Framework for TRSs. All other processors work on the *conditional part* of the rules or (as P_{NQ}) use *active* components (e.g., the rules in $\mathcal{Q}$) which are *missing* in the DP Framework.

Remark 4. Although CTRSs can be *transformed* into TRSs by means of operational-nontermination-preserving transformations (like $\mathcal{U}$ in [9, Definition 7.2.48]), and then existing (narrowing, instantiation) processors of the DP Framework for TRSs (see [3]) could be applied to hopefully obtain similar effects in proofs of operational termination via termination of TRSs, it is unclear (but certainly interesting subject of further research) whether some of our more specific processors P_{SUC}, P_{IC}, P_{SIC}, and P_{NC} could be *simulated* by some of those processors *after* the transformation. Our present view is that this is *unlikely* in most cases. First, it depends on the considered transformation. But even using (a variant of) $\mathcal{U}$ as done by AProVE and VMTL, if this connection were easy, our benchmarks would be closer to those of tools like AProVE which implements most processors and techniques of the DP Framework of TRSs. In contrast, the 2014 edition of the *International Termination Competition*[5] confirms our benchmarks, even with a *time-out* of *five minutes*, instead of the *one minute time-out* used in previous competitions and adopted in our benchmarks, see http://nfa.imn.htwk-leipzig. de/termcomp/show_job_results/5382 for the results of the *TRS Conditional* subcategory. This suggests that the use of *native* processors in the 2D DP Framework is better than transforming CTRS problems into TRS problems and then using the DP Framework for TRSs.

This paper gives strong evidence suggesting that the 2D DP Framework is currently the most powerful technique for proving operational termination of CTRSs. For instance, the CTRSs $\mathcal{R}$ in Examples 1 and 7 cannot be proved operationally terminating by AProVE or VMTL. And $\mathcal{R}$ in Example 4 cannot be proved operationally nonterminating because the transformation used by AProVE and VMTL is *not* complete and does not capture operational nontermination. Furthermore, *neither* these three examples *nor* $\mathcal{R}$ in Example 5 *can* be proved operationally (non)terminating with the processors in [7]. It is true, however, that some of the examples are proved operationally terminating by AProVE or VMTL whereas we are not currently able to provide a proof. And there are examples that cannot be handled by any tool. There is, therefore, room for further

---

[5] See http://www.termination-portal.org/wiki/Termination_Competition/ for the main web site, and http://www.termination-portal.org/wiki/Termination_Competition_2014/ for the 2014 edition, which run from July 19 to 21, 2014.

improvement. In the near future we plan to add more processors to our current implementation. As remarked above, other processors based on transformations or on reducing the proof obligations in the CTRS problems (for instance, by developing a suitable notion of *usable rule* or exploiting *innermost rewriting* [1]) can be added to obtain a more powerful implementation.

Acknoledgements. We thank the referees for their comments and suggestions.

References

1. Arts, T., Giesl, J.: Termination of term rewriting using dependency pairs. Theor. Comput. Sci. **236**(1–2), 133–178 (2000)
2. Alarcón, B., Gutiérrez, R., Lucas, S., Navarro-Marset, R.: Proving termination properties with MU-TERM. In: Johnson, M., Pavlovic, D. (eds.) AMAST 2010. LNCS, vol. 6486, pp. 201–208. Springer, Heidelberg (2011)
3. Giesl, J., Thiemann, R., Schneider-Kamp, P., Falke, S.: Mechanizing and improving dependency pairs. J. Autom. Reasoning **37**(3), 155–203 (2006)
4. Giesl, J., Thiemann, R., Schneider-Kamp, P.: Proving and disproving termination of higher-order functions. In: Gramlich, B. (ed.) FroCos 2005. LNCS (LNAI), vol. 3717, pp. 216–231. Springer, Heidelberg (2005)
5. Giesl, J., Schneider-Kamp, P., Thiemann, R.: AProVE 1.2: automatic termination proofs in the dependency pair framework. In: Furbach, U., Shankar, N. (eds.) IJCAR 2006. LNCS (LNAI), vol. 4130, pp. 281–286. Springer, Heidelberg (2006)
6. Hirokawa, N., Middeldorp, A.: Dependency pairs revisited. In: van Oostrom, V. (ed.) RTA 2004. LNCS, vol. 3091, pp. 249–268. Springer, Heidelberg (2004)
7. Lucas, S., Meseguer, J.: 2D Dependency pairs for proving operational termination of CTRSs. In: Escobar, S. (ed.) WRLA 2014. LNCS, vol. 8663, pp. 195–212. Springer, Heidelberg (2014)
8. Lucas, S., Marché, C., Meseguer, J.: Operational termination of conditional term rewriting systems. Inf. Process. Lett. **95**, 446–453 (2005)
9. Ohlebusch, E.: Advanced Topics in Term Rewriting. Springer, Heidelberg (2002)
10. Schernhammer, F., Gramlich, B.: Characterizing and proving operational termination of deterministic conditional term rewriting systems. J. Logic Algebraic Program. **79**, 659–688 (2010)
11. Schernhammer, F., Gramlich, B.: VMTL–a modular termination laboratory. In: Treinen, R. (ed.) RTA 2009. LNCS, vol. 5595, pp. 285–294. Springer, Heidelberg (2009)

Security

Partial Evaluation for Java Malware Detection

Ranjeet Singh and Andy King(✉)

School of Computing, University of Kent, Kent CT2 7NF, UK
a.m.king@kent.ac.uk

Abstract. The fact that Java is platform independent gives hackers the opportunity to write exploits that can target users on any platform, which has a JVM implementation. To circumvent detection by anti-virus (AV) software, obfuscation techniques are routinely applied to make an exploit more difficult to recognise. Popular obfuscation techniques for Java include string obfuscation and applying reflection to hide method calls; two techniques that can either be used together or independently. This paper shows how to apply partial evaluation to remove these obfuscations and thereby improve AV matching. The paper presents a partial evaluator for Jimple, which is a typed three-address code suitable for optimisation and program analysis, and also demonstrates how the residual Jimple code, when transformed back into Java, improves the detection rates of a number of commercial AV products.

1 Introduction

Java is both portable and architecture-neutral. It is portable because Java code is compiled to JVM byte code for which interpreters exist, not only for the popular desktop operating systems, but for phones and tablets, and as browser plug-ins. It is architecture-neutral because the JVM code runs the same regardless of environment. This presents a huge advantage over languages, such as C/C++, but also poses a major security threat. If an exploit levers a vulnerability in a JVM implementation, it will affect all versions of a JVM that have not closed off that loophole, and well as those users who have not updated their JVM.

JVM vulnerabilities have been used increasingly by criminals in so-called client side attacks, often in conjunction with social engineering tactics. For example, a client-side attack might involve sending a pdf document [14] that is designed to trigger a vulnerability when it is opened by the user in a pdf reader. Alternatively a user might be sent a link to a website which contains a Java applet which exploits a JVM vulnerability [1] to access the user's machine. Client-side attacks provide a way of bypassing a firewall that block ports to users' machines and, are proving to be increasingly popular: last year many JVM exploits were added to the Metasploit package, which is a well-known and widely-used penetration testing platform. This, itself, exacerbates the problem. As well as serving penetration testers and security engineers, a script kiddie and or a skilled blackhat can reuse a JVM vulnerability reported in Metasploit, applying obfuscation so that it is not recognised by even up-to-date AV detection software.

© Springer International Publishing Switzerland 2015
M. Proietti and H. Seki (Eds.): LOPSTR 2014, LNCS 8981, pp. 133–147, 2015.
DOI: 10.1007/978-3-319-17822-6_8

Experimental evidence suggests that commercial AV software use Metasploit as source of popular attack vectors, since exploits from Metasploit are typically detected if they come in an unadulterated form. One can only speculate what techniques an AV vendor actually uses, but detection methods range from entirely static techniques, such as signature matching, to entirely dynamic techniques, in which the execution of the program or script is monitored for suspicious activity. In signature matching, a signature (a hash) is derived, often by decompiling a sample, which is compared against a database of signatures constructed from known malware. Signatures are manually designed to not trigger a false positive which would otherwise quarantine an innocuous file. Dynamic techniques might analyse for common viral activities such as file overwrites and attempts to hide the existence of suspicious files, though it must be said, there are very few academic works that address the classification of Java applets [17].

The essential difference between running a program in an interpreter and partially evaluating it within a partial evaluator is that the latter operates over a partial store in which not all variables have known values; the store determines which parts of the program are executed and which parts are retained in the so-called residual. Partial evaluation warrants consideration in malware detection because it offers a continuum between the entirely static and the entirely dynamic approaches. In particular, one can selectatively execute parts of the program, namely those parts that mask suspicious activity, and then use the residual in AV scanning. This avoids the overhead of full execution while finessing the problem of incomplete data that arises when a program is evaluated without complete knowledge of its environment. On the theoretical side, partial evaluation provides nomenclature (e.g. polyvariance) and techniques (e.g. generalisation) for controlling evaluation and specialisation. On the practical side, partial evaluation seems to be partially appropriate for AV matching because Java exploits are often obfuscated by string obfuscation and by using advanced language features such as reflection. Although reflection is designed for such applications as development environments, debuggers and test harnesses, it can also be applied to hide a method call that is characteristic of the exploit. This paper will investigates how partial evaluation can be used to deobfuscate malicious Java software; it argues that AV scanning can be improved by matching on the residual JVM code, rather than original JVM code itself.

1.1 Contributions

This paper describes how partial evaluation can deobfuscate malicious Java exploits; it revisits partial evaluation from the perspective of Java malware detection which, to our knowledge, is novel. The main contributions are as follows:

- The paper describes the semantics of a partial evaluator for Jimple [20]. Jimple is a typed three-address intermediate representation that is designed to support program analysis and, conveniently, can be decompiled back into Java for AV matching.

```
java.security.Permissions o = new java.security.Permissions();
o.add(new AllPermission());

Class<?> c = Class.forName("java.security.Permissions");
Object o = c.newInstance();
Method m = c.getMethod("add", Permission.class);
m.invoke(o, new AllPermission());
```

Listing 1.1. Method call and its obfuscation taken from CVE-2012-4681

- The paper shows how partial evaluation can be used to remove reflection from Jimple code, as well as superfluous string operations, that can be used to obfuscate malicious Java code.
- The paper describes how partial evaluation can be used in tandem with an abstract domain so as to avoid raising an error when a branch condition cannot be statically resolved [18, Sect. 3.3]. In such a situation, one branch might lead to an environment whose bindings are inconsistent with that generated along another. Rather than abort when inconsistent environments are detected at a point of confluence, we merge the environments into an abstract environment that preserves information from both, so that partial evaluation can continue.

2   Primer on Java Obfuscation

This section will describe techniques that are commonly used to obfuscate Java code to avoid AV detection. The obfuscations detailed below are typically used in combination; it is not as if one obfuscation is more important than another.

2.1   Reflection Obfuscation

An AV filter might check for the invocation of a known vulnerable library function, and to thwart this, malicious applets frequently use reflection to invoke vulnerable methods. This is illustrated by the code in Listing 1.1 which uses the `Class.forName` static method to generate an object c of type `Class`. The c object allows the programmer to access information pertaining to the Java class `java.security.Permissions`, and in particular create an object o of type `java.security.Permissions`. Furthermore, c can be used to create an object m that encapsulates the details of a method call on object o. The invocation is finally realised by applying the `invoke` on m using o as a parameter. This sequence of reflective operation serves to disguise what would otherwise be a direct call to the method `add` on an object of type `Permissions`.

2.2   String Obfuscation

Malicious applets will often assemble a string at run-time from a series of component strings. Alternatively a string can be encoded and then decoded at run-time.

```
public static String getStr(String input) {
  StringBuilder sb = new StringBuilder();

  for(int i = 0; i < input.length(); i++) {
    if(!(input.charAt(i) >= '0' && input.charAt(i) <= '9')) {
      sb.append(input.charAt(i));
    }
  }
  return sb.toString();
}

String str = "1j2a34v5a.s7e8cu9r00i1ty.P3er4m5i6s7s8io9n0s";

Class<?> c = Class.forName(getStr(str));
```

Listing 1.2. String Obfuscation with numeric characters

Either tactic will conceal a string, making it more difficult to recognise class and method names, and thereby improving the chances of outwitting a signature-based AV system. Listing 1.2 gives an example of a string reconstruction method that we found in the wild, in which the string `java.lang.SecurityManager` is packed with numeric characters which are subsequently removed at runtime. Listing 1.3 illustrates an encoder which replaces a letter with the letter 13 letters after it in the alphabet. The encoded strings are then decoded at run-time before they are used to create a handle of type `Class` that can, in turn, be used to instantiate `java.lang.SecurityManager` objects.

```
public static String rot13(String s) {
  StringBuffer sb = new StringBuffer();
  for (int i = 0; i < s.length(); i++) {
    char c = s.charAt(i);
    if        (c >= 'a' && c <= 'm') c += 13;
    else if   (c >= 'A' && c <= 'M') c += 13;
    else if   (c >= 'n' && c <= 'z') c -= 13;
    else if   (c >= 'N' && c <= 'Z') c -= 13;
    sb.append(c);
  }
  return sb.toString();
}

String str = "wnin.frphevgl.Crezvffvbaf";

Class<?> c = Class.forName(rot13(str));
```

Listing 1.3. String obfuscation using the rot13 substitution cipher

2.3   Other Obfuscations

There is also no reason why other obfuscations [6] cannot be used in combination with reflection and string obfuscation. Of these, one of the most prevalent is name obfuscation in which the names of the user-defined class and method names are substituted with fresh names. For example, the name getStr in Listing 1.2 might be replaced with a fresh identifier, so as to mask the invocation of a known decipher method.

3   Partial Evaluation

In this section we outline a partial evaluator for removing string obfuscation and reflection from Jimple code, which is a three address intermediate representation (IR) for the Java programming language and byte code. Jimple is supported by the Soot static analysis framework and, quite apart from its simplicity, Soot provides support for translating between Jimple and Java.

There are two approaches to partial evaluation: online and offline. In the online approach specialisation decisions are made on-the-fly, based on the values of expressions that can be determined at that point in the specialisation process. In the offline approach, the partial evaluator performs binding time analysis, prior to specialisation, so as to classify expressions as static or dynamic, according to whether their values will be fully determined at specialisation time. This classification is then used to control unfolding, so that the specialisation phase is conceptually simple. The online approach, however, mirrors the structure of the interpreter in the partial evaluator, and hence is easier to present (and ultimately justify by abstract interpretation). We therefore follow the online school.

Figures 1 and 2 present some highlights of the partial evaluator, which specialises sequences of Jimple instructions, that are tagged with labels for conditional jumping. The sequel provides a commentary on some representative instructions. In what follows, l denotes a location in memory, x a variable, and v a value. A value is either a primitive value, such as an integer or a boolean, or an object, or $\top$ which is used to indicate the absence of information. An object is considered to be a class name, C, paired with an environment ρ, together denoted $C : \rho$; C is the name of the class from which the object is instantiated and ρ specifies the memory locations where the fields (internal variables) of the object are stored.

The partial evaluator uses an environment ρ and a store σ to record what is known concerning the values of variables. The environment ρ is a mapping from the set of variables to the memory locations, and the store σ is a mapping from locations to values. The partial evaluator is presented as a function $\mathcal{P}[\![S]\!]\langle\rho, \sigma, o\rangle$, which executes the sequence of instructions S in the context of an environment ρ, store σ and current object o.

3.1   Type Declarations

A statement $var\ t\ x$ declares that x is a of type t, where t is either primitive or a user-defined class. Such a declaration is handled by allocating a memory

$$\mathcal{P}[\![var\ t\ x; S]\!]\langle\rho, \sigma, o\rangle =$$
$$v = \mathsf{default}(t)$$
$$l = \mathsf{allocate}(v)$$
$$\rho' = \{x \mapsto l\}$$
$$\sigma' = \{l \mapsto v\}$$
$$emit[\![var\ t\ x]\!]$$
$$\mathcal{P}[\![S]\!]\langle\rho \circ \rho', \sigma \circ \sigma', o\rangle$$

$$\mathcal{P}[\![x := y \oplus z; S]\!]\langle\rho, \sigma, o\rangle =$$
$$\quad \mathsf{if}\ \sigma(\rho(y)) = \top \vee \sigma(\rho(z)) = \top\ \mathsf{then}$$
$$\qquad emit[\![x := y \oplus z]\!]$$
$$\qquad \sigma' = \sigma \circ \{\rho(x) \mapsto \top\}$$
$$\qquad \mathcal{P}[\![S]\!]\langle\rho, \sigma', o\rangle$$
$$\quad \mathsf{else}$$
$$\qquad v = \sigma(\rho(y)) \oplus \sigma(\rho(z))$$
$$\qquad emit[\![x := v]\!]$$
$$\qquad \sigma' = \sigma \circ \{\rho(x) \mapsto v\}$$
$$\qquad \mathcal{P}[\![S]\!]\langle\rho, \sigma', o\rangle$$
$$\quad \mathsf{endif}$$

$$\mathcal{P}[\![x := @this; S]\!]\langle\rho, \sigma, o\rangle =$$
$$emit[\![x := @this]\!]$$
$$\sigma' = \{\rho(x) \mapsto o\}$$
$$\mathcal{P}[\![S]\!]\langle\rho, \sigma \circ \sigma', o\rangle$$

$$\mathcal{P}[\![x := @parameter_i; S]\!]\langle\rho, \sigma, o\rangle =$$
$$emit[\![x := @parameter_i]\!]$$
$$\sigma' = \{\rho(x) \mapsto \sigma(\rho(parameter_i))\}$$
$$\mathcal{P}[\![S]\!]\langle\rho, \sigma \circ \sigma', o\rangle$$

$$\mathcal{P}[\![x := new\ C; S]\!]\langle\rho, \sigma, o\rangle =$$
$$\langle \bar{t}, \bar{f} \rangle = getFields(C)$$
$$\bar{l} = \mathsf{allocates}(\bar{f})$$
$$\bar{v} = \mathsf{defaults}(\bar{t})$$
$$\rho' = \{f_0 \mapsto l_0, \ldots, f_n \mapsto l_n\}$$
$$\sigma' = \{l_0 \mapsto v_0, \ldots, l_n \mapsto v_n, \rho(x) \mapsto C : \rho'\}$$
$$emit[\![x := new\ C]\!]$$
$$\mathcal{P}[\![S]\!]\langle\rho, \sigma \circ \sigma', o\rangle$$

Fig. 1. Outline of partial evaluator: declarations and assignments

location l using the auxiliary $\mathsf{allocate}$ and then updating the environment ρ' to reflect this change. The store is also mutated to map location l to the default value for the type t, which is given by the auxiliary function $\mathsf{default}$. The default values for the primitives types are 0 for *int* and 0 for *boolean*. The default values for object types is *null*.

$$\mathcal{P}[\![return\ x;\ \_]\!]\langle \rho, \sigma, o \rangle =$$
$$\quad \sigma' = \{\rho(return) \mapsto \sigma(\rho(x))\}$$
$$\quad emit[\![return\ x]\!]$$
$$\quad \langle \rho, \sigma \circ \sigma', o \rangle$$

$$\mathcal{P}[\![x\ :=\ \mathsf{virtualinvoke}(obj, m(\bar{t}), \bar{y}); S]\!]\langle \rho, \sigma, o \rangle =$$
$$\quad if\ \sigma(\rho(obj)) = \top\ then$$
$$\qquad emit[\![x\ :=\ \mathsf{virtualinvoke}(obj, m(\bar{t}), \bar{y})]\!]$$
$$\qquad \sigma' = \{\rho(x) \mapsto \top\}$$
$$\qquad \mathcal{P}[\![S]\!]\langle \rho, \sigma \circ \sigma', o \rangle$$
$$\quad else\ if\ \sigma(\rho(obj)) = C : \rho'$$
$$\qquad if\ C = Method \wedge m = invoke$$
$$\qquad\quad if\ \sigma(\rho'(method)) = null\ \mathsf{then}\ error$$
$$\qquad\quad \bar{y}' = \langle y_1, \ldots, y_n \rangle$$
$$\qquad\quad emit[\![x\ :=\ \mathsf{virtualinvoke}(y_0, \sigma(\rho'(method)), \bar{y}')]\!]$$
$$\qquad\quad \sigma' = \{\rho(x) \mapsto \top\}$$
$$\qquad\quad \mathcal{P}[\![S]\!]\langle \rho, \sigma \circ \sigma', o \rangle$$
$$\qquad else$$
$$\qquad\quad \bar{v} = \sigma(\rho(\bar{y}))$$
$$\qquad\quad B = findMethod(C.m(\bar{t}))$$
$$\qquad\quad \bar{l} = allocates(\bar{v})$$
$$\qquad\quad k = allocate(result)$$
$$\qquad\quad \rho'' = \{parameter_0 \mapsto l_0, \ldots, parameter_n \mapsto l_n, result \mapsto k\}$$
$$\qquad\quad \sigma' = \sigma \circ \{l_0 \mapsto v_0, \ldots, l_n \mapsto v_n\}$$
$$\qquad\quad \langle \_, \sigma'', \_\rangle = \mathcal{P}[\![B]\!]\langle \rho'' \circ \rho', \sigma', C : \rho' \rangle$$
$$\qquad\quad \sigma''' = \{\rho(x) \mapsto \sigma''(\rho''(result))\}$$
$$\qquad\quad \mathcal{P}[\![S]\!]\langle \rho, \sigma \circ \sigma''', o \rangle$$

$$\mathcal{P}[\![if\ x\ goto\ l; S]\!]\langle \rho, \sigma, o \rangle =$$
$$\quad if\ \sigma(\rho(x)) = 0\ \mathsf{then}$$
$$\qquad \mathcal{P}[\![S]\!]\langle \rho, \sigma, o \rangle$$
$$\quad else\ if\ \sigma(\rho(y)) = 1\ \mathsf{then}$$
$$\qquad S = lookup(l)$$
$$\qquad \mathcal{P}[\![S]\!]\langle \rho, \sigma, o \rangle$$
$$\quad else\ if\ \sigma(\rho(x)) = \top\ \mathsf{then}$$
$$\qquad n\ =\ lookup(S, P)$$
$$\qquad c\ =\ confluence(l, n)$$
$$\qquad emit[\![if\ n\ goto\ l]\!]$$
$$\qquad T_t\ =\ branch(l, c)$$
$$\qquad T_f\ =\ branch(n, c)$$
$$\qquad \langle \rho, \sigma_t, o \rangle = \mathcal{P}[\![T_t]\!]\langle \rho, \sigma, o \rangle$$
$$\qquad \langle \rho, \sigma_f, o \rangle = \mathcal{P}[\![T_f]\!]\langle \rho, \sigma, o \rangle$$
$$\qquad \sigma' = \{l \mapsto v \mid l \in codomain(\rho) \wedge v = \mathsf{if}\ \sigma_t(l) = \sigma_f(l)\ \mathsf{then}\ \sigma_t(l)\ \mathsf{else}\ \top\}$$
$$\qquad \mathcal{P}[\![P\ drop\ c]\!]\langle \rho, \sigma', o \rangle$$
$$\quad \mathsf{endif}$$

Fig. 2. Outline of partial evaluator: control-flow

3.2 new

A statement $x = new\ C$ instantiates the class C to create an object that is represented by a pair $C : \rho$ where the environment ρ maps the fields of C to memory locations that store their values. Different objects $C : \rho_1$ and $C : \rho_2$ from the same class C map the same field variables to different locations. The auxiliary method getFields retrieves the types and the names of the fields of the class C. The function defaults takes a vector of types t and returns a vector of default values that is used to populate the fields, following the conventions of default.

3.3 Arithmetical Operations

An assignment statement $x := y \oplus z$ can only be statically evaluated if both the variables y and z are bound to known values. In this circumstance the assignment $x := y \oplus z$ is specialised to $x := v$ where v is the value of the expression $y \oplus z$. The store σ' is updated to reflect the new value of x, as the value of x is statically known. Note that the residual includes $x := v$, even though information on x is duplicated in the store σ', so as to permit subsequent statements, with reference x, to be placed in the residual without substituting x with its value v. If there are no statements that reference x then the assignment $x := v$ will be removed by dead variable elimination, which is applied as a post-processing step.

3.4 this and Parameters

In Jimple there is a distinguished variable *this* which stores the current object reference which, in the partial evaluator, is modelled with the current object o, that is passed with the environment and the store. An assignment statement $x := @this$ thus merely updates the location $\rho(x)$ with o.

Also in Jimple, a special variable $parameter_i$ is used to store the location of the ith formal argument of a method call, where the first argument has an index of 0. This is modelled more directly in the partial evaluator, so that an assignment statement $x := @parameter_i$ updates the location $\rho(x)$ with the value of this parameter.

3.5 return and virtualinvoke

The statement $return\ x$ writes the value of x to a special variable return, which is subsequently read by virtualinvoke.

The handling of virtualinvoke$(obj, m(t), y)$ is worthy of special note, both in the way reflective and non-reflective calls are handled. A reflective call arises when the method m coincides with invoke and the object obj is of type Method. The reflective method call is a proxy for the so-called reflected method call. The reflected method call is not applied to the object obj but an object that is prescribed by the first parameter of y. Moreover, the method that is applied to this object is given by obj in a distinguished field that, for our purposes, is called *method*. This field either contains null, indicating that it has been initialised but not been reset, or a string that represents the name of a method that is to be invoked. If the method field is null then an error is issued, otherwise the string stored in the field method is used to generate the residual. Note that the reflected method call is not invoked; merely placed in the residual. Note too that the first argument of virtualinvoke in the residual is y_0 whereas the last is the vector $\boldsymbol{y}'$ which coincides with $\boldsymbol{y}$ with the exception that the first element has been removed. The first argument is the variable name (rather than the object itself) to which the residuated method will ultimately be applied; the third argument is the list of actual arguments that will be passed to the method on its invocation.

In the case of a non-reflected call, the values of the parameters are looked up, and then an auxiliary function findMethod is applied to find a block B which is the entry point into the method of the class C whose signature matches $m(t)$. The function allocates is then called to generate fresh memory locations, one for each actual argument $y_0, \ldots, y_n$. The environment is then extended to map the distinguished variables $parameter_0, \ldots, parameter_n$ to fresh memory locations, so as to store the actual arguments. The partial evaluator is then recursively involved on the block B using $C : \rho'$ as the object. The net effect of the method call is to update the store σ' which is used when evaluating the remaining statements S.

Note that this formulation assumes that method calls are side-effect free. Although this is true for string obfuscation methods that we have seen, for full generality the partial evaluator should be augmented with an alias analysis, in the spirit of side-effect analysis [9], that identifies those variables that are possibly modified by a method call, hence must be reset to $\top$.

3.6 goto

Specialising a conditional jump in Jimple is not conceptually difficult, but is complicated by the way if x goto $l; S$ will drop through to execute the first instruction of the sequence S if the boolean variable x is false. This makes the control-flow more difficult to recover when the value of x is unknown. When x is known to be true, however, the partial evaluator is merely redirected at a block B that is obtained by looking up the sequence whose first instruction is labelled by l. Conversely, if x is known to be false, then partial evaluation immediately proceeds with S.

In the case when x has an undetermined value, the partial evaluator explores both branches until the point of confluence when both branches merge. Then the partial evaluator continues at the merge point, relaxing the store to σ' so that it is consistent with the stores that are derived on both branches. Note that partial evaluation does not halt if the stores are inconsistent; instead it will unify the two stores by replacing any inconsistent assignment for any location with an assignment to $\top$. Note that it is only necessary to unify those locations that are reachable from the variables that are currently in scope.

To realise this approach, an auxiliary function lookup is used to find the position n of the first statement of S in the list P which constitutes the statements

```
//BEFORE
public static void main(String[] args) {
    Main m = new Main();
    String encrypted = "qbFbzrguvat";
    Method method = m.class.getMethod(rot13(encrypted));
    method.invoke(m, null);
}

//AFTER
public static void main(String[] args) {
    Main m = new Main();
    String encrypted = "qbFbzrguvat";
    Method method = m.class.getMethod(rot13(encrypted));
    m.doSomething();
}


//AFTER DEAD VARIABLE ELIMINATION
public static void main(String[] args) {
    Main m = new Main();
    m.doSomething();
}
```

Listing 1.4. Before and after partial evaluation

for the currently executing method. This is the position of the first statement immediately after the conditional branch. Then a function confluence examines the control-flow graph of the method so as to locate the confluence point, of the true and false branches, identified by the index c of S. Both branches are evaluated separately with two copies of the environment, until the confluence point where the two environments are merged. Partial evaluation then resumes at the confluence point, which corresponds to the instruction sequence P drop c, namely, execution is continued at the cth instruction of the sequence P.

3.7   Example

Listing 1.4 gives the before and after for a method call that is obfuscated by reflection and string obfuscation, using the ROT13 simple letter substitution cipher given in Listing 1.3. The residual Jimple code is presented as Java for readability. Completely unfolding the rot13 method call decrypts the string qbFbzrguvat as the string doSomething. This string statically defines the value of the object method, allowing method.invoke(m, null) to be specialised to m.doSomething(), thereby removing the reflective call. Note that the variables encrypted and method cannot be removed without dead variable elimination.

4   Experiments

To assess how partial evaluation can aid in AV matching, a number of known applet malware samples from the Metasploit exploit package [2] were obfuscated using the techniques outlined in Sect. 2. Details of the samples are given in Fig. 3; the samples were chosen entirely at random. So as to assess the effect of partial evaluation against a representative AV tool, we compared the detection rates, with and without partial evaluation, on eight commercial AV products. Together these products cover the majority of the global market, as reported in 2013 [15] and is illustrated in the chart given in Fig. 3. Conveniently, VirusTotal [19] provides a prepackaged interface for submitting malware samples to all of these products, with the exception of Avira, which is why this tool does not appear in our experiments.

Unfortunately, developing a complete partial evaluator for Jimple is a major undertaking, since it is necessary to support the entire Java API and runtime environment, which itself is huge. To side-step this engineering effort, we implemented a partial evaluator in Scala, following the description in Sect. 3, only providing functionality for String, StringBuffer and StringBuilder classes. This was achievable since Java String objects are accessible to Scala. (Scala's parser combinator library also make it is straightforward to engineer a parser for Jimple.) Although other objects could be handled in the same way, we simply took each of these obfuscated CVEs and extracted the Jimple code and methods that manipulated strings. This code was then partially evaluated so as to deobfuscate the string handling. The CVEs were then hand-edited to reflect the residual, and then ran through VirusTotal to check that the effects of obfuscation had been truly annulled. Future implementation work will be to automate the entire process, namely translate the Jimple residual into Java using Soot [8] and then invoke VirusTotal automatically through its public web API.

Table 1 details the detection rates for the AVs given in Fig. 3, without obfuscation, with just string obfuscate, with just reflection obfuscation, and with both obfuscations applied. This gives four experiments in all. It is important to appreciate that the obfuscations used in the fourth experiment include all those obfuscations introduced in the second and third experiments and no more.

The results show that in most cases the AVs detect most of the exploits in their unadulterated form. Exploits CVE-2012-5088 and CVE-2013-2460 go the most undetected, which is possibly because both exploits make extensive use of reflection. It is interesting to see that the product with the highest

CVE	Java Applet Exploit
2012-4681	Remote Code Execution
2012-5076	JAX WS Remote Code Execution
2013-0422	JMX Remote Code Execution
2012-5088	Method Handle Remote Code Execution
2013-2460	Provider Skeleton Insecure Invoke Method

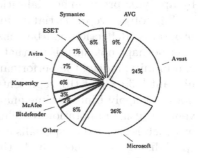

Fig. 3. CVEs and AVs

market share (Microsoft) was unable to detect any of the exploits after string obfuscation, which suggests the removing this obfuscation alone is truly worthwhile. Moreover, after introducing reflection the AV detection count for each exploit drops significantly. Furthermore, applying reflection with string obfuscation is strictly stronger than applying string obfuscation and reflection alone. CVE-2012-4681 represents an anomaly under McAfee since reflection obfuscation impedes detection whereas, bizarrely, using reflection with string obfuscation does not. Interestingly, McAfee classifies this CVE with the message Heuristic.BehavesLike.Java.Suspicious-Dldr.C which suggests that it is using a heuristic behavioural approach which might explain its unpredictability.

Most importantly, applying partial evaluation to the CVEs used in the fourth experiment restores the detection rates to those of the first experiment. Thus detection is improved, without having to redistribute the signature database.

5   Related Work

Although there has been much work in Java security, partial evaluation and reflection, there are few works that concern all three topics. This section provides pointers to the reader for the main works in each of these three separate areas.

One of the very few techniques that has addressed the problem of detecting malicious Java Applets is Jarhead [17]. This recent work uses machine learning to detect malicious Applets based on 42 features which include such things as the number of instructions, the number of functions per class and cyclomatic complexity [13]. Jarhead also uses special features that relate to string obfuscation, such as the number and average length of the strings, and the fraction of strings that contain non-ASCII printable characters. Other features that it applies determine the degree of active code obfuscation, such as the number of times that reflection is used within the code to instantiate objects and invoke methods. Out of a range of classifiers, decision trees are shown to be the most reliable. Our work likewise aspires to be static, though partial evaluation takes this notion to the limit, so as to improve detection rates. Moreover, machine learning introduces the possibility of false negatives and, possibly worse, false positives. Our approach is to scaffold off existing AV products that have been carefully crafted to not trigger false positives, and improve their matching rates by applying program specialisation as a preprocessing step.

The objective of partial evaluation is to remove interpretive overheads from programs. Reflection can be considered to be one such overhead and therefore it is perhaps not surprising that it has attracted interest in the static analysis community; indeed the performance benefit from removing reflection can be significant [16] . Civet [18] represents state-of-the-art in removing Java reflection; it does not apply binding-time analysis (BTA) [3] but relies on programmer intervention, using annotation to delineate static from dynamic data, the correctness of which is checked at specialisation time. Advanced BTAs have been defined for specialising Java reflection [4], though to our knowledge, none have been implemented. We know of no partial evaluator for Jimple, though Soot represents

Table 1. Experimental results

Exploit name	Microsoft	Avast	AVG	Symantec	ESET	Kaspersky	McAfee	Bitdefender
CVE	No obfuscation							
2012-4681	✓	✓	✓	✗	✓	✓	✓	✓
2012-5076	✓	✗	✓	✓	✓	✓	✓	✗
2013-0422	✓	✓	✗	✓	✓	✓	✓	✗
2012-5088	✗	✗	✗	✗	✓	✗	✓	✗
2013-2460	✗	✓	✓	✗	✓	✗	✓	✗
CVE	String obfuscation							
2012-4681	✗	✓	✓	✗	✓	✓	✓	✓
2012-5076	✗	✗	✓	✓	✗	✗	✓	✗
2013-0422	✗	✗	✗	✓	✗	✗	✓	✗
2012-5088	✗	✗	✗	✗	✓	✗	✓	✗
2013-2460	✗	✗	✓	✗	✗	✗	✓	✗
CVE	Reflection obfuscation							
2012-4681	✓	✓	✓	✗	✓	✗	✗	✗
2012-5076	✗	✗	✗	✓	✗	✓	✓	✗
2013-0422	✓	✓	✓	✓	✓	✓	✓	✗
2012-5088	✗	✗	✗	✗	✓	✗	✓	✗
2013-2460	✗	✗	✓	✗	✓	✗	✓	✗
CVE	String and reflection obfuscation							
2012-4681	✗	✓	✓	✗	✗	✗	✓	✗
2012-5076	✗	✗	✗	✓	✗	✗	✓	✗
2013-0422	✗	✗	✗	✓	✗	✗	✓	✗
2012-5088	✗	✗	✗	✗	✓	✗	✓	✗
2013-2460	✗	✗	✓	✗	✗	✗	✓	✗

the ideal environment for developing one [8]. Quite apart from its role in deobfuscation, partial evaluation can also be applied in obfuscation [10]: a modified interpreter, that encapsulates an obfuscation technique, is partially evaluated with respect to the source program to automatically obfuscate the source. Program transformation has been proposed for deobfuscating binary programs [5], by unpacking and removing superfluous jumps and junk, again with the aim of improving AV scanning. This suggest that partial evaluation also has a role in binary analysis, where the aim is to make malware detection more semantic [7].

Reflection presents a challenge for program analysis: quite apart from writes to object fields, reflection can hide calls, and hence mask parts of the call-graph so that an analysis is unsound. Points-to analysis has been suggested [12] as a way of determining the targets of reflective calls which, in effect, traces the flow of strings through the program. This is sufficient for resolving many, but not all calls, hence points-to information is augmented with user-specified annotation so as to statically determine the complete call graph. The use of points-to information represents an advance over using dynamic instrumentation to harvest reflective calls [11] since instrumentation cannot guarantee complete coverage. Partial evaluation likewise traces the flow of strings through the program, though without refining points-to analysis, it is not clear that it has the precision to recover the targets of reflective calls that have been willfully obfuscated with such techniques as a substitution cipher (rot13).

6   Future Work

Termination analysis is a subfield of static analysis within itself and thus far we have not explored how termination can improve unfolding. We simply unfold loops where the loop bound is known at specialisation time. We also deliberately do not unfold recursive methods, though this is a somewhat draconian limitation. Future work will aim to quantify how termination analysis can be applied in an online setting to improve the quality of malware detection.

Although we have not observed this in the wild, there is no reason why reflection cannot be applied to a method that obfuscates a string, such as a decryptor. This would thwart our approach to deobfuscation since the reflected call would be deobfuscated in the residual, but would not actually be evaluated on a given string. Thus we will explore how partial evaluation can be repeatedly applied to handle these multi-layered forms of obfuscation.

We will also examine how partial evaluation can remove less common forms of Java obfuscation such as control flow obfuscation and serialization and deserialization obfusation, the latter appearing to be as amenable to partial evaluation as string obfuscation. In the long term we will combine partial evaluationi with code similarity matching, drawing on techniques from information retrieval.

7   Conclusion

We have presented a partial evaluator for removing string and reflection obfuscation from Java programs, with the aim of improving the detection of malicious Java code. Our work puts partial evaluation in a new light: previous studies have majored on optimisation whereas we argue that partial evaluation has a role in anti-virus matching. To this end, a partial evaluator has been designed for Jimple, which was strength tested on five malware samples from the Metasploit exploit framework, obfuscated using string and reflection obfuscation.

References

1. Rapid 7. Java Applet JMX Remote Code Execution (2013)
2. Rapid 7. Metasploit (2014)
3. Andersen, L.: Binding-time analysis and the taming of C pointers. In: PEPM, pp. 47–58. ACM (1993)
4. Braux, M., Noyé, J.: Towards partially evaluating reflection in Java. In: PEPM, pp. 2–11. ACM (2000)
5. Christodorescu, M., Jha, S., Kinder, J., Katzenbeisser, S., Veith, H.: Software transformations to improve malware detection. J. Comput. Virol. 3(4), 253–265 (2007)

6. Collberg, C., Nagra, J.: Surreptitious Software: Obfuscation, Watermarking, and Tamperproofing for Software Protection. Addison-Wesley, Boston (2009)
7. Dalla Preda, M., Christodorescu, M., Jha, S., Debray, S.: A Semantics-based Approach to Malware Detection. ACM TOPLAS, **30** (2008)
8. Einarsson, A., Nielsen, J.D.: A Survivor's Guide to Java Program Analysis with Soot. Technical report (2008)
9. Flexeder, A., Petter, M., Seidl, H.: Side-effect analysis of assembly code. In: Yahav, E. (ed.) Static Analysis. LNCS, vol. 6887, pp. 77–94. Springer, Heidelberg (2011)
10. Giacobazzi, R., Jones, N.D., Mastroeni, I.: Obfuscation by partial evaluation of distorted interpreters. In: PEPM, pp. 63–72. ACM (2012)
11. Hirzel, M., Diwan, A., Hind, M.: Pointer analysis in the presence of dynamic class loading. In: Odersky, M. (ed.) ECOOP 2004. LNCS, vol. 3086, pp. 96–122. Springer, Heidelberg (2004)
12. Livshits, B., Whaley, J., Lam, M.S.: Reflection analysis for Java. In: Yi, K. (ed.) APLAS 2005. LNCS, vol. 3780, pp. 139–160. Springer, Heidelberg (2005)
13. McCabe, T.J.: A complexity measure. IEEE Trans. Softw. Eng. **2**(4), 308–320 (1976)
14. National Institute of Standards and Technology. Vulnerability Summary for CVE-2013-3346 (2013)
15. OWASP. Metasploit Java Exploit Code Obfuscation and Antivirus Bypass/Evasion (CVE-2012-4681) (2013)
16. Park, J.-G., Lee, A.H.: Removing reflection from Java Programs using partial evaluation. In: Matsuoka, S. (ed.) Reflection 2001. LNCS, vol. 2192, pp. 274–275. Springer, Heidelberg (2001)
17. Schlumberger, J., Kruegel, C., Vigna, G.: Jarhead: analysis and detection of malicious Java applets. In: ACSAC, pp. 249–257. ACM (2012)
18. Shali, A., Cook, W.R.: Hybrid partial evaluation. In: OOPSLA, pp. 375–390. ACM (2011)
19. Sistemas, H.: VirusTotal Analyses Suspicious Files and URLs (2014). https://www.virustotal.com/
20. Valleé Rai, R., Hendren, L.J.: Jimple: Simplifying Java Bytecode for Analyses and Transformations. Technical report TR-1998-4. McGill University (1998)

Access Control and Obligations
in the Category-Based Metamodel:
A Rewrite-Based Semantics

Sandra Alves[1], Anatoli Degtyarev[2], and Maribel Fernández[2]([⊠])

[1] Department of Computer Science, University of Porto, Porto, Portugal
[2] Department of Informatics, King's College London, London WC2R 2LS, UK
Maribel.Fernandez@kcl.ac.uk

Abstract. We define an extension of the category-based access control (CBAC) metamodel to accommodate a general notion of obligation. Since most of the well-known access control models are instances of the CBAC metamodel, we obtain a framework for the study of the interaction between authorisation and obligation, such that properties may be proven of the metamodel that apply to all instances of it. In particular, the extended CBAC metamodel allows security administrators to check whether a policy combining authorisations and obligations is consistent.

Keywords: Security policies · Access control · Obligations · Rewriting

1 Introduction

Access control policies specify which actions users are authorised to perform on protected resources. An authorisation may entail an obligation to perform another action on the same or another resource. Standard languages for the specification of access control policies include also a number of primitives to specify obligations associated with authorisations. For example, within XACML [19], an obligation is a directive from the Policy Decision Point (PDP) to the Policy Enforcement Point (PEP) specifying an action that must be carried out before or after an access is approved.

The notion of obligation helps bridge a gap between requirements and policy enforcement. For example, consider a hospital scenario in which any doctor may be authorised to read the medical record of a patient in an emergency situation, but in that case there is a requirement to inform the patient afterwards. Although access control models deal mainly with authorisations, incorporating obligations facilitates the analysis of compatibility between obligations and authorisations.

A metamodel of access control, which can be specialised for domain-specific applications, has been proposed in [3]. It identifies a core set of principles of access control, abstracting away many of the complexities that are found in

---

This work was partially funded by the European Office of Aerospace Research and Development (EOARD-AFOSR).

M. Proietti and H. Seki (Eds.): LOPSTR 2014, LNCS 8981, pp. 148–163, 2015.
DOI: 10.1007/978-3-319-17822-6_9

specific access control models, in order to simplify the tasks of policy writing and policy analysis. The metamodel focuses on the notion of a *category*, which is a class of entities that share some property. Classic types of groupings used in access control, like a role, a security clearance, a discrete measure of trust, etc., are particular instances of the more general notion of category. In category-based access control (CBAC), permissions are assigned to categories of users, rather than to individual users. Categories can be defined on the basis of user attributes, geographical constraints, resource attributes, etc. In this way, permissions can change in an autonomous way (e.g., when a user attribute changes), unlike, e.g., role-based models [1], which require the intervention of a security administrator.

In this paper, we define an extension of the CBAC metamodel to accommodate a general notion of an *obligation*, obtaining a formal framework for modelling and enforcing access control policies that involve authorisations and obligations. We show examples of application in the context of emergency management. We do not make any specific assumptions on the components of the system. Instead, we aim at defining an abstract model of access control and obligations that can be instantiated in various ways to satisfy specific requirements. To specify dynamic policies involving authorisations and obligations in the metamodel, we adjust the notion of *event* given in [15] and describe a set of core axioms for defining *obligations*.

Summarising, we provide: an axiomatic definition of a *generic framework* for the specification of access control and obligation models, obtained by extending the CBAC metamodel with notions of obligation and duty; a rewrite-based *operational semantics* for the extended metamodel, dealing with *authorisation and obligation* assessment, including mechanisms for the resolution of conflicts between authorisations and obligations; and a rewrite-based technique to prove properties of access control policies involving obligations.

Overview: In Sect. 2, we recall the CBAC metamodel. Section 3 discusses obligations, Sect. 4 presents the extended CBAC metamodel, and Sect. 5 the operational semantics for obligations. In Sect. 6, we discuss related work, and in Sect. 7, conclusions are drawn and further work is suggested.

2   Preliminaries

We assume familiarity with basic notions on first-order logic and term-rewriting systems [2]. We briefly describe below the key concepts underlying the category-based metamodel of access control; see [3] for a detailed description.

Informally, a category is any of several distinct classes or groups to which entities may be assigned. Entities are denoted by constants in a many sorted domain of discourse, including: a countable set $\mathcal{C}$ of categories, denoted $c_0, c_1, \ldots$, a countable set $\mathcal{P}$ of principals, denoted $p_0, p_1, \ldots$ (we assume that principals that request access to resources are pre-authenticated), a countable set $\mathcal{A}$ of *actions*, denoted $a_0, a_1, \ldots$, a countable set $\mathcal{R}$ of *resources*, denoted $r_0, r_1, \ldots$, a finite set $\mathcal{A}uth$ of possible *answers* to access requests (e.g., {grant, deny, undetermined}) and a countable set $\mathcal{S}$ of *situational identifiers* to denote environmental information. More generally, entities can be represented by a data structure (e.g., a principal could be represented by a term $principal(p_i, attributeList)$), but constants will be sufficient for most examples in this paper. A *permission* is a pair (a, r) of an action and a resource, and an *authorisation* is a triple (p, a, r) that associates a permission with a principal. The metamodel includes the following relations to formalise these notions:

- *Principal-category assignment:* $\mathcal{PCA} \subseteq \mathcal{P} \times \mathcal{C}$, such that $(p, c) \in \mathcal{PCA}$ iff a principal $p \in \mathcal{P}$ is assigned to the category $c \in \mathcal{C}$.
- *Permission-category assignment:* $\mathcal{ARCA} \subseteq \mathcal{A} \times \mathcal{R} \times \mathcal{C}$, such that $(a, r, c) \in \mathcal{ARCA}$ iff action $a \in \mathcal{A}$ on resource $r \in \mathcal{R}$ can be performed by the principals assigned to the category $c \in \mathcal{C}$.
- *Authorisations:* $\mathcal{PAR} \subseteq \mathcal{P} \times \mathcal{A} \times \mathcal{R}$, such that $(p, a, r) \in \mathcal{PAR}$ iff a principal $p \in \mathcal{P}$ can perform the action $a \in \mathcal{A}$ on the resource $r \in \mathcal{R}$.

Definition 1 (Axioms). *The relation $\mathcal{PAR}$ satisfies the following core axiom, where we assume that there exists a relationship $\subseteq$ between categories; this can simply be equality, set inclusion (the set of principals assigned to $c \in \mathcal{C}$ is a subset of the set of principals assigned to $c' \in \mathcal{C}$), or an application specific relation.*

$(a1)$ $\forall p \in \mathcal{P}, \forall a \in \mathcal{A}, \forall r \in \mathcal{R},$
$(\exists c, c' \in \mathcal{C}, ((p, c) \in \mathcal{PCA} \wedge c \subseteq c' \wedge (a, r, c') \in \mathcal{ARCA}) \Leftrightarrow (p, a, r) \in \mathcal{PAR})$

Operationally, axiom $(a1)$ can be realised through a set of functions, as shown in [5]. We recall the definition of the function $\mathsf{par}(P, A, R)$ below; it relies on functions pca, which returns the list of categories assigned to a principal, and arca, which returns a list of permissions assigned to a category.

$(a2)$ $\mathsf{par}(P, A, R) \rightarrow if\ (A, R) \in \mathsf{arca}^*(\mathsf{below}(\mathsf{pca}(P)))$ *then* grant *else* deny

The function below computes the set of categories that are subordinate to any of the categories in the list $\mathsf{pca}(P)$. The function $\in$ is a membership operator on lists, grant and deny are answers, and arca^* generalises the function arca to take into account lists of categories.

The axiom $(a1)$, and its algebraic version $(a2)$, state that a request by a principal p to perform the action a on a resource r is authorised if p belongs to a category c such that for some category below c (e.g., c itself) the action a is authorised on r, otherwise the request is denied. There are other alternatives, e.g., considering *undeterminate* as answer if there is not enough information to grant the request. More generally, the relations $\mathcal{BARCA}$ and $\mathcal{BAR}$ were introduced in [6] to explicitly represent prohibitions, that is, to specify that an action is forbidden on a resource; $\mathcal{UNDET}$ was introduced to specify undeterminate answers. These relations obey the following axioms:

$(c1)$ $\forall p \in \mathcal{P}, \forall a \in \mathcal{A}, \forall r \in \mathcal{R},$
$((\exists c \in \mathcal{C}, \exists c' \in \mathcal{C}, (p, c) \in \mathcal{PCA} \wedge c' \subseteq c \wedge (a, r, c') \in \mathcal{BARCA}) \Leftrightarrow$
$(p, a, r) \in \mathcal{BAR})$
$(d1)$ $\forall p \in \mathcal{P}, \forall a \in \mathcal{A}, \forall r \in \mathcal{R},$
$((p, a, r) \notin \mathcal{PAR} \wedge (p, a, r) \notin \mathcal{BAR}) \Leftrightarrow (p, a, r) \in \mathcal{UNDET}$

$(e1)$ $\mathcal{PAR} \cap \mathcal{BAR} = \emptyset$

3   Obligations and Events

Obligations differ from permissions in the sense that, although permissions can be issued but not used, an obligation usually is associated with some mandatory action, which must be performed at a time defined by some temporal constraints or by the occurrence of an event. Fulfilling the obligations may require certain permissions, which can lead to undesirable interactions between permissions and obligations, where the obligations in a policy cannot be fulfilled given the set of assigned permissions in the system at a certain state. If obligations go unfulfilled (that is, become violated), this raises the question of accountability, that is, to whom shall be imputed responsibility for unfulfilled obligations. To address these issues, we will extend the CBAC metamodel in order to be able to specify the assignment of obligations to principals and study the interaction between obligations and permissions. We will define obligations consisting of an action (an atomic or a composed action) on a resource to be performed during an interval specified by an initial event and a terminal event.

Following [16], we consider events as action occurrences, or action happenings, that is, an *event* represents an action that happened in the system. This notion has been used in previous works [7,12,15]; here we distinguish between *event types*, denoted by ge_i (e.g., registration for a course, a fire alarm, etc.) and specific events, denoted by e_i (e.g., the student *Max M.* registered for the Logic course in September 2012, the fire alarm went off in the Strand Building at 5pm on the 30 June 2012). A typing function will be used to classify events.

Definition 2 (Event History and Interval). *An event history, denoted by $h \in \mathcal{H}$, is a sequence of events that may happen in a run of the system being modelled. A subsequence i of h is called an event interval; the first event in i opens the interval and the last one closes it.*

We assume that an event history contains all the relevant events in order to determine, at each point, the set of authorisations and obligations of each principal (i.e., to determine the system state). Events and event types will be specified using a special-purpose language described in Sect. 4.2. We assume that the typing relation associating event types with events is decidable.

Definition 3 (Obligation). *A generic obligation is a tuple (a, r, ge_1, ge_2), where a is an action, r a resource, and ge_1, ge_2 two event types (ge_1 triggers the obligation, and ge_2 ends it). If there is no starting event (resp., no ending event) we write $(a, r, \bot, ge)$ (resp., $(a, r, ge, \bot)$), meaning that the action on the resource must be performed at any point before an event of type ge (resp. at any point after an event of type ge).*

Example 1. Assume that in an organisation, the members of the security team must call the fire-department if a fire alarm is activated, and this must be done before they de-activate the alarm. This obligation could be represented by the tuple $(call, firedept, alarmON, alarmOFF)$;

Models of access control specify *authorisations* by defining the way *permissions* are assigned to principals. Similarly, an obligation model should specify the way *obligations* are assigned to principals, and it should be possible to determine, at each point in the history, which obligations assigned to principals have to be enforced. For the latter, we introduce the notion of a *duty*.

Definition 4 (Duty). *A duty is a tuple (p, a, r, e_1, e_2, h), where p is a principal, a an action, r a resource, e_1, e_2 are two events and h is an event history that includes an interval opened by e_1 and closed by e_2. We replace e_1 (resp. e_2) with $\perp$ if there is no starting (resp. closing) event.*

Unlike access control models, which do not need to check whether the authorised actions are performed or not by the principals, obligation models need to include mechanisms to check whether duties were discharged or not. Specifically, obligation models distinguish four possible states for duties: *invalid* (when at the point a duty is issued the completion point has already passed); *fulfilled* (when the obligation is carried out within the associated interval); *violated* (when it is not carried out within the associated interval) and *pending* (when the obligation has not yet been carried, but the interval is still valid). In some cases, p's duty can be fulfilled by another principal. This is the case in Example 1 above, where all members of the security team have the obligation to call the fire department before deactivating the alarm, but the obligation is fulfilled as soon as one of them makes the call. In order to distinguish both kinds of obligations, we will call *individual obligations* those that have to be fulfilled necessarily by the principal to whom the obligation is assigned, and *collective obligations* those where the obligation is assigned to several principals and can be fulfilled by any of the principals in the group.

4   Obligations in the Category-Based Metamodel

The notion of a category will be used to specify obligations that apply to groups of principals. However, the groupings of principals for authorisations and for obligations are not necessarily the same (for instance, the category UG student is assigned a certain set of permissions, which all UG students enjoy, whereas some UG students belong to the home student category and others to the international student category, with different obligations). We call *permission categories* those used to specify authorisations, and *obligation categories* those used to specify duties. To model obligations and duties we extend the metamodel to include the following sets of entities and relations in addition to the ones defined in Sect. 2:

- Countable sets $\mathcal{E}$ and $\mathcal{GE}$ of events and event types, denoted by $e, e_0, e_1, \dots$ and $ge, ge_0, ge_1, \dots$, respectively.
- A countable set $\mathcal{H}$ of event histories, denoted by $h, h_0, h_1, \dots$.
- *Obligation-category assignment: $\mathcal{OARCA} \subseteq \mathcal{A} \times \mathcal{R} \times \mathcal{GE} \times \mathcal{GE} \times \mathcal{C}$, such that $(a, r, ge_1, ge_2, c) \in \mathcal{OARCA}$ iff the action $a \in \mathcal{A}$ on resource $r \in \mathcal{R}$ must*

be performed in the interval defined by two events of type ge_1, ge_2 by principals assigned to the category $c \in \mathcal{C}$. To accommodate individual and collective obligation assignments, this relation is partitioned into two relations: $\mathcal{OARCA}_\mathcal{I}$ and $\mathcal{OARCA}_\mathcal{C}$. Thus, $(a, r, ge_1, ge_2, c) \in \mathcal{OARCA}_\mathcal{I}$ if *every* member of the category c must perform the action $a \in \mathcal{A}$ on resource $r \in \mathcal{R}$ in the interval defined by ge_1, ge_2, and $(a, r, ge_1, ge_2, c) \in \mathcal{OARCA}_\mathcal{C}$ if *any* member of c must perform the action $a \in \mathcal{A}$ on resource $r \in \mathcal{R}$ in the interval defined by ge_1, ge_2 and it is sufficient that *one* of them does it.

- *Obligation-principal assignment:* $\mathcal{OPAR} \subseteq \mathcal{P} \times \mathcal{A} \times \mathcal{R} \times \mathcal{GE} \times \mathcal{GE}$, such that $(p, a, r, ge_1, ge_2) \in \mathcal{OPAR}$ iff a principal $p \in \mathcal{P}$ must perform the action $a \in \mathcal{A}$ on the resource $r \in \mathcal{R}$ in the interval defined by two events of type ge_1, ge_2. If individual and collective obligations are modelled, then this relation is replaced by two relations: $\mathcal{OPAR}_I \subseteq \mathcal{P} \times \mathcal{A} \times \mathcal{R} \times \mathcal{GE} \times \mathcal{GE}$, defining individual obligations, and $\mathcal{OPAR}_C \subseteq \mathcal{P} \times \mathcal{A} \times \mathcal{R} \times \mathcal{GE} \times \mathcal{GE} \times \mathcal{C}$ for collective obligations, such that $(p, a, r, ge_1, ge_2) \in \mathcal{OPAR}_I$ iff principal $p \in \mathcal{P}$ must perform the action $a \in \mathcal{A}$ on the resource $r \in \mathcal{R}$ in the interval defined by two events of type ge_1, ge_2, and $(p, a, r, ge_1, ge_2, c) \in \mathcal{OPAR}_C$ if principal $p \in \mathcal{P}$ or any other member of c must perform the action $a \in \mathcal{A}$ on the resource $r \in \mathcal{R}$ in the interval defined by two events of type ge_1, ge_2.

- *Duty:* $\mathcal{DPAR} \subseteq \mathcal{P} \times \mathcal{A} \times \mathcal{R} \times \mathcal{E} \times \mathcal{E} \times \mathcal{H}$, such that $(p, a, r, e_1, e_2, h) \in \mathcal{DPAR}$ iff a principal $p \in \mathcal{P}$ must perform the action $a \in \mathcal{A}$ on the resource $r \in \mathcal{R}$ in the interval between the events e_1 and e_2 in the event history h.

  To accommodate individual and collective obligations, this relation is partitioned into $\mathcal{DPAR}_I$ and $\mathcal{DPAR}_C$, similarly to $\mathcal{OPAR}$.

- *Event Typing:* $\mathcal{ET} \subseteq \mathcal{E} \times \mathcal{GE} \times \mathcal{H}$, such that $(e, ge, h) \in \mathcal{ET}$ if the event e is an instance of ge in h. This will be abbreviated as $h \vdash e : ge$.

- *Event Interval:* $\mathcal{EI} \subseteq \mathcal{E} \times \mathcal{E} \times \mathcal{H}$, such that $(e_1, e_2, h) \in \mathcal{EI}$ if the event e_2 closes the interval started by the event e_1 in h.

4.1 Obligation Axioms

In the metamodel, the relations $\mathcal{OPAR}$ and $\mathcal{DPAR}$ are derivable from the others. They satisfy the following core axioms, where we assume that there exists a relationship $\subseteq_o$ between obligation categories; this can simply be equality, set inclusion (the set of principals assigned to $c \in \mathcal{C}$ is a subset of the set of principals assigned to $c' \in \mathcal{C}$), or an application specific relation may be used.

$(o1) \quad \forall p \in \mathcal{P}, \ \forall a \in \mathcal{A}, \ \forall r \in \mathcal{R}, \forall ge_1, ge_2 \in \mathcal{GE},$
$$\big((\exists c, c' \in \mathcal{C}, (p, c) \in \mathcal{PCA} \wedge c \subseteq_o c' \ \wedge (a, r, ge_1, ge_2, c') \in \mathcal{OARCA})$$
$$\Leftrightarrow (p, a, r, ge_1, ge_2) \in \mathcal{OPAR}\big)$$

$(o2) \quad \forall p \in \mathcal{P}, \ \forall a \in \mathcal{A}, \ \forall r \in \mathcal{R}, \forall e_1, e_2 \in \mathcal{E}, \forall h \in \mathcal{H}$
$$\big((\exists ge_1, ge_2 \in \mathcal{GE}, (p, a, r, ge_1, ge_2) \in \mathcal{OPAR},$$
$$h \vdash e_1 : ge_1, h \vdash e_2 : ge_2, (e_1, e_2, h) \in \mathcal{EI})$$
$$\Leftrightarrow (p, a, r, e_1, e_2, h) \in \mathcal{DPAR}\big)$$

The axiom $o1$ is the essence of the category-based metamodel: it specifies that the principals that are members of a category to which the obligation (a, r, ge_1, ge_2) has been assigned have this obligation. The axiom $o2$ shows how to derive duties. The relation $\subseteq_o$ specifies a hierarchy between obligation categories, which does not necessarily correspond to the way permissions are inherited (specified by the relation $\subseteq$ in axiom $(a1)$).

To accommodate collective and individual obligations, the following variants of the axioms should be included.

$(o1_I)$ $\forall p \in \mathcal{P}$, $\forall a \in \mathcal{A}$, $\forall r \in \mathcal{R}, \forall ge_1, ge_2 \in \mathcal{GE}$,
$$((\exists c, c' \in \mathcal{C}, (p, c) \in \mathcal{PCA} \wedge \ c \subseteq_o c' \ \wedge (a, r, ge_1, ge_2, c') \in \mathcal{OARCA_I})$$
$$\Leftrightarrow (p, a, r, ge_1, ge_2) \in \mathcal{OPAR_I})$$
$(o1_C)$ $\forall p \in \mathcal{P}$, $\forall a \in \mathcal{A}$, $\forall r \in \mathcal{R}, \forall ge_1, ge_2 \in \mathcal{GE}, \forall c' \in \mathcal{C}$,
$$((\exists c \in \mathcal{C}, (p, c) \in \mathcal{PCA} \wedge \ c \subseteq_o c' \ \wedge (a, r, ge_1, ge_2, c') \in \mathcal{OARCA_C})$$
$$\Leftrightarrow (p, a, r, ge_1, ge_2, c') \in \mathcal{OPAR_C})$$
$(o1')$ $\mathcal{OPAR_I} \cap \overline{\mathcal{OPAR_C}} = \emptyset$

where $\overline{\mathcal{OPAR_C}}$ represents the projection of the relation which discards the last column (the category). Variants of the axiom $o2$ are defined similarly. Axiom $(o1')$ indicates that the same obligation cannot be both collectivelly and individually fulfilled. Additionally, the relation $\mathcal{OARCA_C}$ should not assign the same obligation to two categories in the $\subseteq_o$ relation, to avoid redundancy.

4.2   Event and Event Type Representation

To consider examples of obligation policies we will describe a possible representation for events and event types.

Actions can be either *elementary* or *compound*, i.e. consisting of sets of (simultaneously happened) elementary actions [12]. We use letters a and e, possibly with indexes, to identify actions and events, respectively. We will use a binary representation for events, introduced in [4] and partly motivated by Davidson's work on event semantics [10]. This choice provides a most flexible representation with negligible computational overheads. In our case, the set of arguments depends not only on the action involved but also on the context where the event description is used.

Definition 5 (Event). *An* event description *is a finite set of ground 2-place facts (atoms) that describe an event, uniquely identified by $e_i, i \in \mathbb{N}$, and which includes two necessary facts, written $happens(e_i, t_j)$ and $act(e_i, a_l)$, and n nonnecessary facts $(n \geq 0)$.*

In [4] an event description includes three *necessary* facts, and n non-necessary facts $(n \geq 0)$. Unlike [4], we use only two types of necessary facts. Their intended meanings may be described as follows: $happens(e_i, t_j)$ means the event e_i happens at time t_j; $act(e_i, a_l)$ means the event e_i involves an action a_l[1]. The events should be positioned in the history in the order in which they happened.

---

[1] We restrict attention to events without a duration.

Sometimes we will consider only the predicate *act*, omitting *happens*. In this case, the history would include both events with specified time and events without time, but whether we use the *happens* predicate or not, the history should reflect the order in which the events happened in the system.

If the event e_i involves the subject s_m, then the corresponding non-necessary fact $subject(e_i, s_m)$ can be added to the event description when we need this fact. In [4], $subject(e_i, s_m)$, would be the third necessary fact. Similarly, if the event e_i involves the resource r_l, then the fact $object(e_i, r_l)$ can be added to the description. And so on. Thus, the event description given by the set $\{happens(e_i, t_j),$ $act(e_i, a_l), subject(e_i, s_m), object(e_i, r_l)\}$ represents a happening e_i at a time t_j of an action a_l performed by a subject s_m on a resource r_l.

To define obligations associated with events, here we also consider event types, denoted by ge_i. As it was noted earlier, in real modelling the set of non-necessary predicates involved in a description of an event is determined not only by the action assigned to this event but also by current context. We define an *event type*, or *generic event*, to consist of the necessary fact indicating an action and a set of predicates over expressions with variables including another necessary predicate *happens/2*.

Definition 6 (Event type and Instance). *A set ge of binary atoms represents an event type, also called generic event, if there exists a substitution σ such that the instantiation of ge with σ, written, geσ is an event description (see Definition 5).*

An event e is an instance of a generic event ge, denoted as e : ge, if there is a substitution σ such that geσ ⊆ e. In other words, if ge subsumes e.

The action of a substitution σ on a generic event *ge* may simply be a syntactic operation (replacing variables by terms), or, depending on the application and the kind of data used to define events, instantiation may require some computation; we call it a *semantic instantiation* in the latter case. We give examples below.

Example 2. The events

$$e_1 = \{happens(e_1, 12.25), act(e_1, activate), object(e_1, alarm)\}$$
$$e_2 = \{happens(e_2, 12.45), act(e_2, deactivate), object(e_2, alarm), subject(e_2, tom)\}$$

are instances, with respective substitutions $\sigma_1 = \{E \mapsto e_1, T \mapsto 12.25\}$ and $\sigma_2 = \{E \mapsto e_2, T \mapsto 12.25, X \mapsto tom\}$, of the generic events

$$alarmON = \{happens(E, T), act(E, activate), object(E, alarm)\}$$
$$alarmOFF = \{happens(E, T + 20), act(E, deactivate), object(E, alarm),$$
$$subject(E, X)\}$$

In the case of e_2 we are using a semantic instantiation function.

Example 3. Assume that in a given university, every international student must have a valid visa before registration day (when documents are checked). We can define a generic event:

$$registration\text{-}day = \{happens(E, T), act(E, open), subject(E, secretary)\}$$

and define categories "home-student" and "international-student", such that $(obtain, visa, \perp, registration\text{-}day, international\text{-}student) \in \mathcal{OARCA}^2$. Here the event type that initiates the obligation is not specified ($\perp$), but the final one is ($registration\text{-}day$). If attendance is required for all students, then $\mathcal{OARCA}$ should contain the tuples: $(attend, reg\text{-}office, reg\text{-}day\text{-}start, reg\text{-}day\text{-}end, home\text{-}student)$ and $(attend, reg\text{-}office, reg\text{-}day\text{-}start, reg\text{-}day\text{-}end, international\text{-}student)$.

Example 4. Consider a hospital, where doctors are allowed to read and write the medical records of the patients in their department. Assume that the policy in place also states that any doctor is allowed to read the medical record of a patient who is in an emergency situation (even if the patient is not in their department) but in that case they or any doctor in their department must declare this access (by creating an entry in the hospital log). Let *patient* be a category consisting of all patients (of a given hospital), and *doctor* be a category consisting of all doctors (of the given hospital). Let $patient(X)$ be a (parameterised) category consisting of all patients in the department X, and $doctor(X)$ be a (parameterised) category consisting of all doctors in the department X, such that for all X, $doctor(X) \subseteq doctor$, i.e., the category $doctor(X)$ inherits all permissions from the category *doctor* and similarly $patient(X) \subseteq patient$. Assume that these categories and hierarchical relation are used both for permissions and for obligations.

To model this scenario, we assume the relations $\mathcal{PCA}$ and $\mathcal{ARCA}$ satisfy the following axioms, where $emerg(bcrd, P)$ is true if an event $brcd$ initiating a cardiac emergency for P has been detected, and no event ending the emergency has been recorded:

$$\forall P, \forall D, \ (P, patient(D)) \in \mathcal{PCA} \ \Rightarrow \ (read, record(P), doctor(D)) \in \mathcal{ARCA}$$
$$\forall P, \forall D, \ (P, patient(D)) \in \mathcal{PCA} \ \wedge \ emerg(bcrd, P) \Rightarrow$$
$$(read, record(P), doctor) \in \mathcal{ARCA}$$

Moreover, we include the following axiom for $\mathcal{OARCA}_C$, where $gen\text{-}read(X, P)$ is a generic event representing the act of $doctor(X)$ reading the medical record of $patient(P)$:

$$\forall P, \forall X, \forall D, \forall D', \ (((P, patient(D)) \in \mathcal{PCA} \wedge (X, doctor(D')) \in \mathcal{PCA} \wedge$$
$$(X, doctor(D)) \notin \mathcal{PCA}) \Rightarrow$$
$$(declare, admin\text{-}log, gen\text{-}read(X, P), \perp, doctor(D')) \in \mathcal{OARCA}_C)$$

5   A Rewrite Semantics for Obligations

Operationally, the axioms $o1$ and $o2$ given in Sect. 4 can be realised through a set of function definitions. In this section we assume all the obligations are individual; the extension to accommodate individual and collective obligations is straightforward. The information contained in the relations $\mathcal{PCA}$ and $\mathcal{OARCA}$ is modelled by the functions pca and oarca, respectively, where pca returns the

---

[2] Or $\mathcal{OARCA}_I$ if we need to distinguish between individual and collective obligations.

list of all the categories assigned to a principal and oarca returns the list of obligations assigned to a category, e.g., defined by the rule:

$$\text{oarca}(c) \rightarrow [(a_1, r_1, ge_1, ge'_1), \ldots, (a_n, r_n, ge_n, ge'_n)].$$

We assume that the function oarca* takes as input a list of obligation categories and computes the list of obligations for all the categories in the list (similar to arca*, see Sect. 2). The function pca was already mentioned in Sect. 2 for authorisations; for efficiency reasons, a separate function opca could be defined to compute obligation categories.

   In addition, we assume that the functions type and interval, specific to the particular system modelled, are also available. The function type implements the typing relation $\mathcal{ET}$ for events, that is, it computes the event type for a given event e in h (taking into account the semantic instantiation relation associated with the events of interest). The function interval implements the relation $\mathcal{EI}$, which links an event e_1 with the event e_2 that closes the interval started by e_1 in h.

Definition 7. *The rewrite-based specification of the axiom* (o1) *in Sect. 4.1 is given by the rewrite rule:*

$$(r1)\ \text{opar}(p, a, r, ge_1, ge_2) \rightarrow (a, r, ge_1, ge_2) \in \text{oarca}^*(\text{obelow}(\text{opca}(p)))$$

where the function $\in$ *is a membership operator on lists, and, as the function name suggests,* obelow *computes the set of categories that are below (w.r.t. the hierarchy defined by the* $\subseteq_o$ *relation) any of the categories given in the list* opca(p). *For example, for a given category* c, *this could be achieved by using a rewrite rule* obelow([c]) $\rightarrow$ [c, c_1, \ldots, c_n]. *Intuitively, this means that if* c' *is below* c, *then* c *inherits all the obligations of* c'.

   The rewrite-based specification of the axiom (o2) *is given by:*

$$(r2)\ \text{duty}(p, a, r, e_1, e_2, h) \rightarrow \text{opar}(p, a, r, \text{type}(e_1, h), \text{type}(e_2, h))\ and$$
$$\text{interval}(e_1, e_2, h)$$

where the functions type *and* interval *are specific to the system modelled, as mentioned above. Additionally, we consider the following function to check the status of obligations for a principal* p *with respect to a history of events:*

$$(r3)\ \textit{eval-obligation}(p, a, r, ge_1, ge_2, h) \rightarrow if\ \text{opar}(p, a, r, ge_1, ge_2)\ then$$
$$\text{append}(\textit{chk-closed}^*(\textit{closed}(ge_1, ge_2, h), p, a, r), \textit{chk-open}^*(\textit{open}(ge_1, ge_2, h), p, a, r))$$
$$else\ [not\text{-}applicable]$$

where the function append *is a standard function that concatenates two lists,* closed *computes the sublists of* h *that start with an event* e_1 *of type* ge_1 *and finish with an event* e_2 *of type* ge_2 *that closes the interval for this obligation, and similarly* open *returns the subhistories of* h *that start with the event* e_1 *of type* ge_1 *and for which there is no event* e_2 *of type* ge_2 *in* h *that closes the interval for this obligation.*

The function chk-closed with inputs h', p, a, r checks whether in the subhistory h' there is an event where the principal p has performed the action a on the resource r, returning a result fulfilled *if that is the case, and* violated *otherwise.*

The function chk-open with inputs h', p, a, r checks whether in the subhistory h' there is an event where the principal p has performed the action a on the resource r, returning a result fulfilled *if that is the case, and* pending *otherwise.*

The functions chk-closed * *and chk-open* * *do the same but for each element of a list of subhistories, returning a list of results.*

According to the previous specification, if the obligation (a, r, ge_1, ge_2) does not concern p then eval-obligation(p, a, r, ge_1, ge_2, h) returns *not-applicable*, otherwise, the function eval-obligation returns a list of results containing the status of p in relation to this obligation according to h. Usually, h will be the event history that is relevant to the scenario being modelled. For instance, it could be the full history of events in the system, or we could restrict ourselves to the events in the last year, or the events that happened during a specific interval. In our model, it is not possible for a duty to be invalid (thanks to axiom $(o2)$). If h is sufficiently long and depending on the events it contains, it is possible that at different points the obligation was fulfilled, violated or pending. This is why the function eval-obligation(p, a, r, ge_1, ge_2, h) returns a list of results.

If the system modelled includes collective and individual obligations, then the functions chk-closed and chk-open should take into account this in order to check if the obligation has been fulfilled.

Example 5. Consider again the hospital scenario mentioned in Example 4. Assume an investigation is taking place to establish if Doctor Tom Smith, who treated patient J. Lewis in an emergency situation occurring in November 2012, but is not J. Lewis's doctor, has fulfilled his obligation with respect to declaring the access to this patient's records. This is a collective obligation which can be discharged by any of the doctors in his department. Accordingly, we assume the functions chk-closed and chk-open perform collective checks.

In this case, we simply evaluate the term

$$\text{eval-obligation}(TomSmith, declare, admin\text{-}log, gen\text{-}read, \bot, h)$$

where $gen\text{-}read = \{act(E, read\text{-}pr(JLewis), sub(E, TomSmith)\}$ and h is the event history that starts on the 1st November 2012 and ends today.

5.1   Analysis of Policies

In previous work, rewriting techniques were used to derive properties of authorisation policies in the metamodel (see, e.g., [5]). Here we apply similar techniques to prove properties of policies that include obligations.

Unicity of Evaluation. The term eval-obligation(p, a, r, ge_1, ge_2, h) should produce *a unique answer*, either indicating that the obligation (a, r, ge_1, ge_2) does not apply to p, or providing the status of the corresponding duty in each of the relevant intervals in h (i.e., compute a list indicating whether it was fulfilled, violated or is still pending in each relevant interval). To prove this property, it is sufficient to prove confluence (which implies the unicity of normal forms) and termination (which implies the existence of normal forms for all terms) of the rewrite system specifying opar, duty and eval-obligation, together with sufficient completeness (which characterises the normal forms). These properties will depend on the specific functions defined for the policy modelled (that is, the specific rules used to compute pca, opca, arca, oarca, etc.), and unfortunately they are generally undecidable. However, sufficient conditions exist and tools such as CiME [9] could be used to help checking properties of the first-order rewrite rules that provide the operational semantics of the policy.

Compatibility. The metamodel is sufficiently expressive to permit the definition of policies where authorisations and obligations co-exist and may depend on each other. In this case, security administrators need to check that obligations and authorisations are compatible: the policy should ensure that every principal has the required permissions in order to fulfill all of its duties. This is not trivial because both the authorisation relation $\mathcal{PAR}$ and the obligations relation $\mathcal{OPAR}$ may be defined using categories whose definition takes into account dynamic conditions (e.g., h). In practice, in order to ensure that only duties that are consistent with authorisations are issued, the semantics could be modified, by adding, as a condition for the assignment of an obligation to a principal in axiom ($o1$), the existence of a corresponding authorisation in $\mathcal{PAR}$. More precisely, consider the relation $\mathcal{OPAR}^+$ defined by adding $(a, r, c) \in \mathcal{PAR}$ to the left-hand side in the axiom ($o1$), i.e. ($o1$) is replaced with ($o1^+$)

$$(o1^+) \; \forall p \in \mathcal{P}, \; \forall a \in \mathcal{A}, \; \forall r \in \mathcal{R}, \forall ge_1, ge_2 \in \mathcal{GE},$$
$$(\exists c, c' \in \mathcal{C}, ((p, c) \in \mathcal{PCA} \wedge \; c \subseteq_o c' \; \wedge (a, r, ge_1, ge_2, c') \in \mathcal{OARCA}$$
$$\wedge (p, a, r) \in \mathcal{PAR}) \Leftrightarrow (p, a, r, ge_1, ge_2) \in \mathcal{OPAR}^+)$$

In the operational semantics, this would boil down to adding in the right-hand side of the rule $r2$ the condition $\mathsf{par}(p, a, r, h, e_1, e_2)$, where the function par computes the authorisation relation within the interval defined by e_1 and e_2 in h. We prefer to follow the separation of concerns principle [11]: instead of adding a condition in ($o1$) to force the compatibility of obligations and authorisations, we axiomatise authorisations and obligations separately, and then check that the authorisations specified are sufficient to enable all the obligations to be fulfilled. Let us first formalise the notion of compatibility.

Definition 8 (Compatibility). *A policy is* compatible *if*

$$\mathcal{OPAR}^+ = \mathcal{OPAR} \tag{1}$$

A policy is weakly compatible *if*

$$\overline{\mathcal{OARCA}} \cap \mathcal{BARCA} = \emptyset \tag{2}$$

where $\overline{\mathcal{OARCA}}$ *is the projection of the relation* $\mathcal{OARCA}$ *on the arguments* a, r, c, *i.e. in notations of relational algebra* $\overline{\mathcal{OARCA}} = \pi_{a,r,c}(\mathcal{OARCA})$.

A policy is strongly compatible *if*

$$\overline{\mathcal{OARCA}} \subseteq \mathcal{ARCA} \qquad (3)$$

Below we summarise the entailment relations between compatibility, strong compatibility and weak compatibility.

Property 1. 1. Strong compatibility implies weak compatibility, $(\mathbf{3} \Rightarrow \mathbf{2})$, under the condition that categories are not empty (i.e., there is at least one principal assigned to each category).
2. Compatibility implies strong compatibility, $(\mathbf{1} \Rightarrow \mathbf{3})$, under the conditions that each principal belongs to a unique category and each category is not empty.
3. Strong compatibility implies compatibility, $(\mathbf{3} \Rightarrow \mathbf{1})$, under the condition that orderings $\subseteq_o$ and $\subseteq$ are the same, more precisely if $\subseteq_o$ implies (is included in) $\subseteq$.

Weak compatibility is, in a sense, minimal among the notions of compatibility defined above since $(\mathbf{2} \not\Rightarrow \mathbf{1})$ and $(\mathbf{2} \not\Rightarrow \mathbf{3})$. Strong compatibility implies weak compatibility, i.e., $(\mathbf{3} \Rightarrow \mathbf{2})$, due to the fact that axiom $(e1)$ (see Sect. 2) implies $\mathcal{ARCA} \cap \mathcal{BARCA} = \emptyset$ if categories are not empty. The proofs of the other implications are omitted due to space constraints. The condition $\subseteq_o$ implies $\subseteq$ means that inheritance of obligations between categories implies inheritance of authorisations. The following sufficient condition for compatibility is a direct consequence of Property 1.

Corollary 1. *A policy is compatible if*

$$\forall c \in \mathcal{C}, \mathsf{subset}(\mathsf{proj} - \mathsf{list}_{ar}(\mathsf{oarca}(c)), \mathsf{arca}(c)) \rightarrow^* \mathsf{True}$$

where we assume $\subseteq_o$ implies $\subseteq$, subset is the function that checks the subset property for sets represented as lists of elements, and proj $-$ list$_{ar}$ is the projection function that projects each element of a list on the first and second components (i.e., action, resource).

Using this result, we can devise a method to automatically prove compatibility of a policy involving authorisations and obligations using a completion tool, such as CiME, to deal with the universal quantification on categories. First, the rewrite system R defining the policy should be proved confluent and terminating by the tool. Then, we add to R the equation $\mathsf{subset}(\mathsf{proj} - \mathsf{list}_{ar}(\mathsf{oarca}(C)), \mathsf{arca}(C)) = \mathsf{True}$. The completion procedure will instantiate C to unify with the left-hand side of the rules defining oarca and arca. If the policy is compatible, the completion procedure terminates successfully without adding any rules to the system. If it is not compatible, the completion procedure will detect an inconsistency $\mathsf{True} = \mathsf{False}$. A similar technique can be used to prove weak compatibility.

6 Related Work

The CBAC model we consider here is an extension of the model defined in [5] for authorisations; all the results regarding the rewriting semantics of authorisations are also valid here, but are not the focus of this paper. Several models dealing with the notion of obligations have been proposed in the literature (see, for example, [8,12–14,17,18,21]), with different features and providing different levels of abstraction in the definition and management of obligations. Some models consider obligations to be fulfilled by the system and therefore never violated, whereas others consider obligations to be fulfilled by users, in which case other questions arise such as how can the system guarantee that the user will be allowed to fulfill the obligations, and the notion of accountability (if a user, with the necessary permissions, does not fulfill its obligations then he becomes accountable).

The framework in [17] deals both with system and user obligations and measures to enforce the fulfillment of obligations. Obligations are triggered by time or events and violated obligations are treated by the system. Our metamodel includes an abstract notion of a principal, which can be a user or the system, and we distinguish between obligations, which are associated with generic events, and duties, which are triggered by events that happen in the system. Time-triggered obligations and duties can be accommodated by defining events corresponding to clock ticks.

The system presented in [14] extends the usage control (UCON) model with obligations and separates system obligations (which are called trusted obligations) from user obligations (called non-trusted). Mechanisms are proposed to deal with the non-trusted obligations. The system does not consider the interaction of obligations with permissions, neither deals with dynamic conditions on obligations, as our metamodel does. In [20], the ABC core model was defined for UCON, dealing with authorisations, obligations and conditions. This approach differs from ours, since it describes not a general metamodel, but instead a family of models depending on the type of authorisations, obligations and conditions that are considered and the necessary components for each model in the family. Our approach also considers authorisations, obligations and conditions, but in a uniform way, and provides a rewrite-based operational semantics for authorisations and obligations.

The approach followed in [8] was to formalise policies (using Horn clauses) trying to minimise the number of actions that need to be performed before allowing an access request. Although the initial system only dealt with systems without violated obligations, this was later extended to handle violation of obligations. This approach was limited in the type of obligations that could be modelled, because it lacked mechanisms to deal with temporal constraints.

In [13] a model is presented for authorisation systems dealing with the notion of accountability; obligations are defined as an action on a resource by a user within a time interval (defined in clock ticks from a predetermined starting point). The monitoring and restoring of accountability was further explored in [21], where a transition relation is defined and conditions to guarantee accountability are established. The notion of obligation defined in these works corresponds to concrete obligations (duties) in our model, and although this is not the focus of this paper, we believe that rewriting can be used to verify the properties of accountability studied in these papers.

A more general model dealing with obligations and its relation to access control and privacy policies was defined in [18]. This model investigates the interaction of obligations and permissions and its undesirable effects such as obligation cascading. We do not deal with privacy policies here, but the category-based model that we present is expressive enough to model the concepts of obligations defined in this work, and rewriting properties can be used to further explore the interplay between permissions and obligations.

Closely related to our work is [12], which considers obligations and authorisations in dynamic systems using a notion of event defined as a pair of a system state and an action. Our notion of event also includes actions, and system states can be included in the event representation. A major difference is that [12] focuses on compliance of events whereas we focus on compatibility properties of policies. Also, we use a rewriting semantics instead of the logic programming approach advocated in [12]. In this sense, our work and [12] are complementary: an answer set semantics could be defined for our policies, and Prolog used to check compatibility provided a tool is available to check termination of the program.

7   Conclusions

We augmented the CBAC metamodel with a general notion of obligation. The framework is expressive enough to deal with most of the features relevant to authorisations and obligations and provides means to reason about them.

Our model distinguishes individual and collective obligations by partitioning the set of obligations into two distinct sets. However, there are some situations where it can be difficult to distinguish between one or the other type. An alternative approach to be investigated in the future is to base this distinction on the definition of the obligation-category assignment relation. That is, although at a certain moment all the principals belonging to a particular category are obliged to perform some action, the definition of the category can depend on the fact of the action not having been performed at the time. In future work we will further develop the notion of event type, and give an operational definition of the typing relation for events. We also wish to explore appropriate mechanisms to deal with compliance and accountability.

References

1. ANSI. RBAC, 2004. INCITS 359–2004
2. Baader, F., Nipkow, T.: Term Rewriting and All That. Cambridge University Press, Cambridge (1998)
3. Barker, S.: The next 700 access control models or a unifying meta-model? In: Proceedings of SACMAT 2009, pp. 187–196. ACM Press (2009)

4. Barker, S., Sergot, M.J., Wijesekera, D.: Status-based access control. ACM Trans. Inf. Syst. Secur. **12**(1), 1–47 (2008)
5. Bertolissi, C., Fernández, M.: Category-based authorisation models: operational semantics and expressive power. In: Massacci, F., Wallach, D., Zannone, N. (eds.) ESSoS 2010. LNCS, vol. 5965, pp. 140–156. Springer, Heidelberg (2010)
6. Bertolissi, C., Fernández, M.: Rewrite specifications of access control policies in distributed environments. In: Cuellar, J., Lopez, J., Barthe, G., Pretschner, A. (eds.) STM 2010. LNCS, vol. 6710, pp. 51–67. Springer, Heidelberg (2011)
7. Bertolissi, C., Fernández, M., Barker, S.: Dynamic event-based access control as term rewriting. In: Barker, S., Ahn, G.-J. (eds.) Data and Applications Security 2007. LNCS, vol. 4602, pp. 195–210. Springer, Heidelberg (2007)
8. Bettini, C., Jajodia, S., Wang, X., Wijesekera, D.: Provisions and obligations in policy rule management. J. Netw. Syst. Manag. **11**(3), 351–372 (2003)
9. Contejean, E., Paskevich, A., Urbain, X., Courtieu, P., Pons, O., Forest, J.: A3pat, an approach for certified automated termination proofs. In: Proceedings of PEPM 2010, pp. 63–72. ACM, New York (2010)
10. Davidson, D.: Essays on Actions and Events. Oxford University Press, Oxford (2001)
11. Dijkstra, E.W.: Selected Writings on Computing - A Personal Perspective. Texts and Monographs in Computer Science. Springer, New York (1982)
12. Gelfond, M., Lobo, J.: Authorization and obligation policies in dynamic systems. In: Garcia de la Banda, M., Pontelli, E. (eds.) ICLP 2008. LNCS, vol. 5366, pp. 22–36. Springer, Heidelberg (2008)
13. Irwin, K., Yu, T., Winsborough, W.H.: On the modeling and analysis of obligations. In: Proceedings of CCS 2006, pp. 134–143. ACM, New York (2006)
14. Katt, B., Zhang, X., Breu, R., Hafner, M., Seifert, J.-P.: A general obligation model and continuity: enhanced policy enforcement engine for usage control. In: Proceedings of SACMAT 2008, pp. 123–132. ACM, New York (2008)
15. Kowalski, R., Sergot, M.: A logic-based calculus of events. New. Gener. Comput. **4**(1), 67–95 (1986)
16. Miller, R., Shanahan, M.: The event calculus in classical logic - alternative axiomatisations. Electron. Trans. Artif. Intell. **3**(A), 77–105 (1999)
17. Mont, M.C., Beato, F.: On parametric obligation policies: enabling privacy-aware information lifecycle management in enterprises. In: POLICY, pp. 51–55 (2007)
18. Ni, Q., Bertino, E., Lobo, J.: An obligation model bridging access control policies and privacy policies. In: Proceedings of SACMAT 2008, pp. 133–142. ACM, New York (2008)
19. OASIS. eXtensible Access Control Markup language (XACML) (2003). http://www.oasis-open.org/xacml/docs/
20. Park, J., Sandhu, R.: The ucon abc usage control model. ACM Trans. Inf. Syst. Secur. **7**(1), 128–174 (2004)
21. Pontual, M., Chowdhury, O., Winsborough, W.H., Yu, T., Irwin, K.: On the management of user obligations. In: Proceedings of SACMAT 2011, pp. 175–184. ACM, New York (2011)

Program Testing and Verification

Concolic Execution and Test Case Generation in Prolog

Germán Vidal[✉]

MiST, DSIC, Universitat Politècnica de València,
Camino de Vera, s/n, 46022 Valencia, Spain
gvidal@dsic.upv.es

Abstract. Symbolic execution extends concrete execution by allowing symbolic input data and then exploring all feasible execution paths. It has been defined and used in the context of many different programming languages and paradigms. A symbolic execution engine is at the heart of many program analysis and transformation techniques, like partial evaluation, test case generation or model checking, to name a few. Despite its relevance, traditional symbolic execution also suffers from several drawbacks. For instance, the search space is usually huge (often infinite) even for the simplest programs. Also, symbolic execution generally computes an overapproximation of the concrete execution space, so that false positives may occur. In this paper, we propose the use of a variant of symbolic execution, called concolic execution, for test case generation in Prolog. Our technique aims at full statement coverage. We argue that this technique computes an underapproximation of the concrete execution space (thus avoiding false positives) and scales up better to medium and large Prolog applications.

1   Introduction

There is a renewed interest in *symbolic execution* [3,9], a well-known technique for program verification, testing, debugging, etc. In contrast to concrete execution, symbolic execution considers that the values of some input data are unknown, i.e., some input parameters $x, y, \ldots$ take *symbolic values* $X, Y, \ldots$ Because of this, symbolic execution is often non-deterministic: at some control statements, we need to follow more than one execution path because the available information does not suffice to determine the validity of a control expression, e.g., symbolic execution may follow both branches of the conditional "if $(x > 0)$ then *exp1* else *exp2*" when the symbolic value X of variable x is not constrained enough to imply neither $x > 0$ nor $\neg(x > 0)$. Symbolic states include a *path condition* that stores the current constraints on symbolic values, i.e., the conditions that must hold to reach a particular execution state.

This work has been partially supported by the EU (FEDER) and the Spanish *Ministerio de Economía y Competitividad (Secretaría de Estado de Investigación, Desarrollo e Innovación)* under grant TIN2013-44742-C4-1-R and by the *Generalitat Valenciana* under grant PROMETEO/2011/052.

M. Proietti and H. Seki (Eds.): LOPSTR 2014, LNCS 8981, pp. 167–181, 2015.
DOI: 10.1007/978-3-319-17822-6_10

E.g., after symbolically executing the above conditional, the derived states for *exp1* and *exp2* would add the conditions $X > 0$ and $X \leq 0$, respectively, to their path conditions.

Traditionally, formal techniques based on symbolic execution have enforced *soundness*: if a symbolic state is reached and its path condition is satisfiable, there must be a concrete execution path that reaches the corresponding concrete state. In contrast, we say that symbolic execution is *complete* when every reachable state in a concrete execution is "covered" by some symbolic state. For the general case of infinite state systems, completeness usually requires some kind of *abstraction* (as in infinite state model checking).

In the context of logic programming, we can find many techniques that use some form of *complete* symbolic execution, like partial evaluation [4,12,13]. However, these overapproximations of the concrete semantics may have a number of drawbacks in the context of testing and debugging. On the one hand, one should define complex subsumption and abstraction operators since the symbolic search space is usually infinite. These abstraction operators, may introduce *false positives*, which is often not acceptable when debugging large applications. On the other hand, because of the complexity of these operators, the associated methods usually do not scale to medium and large applications.

In imperative programming, an alternative approach, called *concolic execution* [6,16], has become popular in the last years. Basically, concolic execution proceeds as follows: first, a concrete execution using random input data is performed. In parallel to the concrete execution, a symbolic execution is also performed, but restricted to the same conditional choices of the concrete execution. Then, by negating one of the constraints in the symbolic execution, new input data are obtained, and the process starts again. Here, only concrete executions are considered and, thus, no false positives are produced. This approach has given rise to a number of powerful and scalable tools in the context of imperative and concurrent programming, like Java Pathfinder [14] and SAGE [7].

In this paper, we present a novel scheme for testing pure Prolog (without negation) based on a notion of concolic execution. To the best of our knowledge, this is the first approach to concolic execution in the context of a declarative programming paradigm.

2   Preliminaries

We assume some familiarity with the standard definitions and notations for logic programs [11]. Nevertheless, in order to make the paper as self-contained as possible, we present in this section the main concepts which are needed to understand our development.

In this work, we consider a first-order language with a fixed vocabulary of predicate symbols, function symbols, and variables denoted by Π, Σ and $\mathcal{V}$, respectively. In the following, we let $\overline{o_n}$ denote the sequence of syntactic objects $o_1, \ldots, o_n$, and we let $\overline{o_{n,m}}$ denote the sequence $o_n, o_{n+1}, \ldots, o_m$. Also, we often use $\overline{o}$ when the number of elements in the sequence is irrelevant. We let $\mathcal{T}(\Sigma, \mathcal{V})$

denote the set of *terms* constructed using symbols from Σ and variables from $\mathcal{V}$. An *atom* has the form $p(\overline{t_n})$ with $p/n \in \Pi$ and $t_i \in \mathcal{T}(\Sigma, \mathcal{V})$ for $i = 1, \ldots, n$. A *goal* is a finite sequence of atoms $A_1, \ldots, A_n$, where the *empty goal* is denoted by *true*. A *clause* has the form $H \to \mathcal{B}$ where H is an atom and $\mathcal{B}$ is a goal (note that we only consider *definite* programs). A logic *program* is a finite sequence of clauses. $\mathcal{V}ar(s)$ denotes the set of variables in the syntactic object s (i.e., s can be a term, an atom, a query, or a clause). A syntactic object s is *ground* if $\mathcal{V}ar(s) = \emptyset$. In this work, we only consider *finite* ground terms.

Substitutions and their operations are defined as usual. In particular, the set $\mathcal{D}om(\sigma) = \{x \in \mathcal{V} \mid \sigma(x) \neq x\}$ is called the *domain* of a substitution σ. We let *id* denote the empty substitution. The application of a substitution θ to a syntactic object s is usually denoted by juxtaposition, i.e., we write $s\theta$ rather than $\theta(s)$. A syntactic object s_1 is *more general* than a syntactic object s_2, denoted $s_1 \leqslant s_2$, if there exists a substitution θ such that $s_2 = s_1\theta$. A *variable renaming* is a substitution that is a bijection on $\mathcal{V}$. Two syntactic objects t_1 and t_2 are *variants* (or equal up to variable renaming), denoted $t_1 \approx t_2$, if $t_1 = t_2\rho$ for some variable renaming ρ. A substitution θ is a unifier of two syntactic objects t_1 and t_2 iff $t_1\theta = t_2\theta$; furthermore, θ is the *most general unifier* of t_1 and t_2, denoted by $mgu(t_1, t_2)$ if, for every other unifier σ of t_1 and t_2, we have that $\theta \leqslant \sigma$.

The notion of *computation rule* $\mathcal{R}$ is used to select an atom within a goal for its evaluation. Given a program P, a goal $\mathcal{G} = A_1, \ldots, A_n$, and a computation rule $\mathcal{R}$, we say that $\mathcal{G} \leadsto_{P,\mathcal{R},\sigma} \mathcal{G}''$ is an *SLD resolution step* for $\mathcal{G}$ with P and $\mathcal{R}$ if

- $\mathcal{R}(\mathcal{G}) = A_i$, $1 \leqslant i \leqslant n$, is the selected atom,
- $H \to \mathcal{B}$ is a renamed apart clause of P (in symbols $H \to \mathcal{B} \lll P$),
- $\sigma = mgu(A, H)$, and
- $\mathcal{G}' \equiv (A_1, \ldots, A_{i-1}, \mathcal{B}, A_{i+1}, \ldots, A_n)\sigma$.

We often omit P, $\mathcal{R}$ and/or σ in the notation of an SLD resolution step when they are clear from the context. An *SLD derivation* is a (finite or infinite) sequence of SLD resolution steps. We often use $\mathcal{G}_0 \leadsto_\theta^* \mathcal{G}_n$ as a shorthand for $\mathcal{G}_0 \leadsto_{\theta_1} \mathcal{G}_1 \leadsto_{\theta_2} \ldots \leadsto_{\theta_n} \mathcal{G}_n$ with $\theta = \theta_n \circ \cdots \circ \theta_1$ (where $\theta = \{\}$ if $n = 0$). An SLD derivation $\mathcal{G} \leadsto_\theta^* \mathcal{G}'$ is *successful* when $\mathcal{G}' = true$; in this case, we say that θ is the *computed answer substitution*. SLD derivations are represented by a (possibly infinite) finitely branching tree.

3 A Deterministic Semantics

In this section, we introduce a deterministic small-step semantics for pure Prolog (without negation). Basically, as we will see in the next section, we need to keep some information through the complete Prolog computation, and the usual semantics based on non-determinism and backtracking is not adequate for this purpose. Therefore, we propose the use of a stack to store alternative execution paths that are tried when a failure is reached. The resulting small-step semantics

$$\text{(unfolding)} \quad \frac{\text{let } \overline{H_n \xleftarrow{\ell_n} \mathcal{B}_n} \ll P \text{ be all the clauses such that}}{\langle \overline{A_m}; \sigma; \mathcal{S} \rangle \xrightarrow{\mathsf{u}(\overline{\ell_n})} \langle (\mathcal{B}_1, \overline{A_{2,m}})\sigma_1; \sigma\sigma_1; [(\ell_2; \sigma\sigma_2; (\mathcal{B}_2, \overline{A_{2,m}})\sigma_2),}$$
$$\cdots,$$
$$(\ell_n; \sigma\sigma_n; (\mathcal{B}_n, \overline{A_{2,m}})\sigma_n)] \mathbin{+\!\!+} \mathcal{S} \rangle$$

$$\text{(backtracking)} \quad \frac{\text{there is no clause } H \xleftarrow{\ell'} \mathcal{B} \ll P \text{ such that } \mathsf{mgu}(A_1, H) \neq \mathit{fail}}{\langle \overline{A_m}; \sigma'; [(\ell; \mathcal{G}; \sigma) | \mathcal{S}] \rangle \xrightarrow{\mathsf{b}(\ell)} \langle \mathcal{G}; \sigma; \mathcal{S} \rangle} \quad (m \geqslant 1)$$

$$\text{(failure)} \quad \frac{\text{there is no clause } H \xleftarrow{\ell} \mathcal{B} \ll P \text{ such that } \mathsf{mgu}(A_1, H) \neq \mathit{fail}}{\langle \overline{A_m}; \sigma; [\,] \rangle \xrightarrow{\mathsf{f}} \langle \mathsf{fail}; \sigma; [\,] \rangle} \quad (m \geqslant 1)$$

Fig. 1. Deterministic small-step semantics

is clearly equivalent to the original one when a depth-first search is considered. Actually, our deterministic semantics is essentially equivalent to (a subset of) the *linear* semantics presented in [17].

In the following, we assume that the program clauses are labeled. In particular, given a program P, we use the notation $H \xleftarrow{\ell} \mathcal{B} \ll P$ to refer to a (renamed apart) labeled clause $H \leftarrow \mathcal{B}$ in P. Labels must be unique. Moreover, we only consider Prolog's left-to-right computation rule, and assume that only the computation of the first answer for the initial goal is relevant (as it is common in practical Prolog applications).

Our semantics deals with *states*, which are defined as follows:

Definition 1 (State). *A* state *is a tuple* $\langle \mathcal{G}; \sigma; \mathcal{S} \rangle$, *where* $\mathcal{G}$ *is a goal,* σ *is a substitution—the (partial) answer computed so far—and* $\mathcal{S}$, *the* stack, *is a (possibly empty) list of tuples* $(\ell; \mathcal{G}'; \sigma')$ *with* ℓ *a clause label,* $\mathcal{G}'$ *a goal and* σ' *a substitution.*

The small-step deterministic semantics is defined as the smallest relation that obeys the labeled transition rules shown in Fig. 1, where $[H|R]$ denotes a list with head H and tail R, and "$+\!\!+$" denotes list concatenation.

Given a goal $\mathcal{G}_0$, the initial state has the form $\langle \mathcal{G}_0; id; [\,] \rangle$. The transition relation is labeled with $\mathsf{u}(\overline{\ell_n})$, denoting an unfolding step with the clauses labeled with $\overline{\ell_n}$, $\mathsf{b}(\ell)$, denoting a backtracking step that tries a clause labeled with ℓ, or f, denoting a failing derivation.

Let us briefly explain the rules of the small-step semantics:

– The unfolding rule proceeds as in standard SLD resolution, but considers all matching clauses, so that all SLD resolution steps are performed in one go. The first unfolding step is used to replace the goal component of the state, while the remaining ones (if any) are added on top of the stack (thus we mimic the usual depth-first search of Prolog). Here, the labels of the clauses and the partial computed answers are also stored in the stack in order to recover this information when a backtracking step is performed.

- The backtracking rule applies when no further unfolding is possible and the goal component is not *true* (the empty goal). In this case, we discard the current goal and consider the first goal in the stack, extracting the clause label and the partial answer that are needed for the transition step.
- Finally, the failure rule is used to terminate a computation that reaches a goal in which the selected atom does not match any rule and, moreover, there are no alternatives in the stack.

A *successful* computation has the form $\langle \mathcal{G}_0; id; [\,] \rangle \overset{s_1}{\to} \langle \mathcal{G}_1; \sigma_1; \mathcal{S}_1 \rangle \overset{s_2}{\to} \ldots \overset{s_n}{\to} \langle true; \sigma_n; \mathcal{S}_n \rangle$, where σ_n (restricted to the variables of $\mathcal{G}_0$) is the computed answer substitution. A *failing* computation has the form $\langle \mathcal{G}_0; id; [\,] \rangle \overset{s_1}{\to} \langle \mathcal{G}_1; \sigma_1; \mathcal{S}_1 \rangle \overset{s_2}{\to} \ldots \overset{s_n}{\to} \langle \mathcal{G}_n; \sigma_n; \mathcal{S}_n \rangle \overset{f}{\to}_{id} \langle fail; \sigma_n; [\,] \rangle$; we keep σ_n in the last state since it might be useful for analyzing finite failure derivations.

Now, we introduce the following notion of execution *trace*, that will be used in the next section to steer the symbolic execution.

Definition 2 (Trace). Let $\langle \mathcal{G}_0; id; \mathcal{S}_0 \rangle \overset{s_1}{\to} \langle \mathcal{G}_1; \sigma_1; \mathcal{S}_1 \rangle \overset{s_2}{\to} \ldots \overset{s_n}{\to} \langle \mathcal{G}_n; \sigma_n; \mathcal{S}_n \rangle$ be a computation. The associated trace is the list $[s_1, s_2, \ldots, s_n]$, where each s_i is either of the form $u(\overline{\ell_m})$, $u(\ell)$ or f.

Example 1. Consider the rev_acc_type program to reverse a list using an accumulator and also checking the type of the input parameter (from the DPPD library [10]), extended with predicates main, length, and foo:

```
(1) main(L,N,R) :-          (5) is_list([]).
        length(L,N),        (6) is_list([_H|T]) :-
        rev(L,[],R),            is_list(T).
        foo(a).
(2) main(_L,_N,error).
                            (7) length([],0).
(3) rev([],A,A).            (8) length([_H|R],s(N)) :-
(4) rev([H|T],Acc,Res) :-      length(R,N).
        is_list(Acc),
        rev(T,[H|Acc],Res). (9) foo(b).
```

Here, we use natural numbers as clause labels. Predicate main considers two cases: if the input list L has length N (the length is represented using natural numbers built from 0 and s(_) to avoid the use of built-ins), the reverse of L is computed; otherwise, we assume that an error occurs. The computation for the initial goal main([a, b], s(s(0)), R) is shown in Fig. 2, where only the relevant computed substitutions are shown. The trace associated to the computation is

$$[u(1,2), u(8), u(8), u(7), u(4), u(5), u(4), u(6), u(5), u(3), b(2)]$$

$\langle \mathtt{main}([\mathtt{a}, \mathtt{b}], \mathtt{s}(\mathtt{s}(0)), \mathtt{R}); \; id; \; [\,] \rangle$

$\overset{u(1,2)}{\rightarrow} \langle \mathtt{length}([\mathtt{a}, \mathtt{b}], \mathtt{s}(\mathtt{s}(0))), \mathtt{rev}([\mathtt{a}, \mathtt{b}], [\,], \mathtt{R}), \mathtt{foo}(\mathtt{a}); \; id; \; [(2; \{\mathtt{R}/\mathtt{error}\}; \mathtt{true})] \rangle$

$\overset{u(8)}{\rightarrow} \langle \mathtt{length}([\mathtt{b}], \mathtt{s}(0)), \mathtt{rev}([\mathtt{a}, \mathtt{b}], [\,], \mathtt{R}), \mathtt{foo}(\mathtt{a}); \; id; \; [(2; \{\mathtt{R}/\mathtt{error}\}; \mathtt{true})] \rangle$

$\overset{u(8)}{\rightarrow} \langle \mathtt{length}([\,], 0), \mathtt{rev}([\mathtt{a}, \mathtt{b}], [\,], \mathtt{R}), \mathtt{foo}(\mathtt{a}); \; id; \; [(2; \{\mathtt{R}/\mathtt{error}\}; \mathtt{true})] \rangle$

$\overset{u(7)}{\rightarrow} \langle \mathtt{rev}([\mathtt{a}, \mathtt{b}], [\,], \mathtt{R}), \mathtt{foo}(\mathtt{a}); \; id; \; [2; (\{\mathtt{R}/\mathtt{error}\}; \mathtt{true})] \rangle$

$\overset{u(4)}{\rightarrow} \langle \mathtt{is\_list}([\,]), \mathtt{rev}([\mathtt{b}], [\mathtt{a}], \mathtt{R}), \mathtt{foo}(\mathtt{a}); \; id; \; [(2; \{\mathtt{R}/\mathtt{error}\}; \mathtt{true})] \rangle$

$\overset{u(5)}{\rightarrow} \langle \mathtt{rev}([\mathtt{b}], [\mathtt{a}], \mathtt{R}), \mathtt{foo}(\mathtt{a}); \; id; \; [(2; \{\mathtt{R}/\mathtt{error}\}; \mathtt{true})] \rangle$

$\overset{u(4)}{\rightarrow} \langle \mathtt{is\_list}([\mathtt{a}]), \mathtt{rev}([\,], [\mathtt{b}, \mathtt{a}], \mathtt{R}), \mathtt{foo}(\mathtt{a}); \; id; \; [(2; \{\mathtt{R}/\mathtt{error}\}; \mathtt{true})] \rangle$

$\overset{u(6)}{\rightarrow} \langle \mathtt{is\_list}([\,]), \mathtt{rev}([\,], [\mathtt{b}, \mathtt{a}], \mathtt{R}), \mathtt{foo}(\mathtt{a}); \; id; \; [(2; \{\mathtt{R}/\mathtt{error}\}; \mathtt{true})] \rangle$

$\overset{u(5)}{\rightarrow} \langle \mathtt{rev}([\,], [\mathtt{b}, \mathtt{a}], \mathtt{R}), \mathtt{foo}(\mathtt{a}); \; id; \; [(2; \{\mathtt{R}/\mathtt{error}\}; \mathtt{true})] \rangle$

$\overset{u(3)}{\rightarrow} \langle \mathtt{foo}(\mathtt{a}); \; \{\mathtt{R}/[\mathtt{b}, \mathtt{a}]\}; \; [(2; \{\mathtt{R}/\mathtt{error}\}; \mathtt{true})] \rangle$

$\overset{b(2)}{\rightarrow} \langle \mathtt{true}; \; \{\mathtt{R}/\mathtt{error}\}; \; [\,] \rangle$

Fig. 2. Successful computation for $\mathtt{main}([\mathtt{a}, \mathtt{b}], \mathtt{s}(\mathtt{s}(0)), \mathtt{R})$.

4 Concolic Execution

In this section, we introduce the semantics of concolic execution. Essentially, it deals with symbolic input data (free variables in our context), as in standard symbolic execution, but is driven by a concrete execution. Often, a single algorithm mixing both concrete and symbolic execution is introduced. In contrast, for clarity, we prefer to keep both calculi independent: the concrete semantics produces a trace, which is then used to steer the symbolic execution.[1]

The symbolic states for concolic execution are defined as follows:

Definition 3 (Symbolic State). *A symbolic state is a tuple* $\langle \tau; \mathcal{L}; \mathcal{G}; \sigma; \mathcal{S}; T \rangle$, *where*

- τ *is a computation trace,*
- $\mathcal{L}$ *is a list of clause labels (namely, a stack that keeps track of the current clause environment),[2]*
- $\mathcal{G}$ *is a goal,*
- σ *is the partial answer computed so far (a substitution),*
- $\mathcal{S}$ *is a (possibly empty) list of tuples,* $(\ell; \mathcal{L}'; \sigma; \mathcal{G}')$, *where* $\mathcal{L}'$ *is also a list of clause labels, and*
- T *is a set of clause labels (the labels of those clauses not yet completely evaluated).*

---

[1] Nevertheless, an implementation of this technique may as well combine both calculi into a single algorithm to improve efficiency.

[2] The usefulness of keeping the clause stack will become clear in the next section. Basically, it is needed to know which other clauses can be completely evaluated when a given clause—the one that is on top of the stack—is completely evaluated.

$$\text{(unfolding)} \quad \frac{\begin{array}{c} \text{if } H_n \xleftarrow{\ell_n} \mathcal{B}_n \ll P \text{ with } \mathrm{mgu}(A_1, H_i) = \sigma_i, i = 1, \ldots, n \text{ and} \\ H'_k \xleftarrow{\ell'_k} \mathcal{B}'_k \ll P \text{ with } \mathrm{mgu}(A_1, H'_j) = \sigma'_j, \ j = 1, \ldots, k \\ \text{and } \{\overline{\ell_n}\} \cap \{\overline{\ell'_k}\} = \{\,\} \end{array}}{\langle [\mathsf{u}(\overline{\ell_n}) | \tau]; \mathcal{L}; \overline{A_m}; \sigma; \mathcal{S}; T \rangle \xrightarrow{\{\mathsf{c}(\ell'_k, \sigma'_k)\}}}$$

$$\begin{array}{c} \langle \tau; [\ell_1 | \mathcal{L}]; \\ (\mathcal{B}_1, \mathsf{e}(\ell_1), \overline{A_{2,m}})\sigma_1; \ \sigma\sigma_1; \\ [(\ell_2; [\ell_2 | \mathcal{L}]; \sigma\sigma_2; (\mathcal{B}_2, \mathsf{e}(\ell_2), \overline{A_{2,m}})\sigma_2), \\ \ldots, \\ (\ell_n; [\ell_n | \mathcal{L}]; \sigma\sigma_n; (\mathcal{B}_n, \mathsf{e}(\ell_n), \overline{A_{2,m}})\sigma_n)] \!+\!\! + \mathcal{S}; T \rangle \end{array}$$

$$\text{(exit)} \quad \frac{A_1 = \mathsf{e}(\ell)}{\langle \tau; [\ell | \mathcal{L}]; \overline{A_m}; \sigma; \mathcal{S}; T \rangle \xrightarrow{\{\}} \langle \tau; \mathcal{L}; \overline{A_{2,m}}; \sigma; \mathcal{S}; T \setminus \{\ell\} \rangle}$$

$$\text{(backtracking)} \quad \frac{A_1 \neq \mathsf{e}(\,), \ \overline{H_n \xleftarrow{\ell_n} \mathcal{B}_n} \ll P \text{ s.t. } \mathrm{mgu}(A_1, H_i) = \sigma \neq \mathit{fail}, \ i = 1, \ldots, n}{\langle [\mathsf{b}(\ell) | \tau]; \mathcal{L}; \overline{A_m}; \sigma; [(\ell; \mathcal{L}'; \sigma'; \mathcal{G}) | \mathcal{S}]; T \rangle \xrightarrow{\{\mathsf{c}(\ell_n, \sigma_n)\}} \langle \tau; \mathcal{L}'; \mathcal{G}; \sigma'; \mathcal{S}; T \rangle} \quad (m \geqslant 1)$$

$$\text{(failure)} \quad \frac{\text{if } \overline{H_n \xleftarrow{\ell_n} \mathcal{B}_n} \ll P \text{ such that } \mathrm{mgu}(A_1, H_i) = \sigma \neq \mathit{fail}, \ i - 1, \ldots, n}{\langle [\mathsf{f} | \tau]; \mathcal{L}; \overline{A_m}; \sigma; [\,]; T \rangle \xrightarrow{\{\mathsf{c}(\ell_n, \sigma_n)\}} \langle \tau; \mathcal{L}; \mathsf{fail}; \sigma; [\,]; T \rangle} \quad (m \geqslant 1)$$

Fig. 3. Concolic execution semantics

The concolic execution semantics is defined as the smallest relation that obeys the labeled transition rules shown in Fig. 3. Given a trace τ, the initial symbolic state has the form $\langle \tau; [\,]; \mathcal{G}_0; id; [\,]; T \rangle$, where $\mathcal{G}_0$ is a goal with the same predicates as in the concrete execution, but with fresh variables as arguments, and T is a set with the labels of all program clauses. The transition relation is labeled with a (possibly empty) list of terms of the form $\mathsf{c}(\ell, \theta)$, which denote possible alternatives for unfolding that concrete execution did not consider. Missing alternatives will be used to generate new input data that explore different execution paths.

Let us briefly explain the rules of concolic execution semantics:

- The unfolding rule follows the trace of the concrete execution and applies the same unfolding step. Here, a call of the form $\mathsf{e}(\ell)$ is added to the end of the clause bodies to mark when the clauses are completely evaluated. This is required in our context since we only consider that a clause is *covered* when all body atoms are successfully executed.[3] This will be a useful information

---

[3] Observe that other, more relaxed, notions of clause covering are possible; e.g., consider that a clause is covered as soon as the clause is used in an unfolding step. Also, see [2] for a more declarative notion of test coverage.

for test case generation, as we will see in the next section. Moreover, in this rule, we label the step with the information regarding the remaining clauses whose head unifies with the selected atom (and did not match with it in the concrete execution). Finally, we add ℓ_1 to the stack of clause labels (the current environment).

- The exit rule applies when the selected atom has the form $e(\ell)$. In this case, we remove ℓ from the top of the environment stack, and also delete ℓ from the set of clause labels T (i.e., clause ℓ has been completely evaluated).
- The backtracking and failure rules proceed analogously to the unfolding rule by labeling the step with the information regarding the additional clauses whose head unify with the selected atom (if any).

When the set labeling rules unfolding, backtracking and failure is not empty, we have identified situations in which the symbolic state can follow an execution path that is not possible with the concrete goal. Therefore, they allow us to construct new input data for the initial goal so that a different execution path is followed.

Let us now show a simple computation with the concolic execution semantics. We postpone to the next section the algorithm for test case generation.

Example 2. Consider the following simple program:

```
(1)   p(X) :- q(X),r(X).
(2)   q(X) :- s(X).
(3)   s(a).
(4)   s(b).
(5)   r(b).
```

where we again consider natural numbers as clause labels. The concrete execution for the initial goal $p(a)$ is as follows:[4]

$$\langle p(a); id; [] \rangle \xrightarrow{u(1)} \langle q(a), r(a); id; [] \rangle \xrightarrow{u(2)} \langle s(a), r(a); id; [] \rangle \xrightarrow{u(3)} \langle r(a); id; [] \rangle \xrightarrow{f} \langle \mathsf{fail}; id; [] \rangle$$

Therefore, its associated trace is $\tau = [u(1), u(2), u(3), f]$. Now, for concolic execution, we consider the trace τ and the initial goal $p(X)$. The concolic execution is shown in Fig. 4. As can be seen, the execution of clauses 1, 4 and 5 has not been completed. Moreover, we can observe that there was only one missing alternative when unfolding $s(X)$. In the next section, we show how this information can be used for test case generation.

5   Test Case Generation

In this section, we present an algorithm for test case generation using concolic execution. In contrast to previous approaches for Prolog testing, our technique

---

[4] In the examples, we restrict the (partial) computed answers to the variables of the initial goal.

$$\langle[u(1),u(2),u(3),f];[\,]; \quad p(X); \qquad\qquad id; \qquad [\,];[1,2,3,4,5]\rangle$$
$$\underset{\curvearrowright}{\Downarrow} \qquad \langle[u(2),u(3),f];[1]; \quad q(X),r(X),e(1); \quad id; \qquad [\,];[1,2,3,4,5]\rangle$$
$$\underset{\curvearrowright}{\Downarrow}$$
$$\underset{\{c(4,\{X/b\})\}}{\curvearrowright} \qquad \langle[u(3),f];[2,1]; \quad s(X),e(2),r(X),e(1); \; id; \quad [\,];[1,2,3,4,5]\rangle$$
$$\langle[f];[3,2,1]; \quad e(3),e(2),r(a),e(1); \; \{X/a\};[\,];[1,2,3,4,5]\rangle$$
$$\underset{\curvearrowright}{\Downarrow} \qquad\qquad \langle[f];[2,1]; \quad e(2),r(a),e(1); \qquad \{X/a\};[\,];[1,2,4,5]\rangle$$
$$\underset{\curvearrowright}{\Downarrow} \qquad\qquad \langle[f];[1]; \quad r(a),e(1); \qquad\qquad \{X/a\};[\,];[1,4,5]\rangle$$
$$\underset{\curvearrowright}{\Downarrow} \qquad\qquad \langle[\,];[1]; \quad fail; \qquad\qquad\qquad \{X/a\};[\,];[1,4,5]\rangle$$

Fig. 4. Concolic execution for $[u(1),u(2),u(3),f]$ and $p(X)$

considers an underapproximation, i.e., only actual executions are considered (since there is no abstraction involved). Therefore, no false positives may occur. If a test case shows an error, this is an actual error in the considered program.

5.1 The Algorithm

In this section, we assume that the program contains a single predicate that starts the execution, which we denote with *main*. This is not unusual for real applications. Moreover, we consider a particular *mode* for *main*,[5] where $in(main/n) = \{i_1, \ldots, i_m\}$ denotes the set of input parameters of *main*.

The algorithm for test case generation proceeds as follows:

1. First, a random goal of the form $main(\overline{t_n})$ is produced, where at least the input arguments (according to $in(main/n)$) must be ground.
2. Now, we use the concrete semantics to execute the goal $main(\overline{t_n})$, thus obtaining an associated trace τ. We assume that this execution terminates, which is reasonable since the input arguments are ground. In practice, one can use a timeout and report a warning when the execution takes more time.
3. Then, we use concolic execution to run an initial symbolic state of the form

$$\langle \tau; [\,]; main(\overline{X_n}); id; [\,]; T \rangle$$

where T is a set with the labels of all program clauses. Since the concrete execution was finite, so is the concolic execution (since it performs exactly the same steps). Let us consider that it has the following form:

$$\langle \tau_0; \mathcal{L}_0; \mathcal{G}_0; \sigma_0; \mathcal{S}_0; T_0 \rangle \overset{c_1}{\rightsquigarrow} \ldots \overset{c_m}{\rightsquigarrow} \langle \tau_m; \mathcal{L}_m; \mathcal{G}_m; \sigma_m; \mathcal{S}_m; T_m \rangle$$

where $\tau_0 = \tau$, $\mathcal{L}_0 = [\,]$, $\mathcal{G}_0 = main(\overline{X_n})$, $\sigma_0 = id$, $\mathcal{S}_0 = [\,]$, $T_0 = T$, and $\mathcal{G}_m$ is either *true* or fail.

---

[5] Extending our approach to multiple modes would not be difficult, but would introduce another source of nondeterminism when grounding an input goal.

4. Now, we check the value of T_m. If $T_m = \{\}$, the algorithm terminates since all clauses have been completely executed. Otherwise, we identify the *last* state $\langle \tau_i; \mathcal{L}_i; \mathcal{G}_i; \sigma_i; \mathcal{S}_i; T_i \rangle$ in the above concolic execution such that
   - the previous transition $\overset{c_i}{\leadsto}$ is labeled with $c_i = \{c(\ell_k, \theta_k)\}$, $k > 0$, and
   - there exists $j \in \{1, \ldots, k\}$ such that either $\ell_j \in T_m$ or $\mathcal{L}_i$ contains (not necessarily in a top position) some labels from T_m; the reason to also consider the labels from $\mathcal{L}_i$ is that considering an alternative clause may help to complete the execution of *all* the clauses in the current clause stack. Either way, we choose a clause j in a non-deterministic way.

Therefore, we have a prefix of the complete concolic execution of the form:

$$\langle \tau_0; \mathcal{L}_0; \mathcal{G}_0; \sigma_0; \mathcal{S}_0; T_0 \rangle \overset{c_1}{\leadsto} \ldots \overset{c_{i-1}}{\leadsto} \langle \tau_{i-1}; \mathcal{L}_{i-1}; \mathcal{G}_{i-1}; \sigma_{i-1}; \mathcal{S}_{i-1}; T_{i-1} \rangle$$
$$\overset{c_i}{\leadsto} \langle \tau_i; \mathcal{L}_i; \mathcal{G}_i; \sigma_i; \mathcal{S}_i; T_i \rangle$$

and we are interested in the (possibly) partial computed answer $\sigma_{i-1}\theta_j$, since it will allow us to explore a different execution path, possibly covering some more program clauses.

Hence, we have a second test case: $\mathcal{G}'_0 = main(\overline{X_n})\sigma_{i-1}\theta_j\gamma$, where γ is a substitution that is only aimed at grounding the input parameters $in(main/n)$ of *main* using arbitrary values (of the right type, preferably minimal ones).

5. Finally, we consider the initial state $\langle \mathcal{G}'_0; id; [\,] \rangle$ and obtain a new trace using the concrete execution semantics τ', so that a new initial symbolic state is defined as follows: $\langle \tau'; [\,]; main(\overline{X_n}); id; [\,]; T_m \rangle$ and the process starts again (i.e., we jump again to step 3). Observe that the initial state includes the set of clause labels T_m obtained in the last state of the previous concolic execution, in order to avoid producing new tests for clauses that are already covered by some previous test case.

Let us now illustrate the complete test case generation process with an example.

5.2 Test Case Generation in Practice

In this section, we illustrate the generation of test cases using a slight modification of the program in Example 1:

```
(1) main(L,N,R) :-          (5) is_list([]).
        length(L,N),         (6) is_list([_H|T]) :-
        rev(L,[],R).              is_list(T).
(2) main(_L,_N,error).

                             (7) length([],0).
(3) rev([],A,A).             (8) length([_H|R],s(N)) :-
(4) rev([H|T],Acc,Res) :-        length(R,N).
        is_list(Acc),
        rev(T,[H|Acc],Res).
```

Observe that, in this example, using a random generation of test cases would be useless since the length of the generated list and the second argument would

$$\langle [u(1,2), u(8), b(2)]; [\,]; \quad \text{main}(L,N,R); \qquad id; \qquad [\,]; \{1,\ldots,8\}\rangle$$

$$\overset{\{\}}{\rightarrow} \qquad \langle [u(8), b(2)]; [1]; \quad \begin{pmatrix} \text{length}(L,N), \\ \text{rev}(L,[\,],R), e(1) \end{pmatrix}; \quad id; \qquad S; \{1,\ldots,8\}\rangle$$

$$\overset{\{c(7,\{L/[\,]],N/0\})\}}{\rightarrow} \quad \langle [b(2)]; [8,1]; \begin{pmatrix} \text{length}(L',N'), e(8), \\ \text{rev}([X|L'],[\,],R), e(1) \end{pmatrix}; \{L/[X|L'], N/s(N')\}; S; \{1,\ldots,8\}\rangle$$

$$\overset{\{c(7,\{L'/[\,]],N'/0\}),\ldots\}}{\rightarrow} \quad \langle [\,]; [2]; \quad e(2); \qquad\qquad \{R/\text{error}\}; \qquad [\,]; \{1,\ldots,8\}\rangle$$

$$\overset{\{\}}{\rightarrow} \qquad \langle [\,]; [\,]; \quad \text{true}; \qquad\qquad \{R/\text{error}\}; \qquad [\,]; \{1,3,\ldots,8\}\rangle$$

with $S = [(2; [2]; \{R/\text{error}\}; e(2))];$

Fig. 5. Concolic execution for $\langle [u(1,2), u(8), b(2)]; [\,]; \text{main}(L,N,R); id; [\,]; \{1,2,\ldots,8\}\rangle$

hardly coincide. Also, using standard symbolic execution might be difficult too since the search space is infinite and, moreover, due to the use of predicate rev that includes an accumulating parameter, there are goals that are not instances of any previous goal, thus requiring some powerful abstraction operators.

Using concolic execution, though, we can easily generate appropriate test cases.

First Iteration. We start with a random initial goal, e.g., $\text{main}([a,b], s(0), R)$, where the input arguments $[1,2]$ are assumed ground. The associated concrete execution is the following:

$$\langle \text{main}([a,b], s(0), R); \, id; \, [\,]\rangle$$
$$\overset{u(1,2)}{\rightarrow} \langle \text{length}([a,b], s(0)), \text{rev}([a,b],[\,],R); \, id; \, [(2; \{R/\text{error}\}; \text{true})]\rangle$$
$$\overset{u(8)}{\rightarrow} \langle \text{length}([b],0), \text{rev}([a,b],[\,],R); \, id; \, [(2; \{R/\text{error}\}; \text{true})]\rangle$$
$$\overset{b(2)}{\rightarrow} \langle \text{true}; \, \{R/\text{error}\}; \, [\,]\rangle$$

and its associated trace is thus $\tau = [u(1,2), u(8), b(2)]$.

Now, we use concolic execution and produce the computation shown in Fig. 5. Therefore, by executing $\text{main}([a,b], s(0), R)$ only clause (2) is completely evaluated. According to the previous algorithm for test case generation, we now consider the following prefix of the concolic execution:

$$\langle [u(1,2), u(8), b(2)]; [\,]; \text{main}(L,N,R); id; [\,]; \{1,\ldots,8\}\rangle \overset{\{\}}{\rightarrow} \ldots \overset{\{c(7,\{L'/[\,]],N'/0\})\}}{\rightarrow} \langle \ldots \rangle$$

and the associated substitution $\{L/[X], N/s(0)\}$.

Second Iteration. Now, we consider the goal $\text{main}([X], s(0), R)$. Since the first two arguments must be ground, as mentioned before, we apply a minimal grounding substitution and get, e.g., $\text{main}([a], s(0), R)$. The concrete execution, which is shown in Fig. 6, computes the following trace:

$$\tau'' = [u(1,2), u(8), u(7), u(4), u(5), u(3)]$$

$\langle \mathtt{main}([a], s(0), R); id; [\,] \rangle$

$\overset{u(1,2)}{\longrightarrow} \langle \mathtt{length}([a], s(0)), \mathtt{rev}([a], [\,], R); id; [(2; \{R/\mathtt{error}\}; \mathtt{true})] \rangle$

$\overset{u(8)}{\longrightarrow} \langle \mathtt{length}([\,], 0), \mathtt{rev}([a], [\,], R); id; [(2; \{R/\mathtt{error}\}; \mathtt{true})] \rangle$

$\overset{u(7)}{\longrightarrow} \langle \mathtt{rev}([a], [\,], R); id; [2; (\{R/\mathtt{error}\}; \mathtt{true})] \rangle$

$\overset{u(4)}{\longrightarrow} \langle \mathtt{is\_list}([\,]), \mathtt{rev}([\,], [a], R); id; [(2; \{R/\mathtt{error}\}; \mathtt{true})] \rangle$

$\overset{u(5)}{\longrightarrow} \langle \mathtt{rev}([\,], [a], R); id; [(2; \{R/\mathtt{error}\}; \mathtt{true})] \rangle$

$\overset{u(3)}{\longrightarrow} \langle \mathtt{true}; \{R/[a]\}; [(2; \{R/\mathtt{error}\}; \mathtt{true})] \rangle$

Fig. 6. Concrete execution for $\mathtt{main}([a], s(0)), R$.

Then, we use concolic execution again as shown in Fig. 7. Therefore, according to the algorithm, we consider the following prefix of the concolic execution:

$$\langle [u(1,2), u(8), \dots]; [\,]; \mathtt{main}(L, N, R); [\,]; \{4, 5, 6, 8\} \rangle \overset{\{\}}{\longrightarrow} \langle \dots \rangle \overset{\{c(7, \theta_1)\}}{\longrightarrow} \langle \dots \rangle \overset{\{c(8, \theta_2)\}}{\longrightarrow} \langle \dots \rangle$$

with the associated substitution $\sigma_1 \theta_2 = \{L/[X, Y|L''], N/s(s(N''))\}$.

Third (and Last) Iteration. As in the previous case, the instantiated goal, $\mathtt{main}([X, Y|L''], s(s(N'')), R)$, is not ground enough according to its input mode and, thus, we apply a minimal grounding substitution. In this case, we get the initial goal $\mathtt{main}([a, b], s(s(0)), R)$. Here, the concrete execution is basically the same shown in Fig. 2, except for the last (backtracking) step. Therefore, the associated trace is

$$\tau''' = [u(1, 2), \; u(8), \; u(8), \; u(7), \; u(4), \; u(5), \; u(4), \; u(6), \; u(5), \; u(3)]$$

Now, concolic execution from the initial state

$$\langle [u(1,2), u(8), u(8), u(7), u(4), u(5), u(4), u(6), u(5), u(3)]; [\,]; \mathtt{main}(L, N, R); id; [\,]; \{6\} \rangle$$

proceeds similarly to the derivation shown in Fig. 7, but now clause (6) is also completely evaluated, which means that the algorithm terminates successfully.

To summarize, concolic testing generated four test cases:

$$\mathtt{main}([a, b], s(0), R) \qquad \mathtt{main}([a], s(0), R)$$
$$\mathtt{main}([\,], 0, R) \qquad \mathtt{main}([a, b], s(s(0)), R)$$

which suffice to cover the complete evaluation of all program clauses.

In general, when the test case generation algorithm terminates, concolic testing is *sound* (i.e., there are no false positives since only concrete executions are considered) and *complete* (in the sense that all clauses are completely

$$\langle [u(1,2), u(8), \ldots]; [\,]; \quad \mathtt{main(L, N, R)}; \quad id; \quad [\,]; \{4,5,6,8\}\rangle$$

$\xrightarrow{\{\}}$ $\langle [u(8), u(7), \ldots]; [1]; \quad \begin{pmatrix} \mathtt{length(L,N),} \\ \mathtt{rev(L,[\,],R), e(1)} \end{pmatrix}; \quad id; \quad \mathcal{S}; \{4,5,6,8\}\rangle$

$\xrightarrow{\{c(7,\theta_1)\}}$ $\langle [u(7), u(4), \ldots]; [8,1]; \quad \begin{pmatrix} \mathtt{length(L',N'), e(8),} \\ \mathtt{rev([X|L'],[\,],R), e(1)} \end{pmatrix}; \quad \sigma_1; \quad \mathcal{S}; \{4,5,6,8\}\rangle$

$\xrightarrow{\{c(8,\theta_2)\}}$ $\langle [u(4), u(5), u(3)]; [7,8,1]; \begin{pmatrix} \mathtt{e(7), e(8),} \\ \mathtt{rev([X|L'],[\,],R), e(1)} \end{pmatrix}; \quad \sigma_1\sigma_2; \mathcal{S}; \{4,5,6,8\}\rangle$

$\xrightarrow{\{\}}$ $\langle [u(4), u(5), u(3)]; [8,1]; \quad \mathtt{e(8), rev([X|L'],[\,],R), e(1)}; \quad \sigma_1\sigma_2; \mathcal{S}; \{4,5,6,8\}\rangle$

$\xrightarrow{\{\}}$ $\langle [u(4), u(5), u(3)]; [1]; \quad \mathtt{rev([X|L'],[\,],R), e(1)}; \quad \sigma_1\sigma_2; \mathcal{S}; \{4,5,6\}\rangle$

$\xrightarrow{\{\}}$ $\langle [u(5), u(3)]; [4,1]; \quad \begin{pmatrix} \mathtt{is\_list([\,]), rev(L',[X],R),} \\ \mathtt{e(4), e(1)} \end{pmatrix}; \sigma_1\sigma_2; \mathcal{S}; \{4,5,6\}\rangle$

$\xrightarrow{\{\}}$ $\langle [u(3)]; [5,4,1]; \quad \mathtt{e(5), rev(L',[X],R), e(4), e(1)}; \quad \sigma_1\sigma_2; \mathcal{S}; \{4,5,6\}\rangle$

$\xrightarrow{\{\}}$ $\langle [u(3)]; [4,1]; \quad \mathtt{rev(L',[X],R), e(4), e(1)}; \quad \sigma_1\sigma_2; \mathcal{S}; \{4,6\}\rangle$

$\xrightarrow{\{\}}$ $\langle [\,]; [3,4,1]; \quad \mathtt{e(3), e(4), e(1)}; \quad \sigma_1\sigma_2; \mathcal{S}; \{4,6\}\rangle$

$\xrightarrow{\{\}}$ $\langle [\,]; [4,1]; \quad \mathtt{e(4), e(1)}; \quad \sigma_1\sigma_2; \mathcal{S}; \{4,6\}\rangle$

$\xrightarrow{\{\}}$ $\langle [\,]; [1]; \quad \mathtt{e(1)}; \quad \sigma_1\sigma_2; \mathcal{S}; \{6\}\rangle$

$\xrightarrow{\{\}}$ $\langle [\,]; [\,]; \quad \mathtt{true}; \quad \sigma_1\sigma_2; \mathcal{S}; \{6\}\rangle$

with $\mathcal{S} = [(2; [2]; \{\mathtt{R/error}\}; \mathtt{e(2)})];$

Fig. 7. Concolic execution for $\langle [u(1,2), u(8), \ldots]; id; [\,]; \mathtt{main(L, N, R)}; [\,]; \{4,5,6,8\}\rangle$

evaluated when using the computed test cases, i.e., we get a 100% coverage). When the process is stopped (e.g., because it does not terminate or takes too much time), our test case generation is only sound. Note that this contrasts with other approaches to test case generation in Prolog (and CLP), e.g., [8,15], where full coverage is not considered. In [1], however, the authors consider a refined notion of coverage criterion: a pair $\langle TC, SC \rangle$, where TC is a termination criterion (e.g., the maximum number of recursive calls to a predicate allowed in symbolic execution) and SC is the selection criterion, used to determine which test cases must be produced. In particular, using the SC program-points(P), where P includes all program points, amounts to the statement coverage that we consider in this paper. On the other hand, the authors have introduced a refinement for driving symbolic execution by means of "trace terms" (terms representing the shape of a particular subset of the SLD search space). The idea behind this technique is closer to that of concolic execution, although they do not have the possibility of using concrete data in symbolic executions (which is one of the main advantages of concolic execution).

6 Concluding Remarks and Future Work

We have introduced a novel approach to Prolog testing and debugging. The so called concolic execution that mixes concrete and symbolic execution has been shown quite successful in other programming paradigms, especially when dealing with large applications. Therefore, it might have a great potential for Prolog testing, too.

In this paper, we have only considered a simple form of coverage, *statement* coverage, and a limited scenario: pure Prolog without negation. Nevertheless, the main distinctive features of the Prolog programming language—i.e., unification, non-determinism and backtracking—are present here, so adapting the standard concolic execution approach [6,16] to Prolog was not trivial. The challenge, now, is experimentally verifying the effectiveness and scalability of our approach with real Prolog programs. For this purpose, though, we first need to extend concolic execution to deal with negation, built-in's, extra-logical features, etc. For this purpose, we will consider the linear operational semantics of [17], and its symbolic version [5], as a promising starting point.

Acknowledgements. The author gratefully acknowledges the anonymous referees and the participants of LOPSTR 2014 for many useful comments and suggestions. I would also like to thank Fred Mesnard and Etienne Payet for their remarks to improve the paper.

References

1. Albert, E., Arenas, P., Gómez-Zamalloa, M., Rojas, J.M.: Test case generation by symbolic execution: basic concepts, a CLP-based instance, and actor-based concurrency. In: Bernardo, M., Damiani, F., Hähnle, R., Johnsen, E.B., Schaefer, I. (eds.) SFM 2014. LNCS, vol. 8483, pp. 263–309. Springer, Heidelberg (2014)
2. Belli, F., Jack, O.: Implementation-based analysis and testing of Prolog programs. In: ISSTA, pp. 70–80. ACM (1993)
3. Clarke, L.A.: A program testing system. In: Proceedings of the 1976 Annual Conference (ACM'76), Houston, pp. 488–491 (1976)
4. De Schreye, D., Glück, R., Jørgensen, J., Leuschel, M., Martens, B., Sørensen, M.H.: Conjunctive partial deduction: foundations, control, algorithms, and experiments. J. Log. Program. 41(2&3), 231–277 (1999)
5. Giesl, J., Ströder, T., Schneider-Kamp, P., Emmes, F., Fuhs, C.: Symbolic evaluation graphs and term rewriting: a general methodology for analyzing logic programs. In: PPDP'12, pp. 1–12. ACM (2012)
6. Godefroid, P., Klarlund, N., Sen, K.: DART: directed automated random testing. In: Proceedings of PLDI'05, pp. 213–223. ACM (2005)
7. Godefroid, P., Levin, M.Y., Molnar, D.A.: Sage: whitebox fuzzing for security testing. Commun. ACM 55(3), 40–44 (2012)
8. Gómez-Zamalloa, M., Albert, E., Puebla, G.: Test case generation for object-oriented imperative languages in CLP. TPLP 10(4–6), 659–674 (2010)
9. King, J.C.: Symbolic execution and program testing. Commun. ACM 19(7), 385–394 (1976)

10. Leuschel, M.: The DPPD (Dozens of Problems for Partial Deduction) Library of Benchmarks. http://www.ecs.soton.ac.uk/mal/systems/dppd.html (2007)
11. Lloyd, J.W.: Foundations of Logic Programming, 2nd edn. Springer, Berlin (1987)
12. Lloyd, J.W., Shepherdson, J.C.: Partial evaluation in logic programming. J. Log. Program. **11**, 217–242 (1991)
13. Martens, B., Gallagher, J.: Ensuring global termination of partial deduction while allowing flexible polyvariance. In: Proceedings of ICLP'95, pp. 597–611. MIT Press (1995)
14. Pasareanu, C.S., Rungta, N.: Symbolic PathFinder: symbolic execution of Java bytecode. In: Pecheur, C., Andrews, J., Di Nitto, E. (eds.) ASE, pp. 179–180. ACM (2010)
15. Rojas, J.M., Gómez-Zamalloa, M.: A framework for guided test case generation in constraint logic programming. In: Albert, E. (ed.) Proceedings of LOPSTR. LNCS, vol. 7844, pp. 176–193. Springer, Heidelberg (2013)
16. Sen, K., Marinov, D., Agha, G.: CUTE: a concolic unit testing engine for C. In: Proceedings of ESEC/SIGSOFT FSE 2005, pp. 263–272. ACM (2005)
17. Ströder, T., Emmes, F., Schneider-Kamp, P., Giesl, J., Fuhs, C.: A linear operational semantics for termination and complexity analysis of ISO Prolog. In: Vidal, G. (ed.) LOPSTR'11. LNCS, vol. 7225, pp. 237–252. Springer, Heidelberg (2012)

Liveness Properties in CafeOBJ – A Case Study for Meta-Level Specifications

Norbert Preining$^{(\boxtimes)}$, Kazuhiro Ogata, and Kokichi Futatsugi

Japan Advanced Institute of Science and Technology, Research Center
for Software Verification, Nomi, Ishikawa, Japan
{preining,ogata,futatsugi}@jaist.ac.jp

Abstract. We provide an innovative development of algebraic specifications and proof scores in CAFEOBJ by extending a base specification
to the meta-level that includes infinite transition sequences. The infinite
transition sequences are modeled using behavioral specifications with
hidden sort, and make it possible to prove safety and liveness properties
in a uniform way.

As an example of the development, we present a specification of
Dijkstra's binary semaphore, a protocol to guarantee exclusive access to
a resource. For this protocol we will give three different properties, one
being the mutual exclusion (or safety) property, and two more regarding
different forms of liveness, which we call progress property and entrance
property. These three properties are verified in a computationally uniform way (by term rewriting) based on the new development.

Besides being a case study of modeling meta-properties in CAFEOBJ,
we provide an initial characterization of strength of various properties.
Furthermore, this method can serve as a blue-print for other specifications, in particular those based on Abstract State System (ASSs).

Keywords: Algebraic specification · Liveness · CAFEOBJ · Verification

1 Introduction

QLOCK, an abstract version of Dijkstra's binary semaphore, is a protocol to guarantee exclusive access to a resource. Besides the initial specification and verification in CAFEOBJ(see for example [6]), it saw implementations in COQ [12] and
MAUDE [14]. Most of these specifications only consider safety properties, in the
current case the mutual exclusion property, that no two agents will have access to
the resource at the same time. However, liveness properties are normally left open.
These properties ensure that 'there is progress'. In our particular case, they ensure
that agents do not block out other agents from acquiring access to the resource.

We are using CAFEOBJ as specification and verification language. CAFEOBJ
is a many- and order-sorted algebraic specification language from the OBJ family,

---

This work was supported in part by Grant-in-Aid for Scientific Research (S) 23220002
from Japan Society for the Promotion of Science (JSPS).

M. Proietti and H. Seki (Eds.): LOPSTR 2014, LNCS 8981, pp. 182–198, 2015.
DOI: 10.1007/978-3-319-17822-6_11

related to languages like CASL and MAUDE. CAFEOBJ allows us to have both the specification and the verification in the same language. It is based on powerful logical foundations (order-sorted algebra, hidden algebra, and rewriting logic) with an executable semantics [8,10,11].

The particular interest of the current development is two-fold: Firstly, it extends base specifications in order-sorted and rewriting logics to a meta-level, which requires behavioral logic, thus using the three logics together to achieve the proofs. Secondly, we use a search predicate and covering state patterns that allow us to prove the validity of a property over all possible one-step transitions, by which safety and liveness properties in the base and meta-level can be proven.

1.1 Related Work

Our work is closely related in spirit to [1, 2], where the authors discuss verification and model checking of temporal properties over infinite-state transition systems. Both works discuss variants of a mutual exclusion protocol, but while the main focus of their work is on model checking, we target theorem proving. Furthermore, they use rewriting logic and narrowing, while we are employing behavioral logic to represent infinite data structures. On the other hand, Goguen and Lin [9] use behavioral algebra to specify and verify properties on the Alternating Bit Protocol, but they do not use rewriting logic.

Many of the works done on Unity bear resemblance and relation with our work. Our methodology is closely related to concepts of UNITY. Theorem proving over UNITY using the Larch prover is the target of [5], while mechanization of UNITY in ISABELLE is discussed in [15].

Although the approach taken in [3] is similar to ours, there only the *progress property* (in our words) is discussed, while fairness based on or similar to our concept of fairness of execution sequences allows for stronger properties like the entrance property.

While all the above (and more) related works often deal with similar concepts, we believe that it is the first time that behavioral logic, rewriting logic, and order-sorted logic, are used together for system specification and treatment of liveness properties. This is also the reason why the *entrance property* introduced here has not been discussed hitherto.

1.2 Layout of the Article

In Sect. 2 we introduce the QLOCK protocol and various properties for verification, give a short introduction to the CAFEOBJ language, and provide the base specification onto which the current work is building. This section also discusses briefly the proof method by induction and exhaustive search.

In Sect. 3 we extend the base specification to include infinite transition sequences. Here we also discuss the methodology of meta-modeling.

In Sect. 4 we provide (parts) of the proof score verifying the three properties.

In the final section we provide a discussion of the approach with respect to applicability to different problems, and conclude with future research directions.

2   The QLOCK Protocol

The QLOCK protocol regulates access of an arbitrary number of agents to a resource by providing a queue (first-in-first-out list). Agents start in the *remainder section*, henceforth indicated by rs. The mode of operation is regulated by the following set of rules:

- If an agent wants to use the resource, it puts a unique identifier into the queue, and by this it transitions into the *waiting section* (ws).
- In the waiting section, an agent checks the top of the queue. If it is the agent's unique identifier, the agent transitions into the *critical section* (cs), during which the agent can use the resource.
- After having finished with the resource usage, the agent removes the head of the queue and transitions back into the remainder section (rs).

See Fig. 1 for a schematic flow diagram.

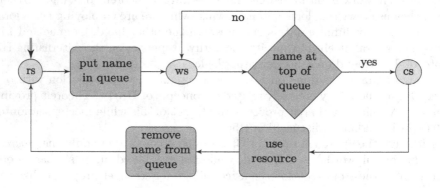

Fig. 1. Flow diagram of QLOCK protocol

2.1   Verification Properties

The basic safety property of QLOCK is the *mutual exclusion property* (mp):

Property 1 (mutual exclusion property). At any time, at most one agent is in the critical section.

While this is the most important property for safety concerns, it does not guarantee that an agent wanting to use the resource ever gets the chance to use it (for example, in case of denial-of-service attack to a server). To guarantee this, we define two liveness properties: The first concerns the transition from ws to cs, and is called the *progress property* (pp). This property has already been discussed in [13] as *lockout freedom property*.

Property 2 (progress property). An agent that has entered into the waiting section (ws), i.e., has put his unique identifier into the queue, will eventually transition into the critical section (cs), i.e., progress to the top of the queue and gain access to the resource.

The last one concerns the transition from the remainder section rs to the waiting section ws, called *entrance property* (ep):

Property 3 (entrance property). An agent will eventually transition into the queue, i.e., from the remainder section to the waiting section.

Although it might sound counter-intuitive that the entrance property should hold for each agent at all times, we believe that there are good reasons to consider this property: This is motivated by the fact that given any finite run, i.e., finite transition sequence, of the QLOCK protocol, we can always extend it to an infinite and fair transition sequence (see later sections for details). Thus, what we are actually proving is that for each agent, either the entrance property holds, or the execution terminates before the agent had a chance. In circumstances of long-running services (like most client-server interaction where a server is practically never stopped), this ensures that agents, or clients, will – given long enough execution time – eventually be served.

As we will see later on, to prove this property we need additional assumptions, in particular fairness, see Sect. 3.2, which makes it conceptually different from the first two properties. To continue with the analogy set forth above, the schedulers used in most operating systems or network hubs try to create a fair execution sequence by using round-robin or similar techniques [20, 21].

2.2   Short Introduction to CafeOBJ

Although we cannot give a full introduction to the CAFEOBJ language, to aid readers unfamiliar with it we give a short introduction. Users acquainted with MAUDE can safely skip this section, as syntaxes of the two languages are very similar.

CAFEOBJ is an algebraic specification language, thus the unit of specification is an algebra, in particular an order-sorted algebra. To specify an algebra, the following information have to be given:

Signature. Similar to normal algebras, a signature consists of operators and their arities. In the multi-sorted setting we are working in, this means that sorts have to be defined, and for each operator (or function) the number and sorts of the arguments and result have to be specified.

Axioms. They provide an equational theory over the (many-sorted) signature defined above. These axioms (or equations) make up the core of any algebraic specification.

We will demonstrate these concepts on a simple definition of natural numbers with successor and addition, but no comparison or subtraction:

```
1 mod! SIMPLE-NAT {
2   signature {
3     [ Zero NzNat < Nat ]
4     op 0 : -> Zero {constr}
5     op s : Nat -> NzNat {constr}
6     op _+_ : Nat Nat -> Nat
7   }
8   axioms {
9     vars N N' : Nat
10    eq 0 + N = N .
11    eq s(N) + N' = s(N + N') .
12  }
13 }
```

Line 1 begins the specification of the algebra called SIMPLE-NAT with initial semantics (indicated by the ! after mod, which can be replaced with * to indicate loose semantics). The body of the specification consists of two blocks. Lines 2-7 define the signature, lines 8-12 the axioms. In line 3 the sorts and their order are introduced by defining a partial order of sorts between brackets. In case of hidden sorts (of behavioural algebras) we use *[...]*. In this above case there are three sorts, Zero, NzZero, and Nat. The order relation expresses that the former two are a subsort of Nat, but does not specify a relation between themselves. Lines 4-6 give three operators, the constant 0 of sort Zero, the successor function s, and the addition, written in infix notation. Two of the operators are furthermore tagged as *constructors* of the algebra. Note that in the specification of operators, the _ represents the places of arguments.

The second block defines the equations by first declaring two variables of sort Nat. Axioms are introduced with the eq keyword, and the left- and right-hand side of the equation are separated by =. Thus, the two axioms provided here provide the default inductive definition of addition by the successor function.

Not exhibited here, but used later in the code is the syntax protecting(...), which imports all the sorts and axioms of another module, but does not allow to alter them.

In the following, the signature and axioms block declaration will be dropped, as they are not necessary.

2.3   Base Specification

We are building upon a previously obtained specification and verification of QLOCK [6]. Although we are providing the complete source of the specification part, we cannot, due to space limits, include the full verification part. The reader is referred to the full code at [17].

The basic idea of the following specification is to use the natural abstraction of QLOCK and its transitions as an Abstract State System (Ass). The usual steps in providing an algebraic specification in this setting are:

- Define proper abstraction as modules/algebras of the *players* (e.g., agent, queue) in the protocol.

- Model the complete state as a module/algebra.
- Use transitions between terms of the state algebra to describe transitions in the protocol.
- Specify (and later verify) properties on states (and transitions) to ensure safety and liveness.

Let us start with the most basic item, namely modeling the sections an agent can be in. For literals, i.e., elements of the sort LabelLt, which is a sub-sort of the sort Label, we define identity via the syntactical identity. The second module specification defines an agent, or more specifically, the algebra of agent identifiers AID, without any further axioms, which implies that identifiers are considered different if they are syntactically different.

```
mod! LABEL { [ LabelLt < Label ]
  vars L1 L2 : LabelLt .
  ops rs ws cs : -> LabelLt {constr} .
  eq (L1 = L2) = (L1 == L2) . }
mod* AID { [ Aid ] }
```

In the next step we model a queue, a first-in-first-out storage. Note in the following code, that CAFEOBJ allows for parametrized modules. In the present case the parameter X has no further requirements, which is expressed by the fact that it only needs to belong to the trivial algebra. Another important point to note is that we are using associative constructors, which allows us to freely use any way of parenthesizing. Similarly, we introduce a module for parametrized sets, where we use an associative and commutative constructor.

```
mod! QUEUE (X :: TRIV) { [ Elt.X < Qu ]
  vars Q Q1 Q2 : Qu . vars E E1 E2 : Elt .
  op empQ : -> Qu {constr} .
  op (_&_) : Qu Qu -> Qu {constr assoc id: empQ} .
  eq (empQ = (E & Q)) = false .
  eq ((E1 & Q1) = (E2 & Q2)) = ((E1 = E2) and (Q1 = Q2)) . }
mod! SET(X :: TRIV) { [ Elt.X < Set ]
  vars E : Elt .
  op empty : -> Set {constr} .
  op (_ _) : Set Set -> Set {constr assoc comm id: empty} .
  eq E E = E . }
```

Concerning agents, we model them as terms of an algebra of *agent observers* which associates agent identifiers with labels, expressing the fact that the agent is in the current state. More formally, the meaning of the term lb[A]:S is that the agent A is in section S:

```
mod! AOB {protecting(LABEL) protecting(AID) [ Aob ]
  op (lb[_]:_) : Aid Label -> Aob {constr} . }
```

In the final step we instantiate the parametrized queue with agent ids, and define the state algebra as a pair of one queue and an arbitrary set of agent observers. Note that the pairing is done by the syntax l $ r, CAFEOBJ allows nearly arbitrary syntax:

```
mod! AID-QUEUE { protecting( QUEUE(AID{sort Elt -> Aid}) ) }
mod! STATE{ protecting(AID-QUEUE)
  protecting(SET(AOB{sort Elt -> Aob})*{sort Set -> Aobs})
  [State] op _$_ : Qu Aobs -> State {constr} . }
```

With this we have given a complete definition of the state algebra, but the dynamic aspect of the protocol has been left out till now. We are now providing transition rules over states to express this dynamic aspect. In the following code segments, the two states of the transition are aligned, and changing parts are indicated with a bold font. The three transitions are WaitTrans, where an agent transitions from rs to ws, TryTrans, where an agent tries to enter cs, and ExitTrans, where an agent leaves the critical state:

```
mod! WaitTrans { protecting(STATE) .
  var Q : Qu . var A : Aid . var AS : Aobs .
  trans[wt]: (Q        $ ((lb[A]: rs) AS))
        => ((Q & A ) $ ((lb[A]: ws) AS)) . }
mod! TryTrans { protecting(STATE) .
  var Q : Qu . var A : Aid . var AS : Aobs .
  trans[ty]: ((A & Q) $ ((lb[A]: ws) AS))
        => ((A & Q) $ ((lb[A]: cs) AS)) .      }
mod! ExitTrans { protecting(STATE) .
  var Q : Qu . vars A1 A2 : Aid . var AS : Aobs .
  trans[ex]: ((A1 & Q) $ ((lb[A2]: cs) AS))
        => (    Q  $ ((lb[A2]: rs) AS)) . }
```

Based on the above specification, it is possible to provide a *proof score*, i.e., a program in CAFEOBJ, that verifies the mutual exclusion property mp. As usual with proofs by induction over the reachable states (see below), the target property by itself does not suffice to work as inductive property, making the introduction of further properties on states necessary. Example properties that have to be used are uniqueness properties (e.g., the same agent identifier cannot appear several times in the queue) or initial state properties (e.g., the queue is empty at the beginning). Obtaining an inductive property (set of properties) is one of the challenging aspects of verifications, and requires an iterative and interactive approach. Readers interested in the details are referred to the code at [17].

We conclude this section with a short discussion on the verification methodology applied here. Details concerning this verification technique and a more generalized methodology will be presented at [18] and the upcoming proceeding.

2.4 Verification by Induction and Exhaustive Search

Verification of properties of an Ass is often done by induction on reachable states, more specifically by induction over the length of transition sequences from initial states to reachable states. That is, we show that a certain property (invprop) holds in the set of initial states, characterized by init. Furthermore, as we proceed through transitions (state changes), the invprop is preserved.

But to show liveness properties, considering only invariant properties on states is not enough. We thus use an extended method that does inductive proofs on the reachable state space, and in parallel proves properties (transprop) on all transitions between reachable states. To be a bit more specific, assume that $S \Rightarrow S'$ is a transition from one state (state term, state pattern) S to a state S'. We show that if invprop(S) holds, then also invprop(S') (the induction on reachable states), but also that for this transition transprop(S, S') holds.

Both of these are done with CAFEOBJ's built-in search predicate (see Sect. 4.2), which exhaustively searches and tests all possible transitions from a given state (pattern). The concepts introduced here are an extension and generalization of *transition invariants* [16], details can be found in [7].

In the CAFEOBJ setting, which means rewrite-based, we have to ensure that both of the following implications reduce to True:

$$\text{init}(S) \rightarrow \text{invprop}(S)$$
$$\text{invprop}(S) \rightarrow \text{invprop}(S') \qquad \text{where } S \rightarrow S' \text{ is a transition}$$

where S and S' are states (state terms) describing the pre- and post-transition states, respectively. This has to be checked for all possible transitions available in the specification.

If this can be achieved, we can be sure that in all *reachable* states, i.e., those that can actually occur when starting from an initial state, the required property invprop holds.

3   Extended Specification

The starting point of the following discussion is the question of how to verify liveness properties. Initial work by the first author led to a specification which kept track of the waiting time in the queue. Combined with the assumption that there are only finitely (but arbitrary) many agents, we could give a proof score not only for the mutual exclusion property mp, but also for the progress property pp. This work was extended by the third author to the currently used base specification.

To verify the last property, ep, operational considerations alone do not suffice. On the level of observers, we cannot guarantee that an agent will ever enter the queue, since we have no control over which transitions are executed by the system. To discuss (verify) this property, we have to assume a certain meta-level property, in this case the fairness of the transition sequence. A similar approach has been taken in [9] for the Alternating Bit Protocol, where fair *event mark streams* are considered.

3.1   Fairness

The concept of fairness we are employing here is based on the mathematically most general concept:

Definition 1 (Fairness). *A sequence of transitions S is called* fair, *if every* finite *sequence of transitions appears as sub-sequence of S.*

A necessary consequence of this definition is that every fair sequence is infinite.

Relation to Other Concepts of Fairness. The methodology of using Ass in CAFEOBJ has been strongly influenced by UNITY [4], which builds upon a semantics similar to Ass of sequences of states and temporal operators. It provides an operator *ensures*, which can be used to model fairness via a measure function.

Another approach to the concept of fairness is taken by LTL logic [19], where two types of fairness, strong and weak, are considered, referring to *enabled* and *applied* state of transitions.

$$\text{weak} \qquad \Diamond\Box\, enabled(t) \rightarrow \Box\Diamond\, applied(t)$$
$$\text{strong} \qquad \Box\Diamond\, enabled(t) \rightarrow \Box\Diamond\, applied(t)$$

where *enabled* and *applied* are properties on transition instances. In the particular case we are considering, *enabled* is always true, as we can execute every instance of a transition at any time, due to the fact that the wait-state transition can be applied even if the agent is not at the top of the queue. Fairness in this case means that, at some point every transition will be applied.

Both concepts can be represented in suitable way by the definition of fairness, or in other words, the definition used in this setting (every finite sequence is subsequence) subsumes these two concepts.

3.2   Transition Sequence

As mentioned above, modeling fairness requires recurring to a meta-assumption, namely that the sequence of transitions is fair, i.e., every instance of a transition appears infinitely often in the sequence. In our case we wanted to have a formalization of this meta-assumption that can be expressed with the rewriting logic of CAFEOBJ.

The approach we took models transition sequences using behavioral specification with hidden algebra [9], often used to express infinite entities. Note that we are modeling the transition sequence by an infinite stream of agent identifiers, since the agent uniquely defines the instance of transition to be used, depending on the current state of the agent. This is a fortunate consequence of the modeling scheme at the base level, where, if we pick an agent, we can always apply the transition that is uniquely defined by that agent. Translated into the language of the above mentioned LTL logic it means that all transitions are permanently enabled.

```
mod* TRANSSEQ { protecting(AID)
  *[ TransSeq ]*
  op (_&_) : Aid TransSeq -> TransSeq . }
```

The transition sequence is then used to model a *meta-state*, i.e., the combination of the original state of the system as specified in the base case, together with the list of upcoming transitions:

```
mod! METASTATE { protecting(STATE + ... ) [MetaState]
  op _^_ : State TransSeq -> MetaState . ... }
```

The dots at the end of the definition refer to a long list of functions on meta-states, we will return to this later on.

In the same way, transitions from the base case are lifted to the meta-level. Here we have to ensure that the semantics of transition sequences and the transition in the original (non-meta level) system do not digress. That means first and foremost, that only the agent at the top of the transition sequence can be involved in a transition.

Let us consider the first transition WaitTrans:

```
mod! MWT {protecting(METASTATE) var Q : Queue .
   var A : Aid . var AS : Aobs . var T : TransSeq .
   trans[meta-wt]:
      ( (Q        $ ((lb[ A ]: rs) AS)) ^ (A & T))
   => ( (Q & A) $ ((lb[ A ]: ws) AS)  ^      T) .
}
```

Due to the structural definition of the meta-transition, we see that it can only be applied under the following conditions that

- the agent is at the top of the transition sequence, and
- the agent is in remainder section rs.

The next transition is TryTrans, where an agent checks whether it is at the top of the queue, and if yes, enters into the critical section. Adding the meta-level requires only that the agent is also at the head of the transition sequence. This transition uses the built-in operator if_then_else_fi, because we want to destructively use up the top element of the transition sequence to ensure that no empty transitions appear.

```
mod!  MTY {pr(METASTATE) var Q : Queue .
   vars A B : Aid . var AS : Aobs . var T : TransSeq .
   trans[meta-ty]:
         (((B & Q) $ ((lb[ A ]: ws) AS)) ^ (A & T))
      => if (A = B) then
            (((A & Q) $ ((lb[ A ]: cs) AS))  ^      T)
         else
            (((B & Q) $ ((lb[ A ]: ws) AS))  ^      T)
         fi .
}
```

The final transition is ExitTrans, where an agent returns into the remainder section:

```
mod! MEX {pr(METASTATE) var Q : Queue .
  var A : Aid . var AS : Aobs . var T : TransSeq .
  trans [meta-ex]:
          (((A & Q) $ ((lb[ A ]: cs) AS)) ^ (A & T))
      =>  ((     Q  $ ((lb[ A ]: rs) AS)) ^      T) .
}
```

Relation between meta-transition and transition. Comparing the transition of the base system (see listing on p. 7), we see that in the meta-transition the part to the left of ^, the state of the base system, behaves exactly like the corresponding part in the base transition. The only difference is that an additional guard is added, namely that the active agent has to be also at the top of the transition sequence to the right of the ^ marker. This can be considered a general procedure of meta-level reflection.

Combining all these algebras provides the specification of the meta system:

```
mod! METAQLOCKsys{ pr(MWT + MTY + MEX) }
```

As mentioned above, to express the fairness condition, we recur to an equivalent definition, namely that every finite sequence of agent identifiers can be found as a sub-sequence of the transition sequence, rephrased here in an indirect way in the sense that it cannot happen that we cannot find a (finite) sequence of agent identifiers in a transition sequence:

```
eq ( find ( Q, T ) = empQ ) = false .
```

3.3   Waiting Times

The axiom shown above uses the function **find**, which has been defined in the algebra **METASTATE**. We mentioned that several additional functions are defined, too. These functions are necessary to compute the *waiting time* for each agent.

Waiting time here refers to the number of meta-transitions until a particular agent is actually *changing* its section. Here complications arise due to the trial transition from ws to cs (by means of the **if_then_else_fi** usage), where an agent might find itself still in ws due to not being at the head of the queue. This has to be considered while computing waiting times for each agent.

Closer inspection of the transition system provides the following values for waiting times:

For agents in rs *and* cs. The waiting time is determined by the next occurrence of its agent id in the transition sequence, since there are no further requirements. In CAFEOBJ notation, the length of the following sequence:

```
find( A, T )
```

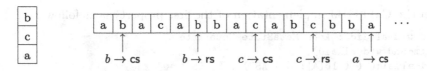

Fig. 2. Computing of waiting time from ws to cs

For agents in ws. Here we have to ensure that the agent advances to the top of the queue. Thus, each agent higher up in the queue has to appear two times in the transition sequence, once for entering the critical section, and once for leaving it. Let Q_A be the part of the queue that is above (higher up) the agent id a. Then the waiting time for a would be determined by doubling Q_A, then searching the transition sequence first for the doubled Q_A, and finally searching for the next appearance of a. Consider for example the state represented in Fig. 2. Assume that initially the queue contains the three agent identifiers b, c, and a (in this order), and all of them are initially in rs. To see when the a at the bottom of the queue can transition into cs, we first have to bring b into cs, which happens at position 2. After that there is a series of trial transitions without success (a, c, a) until another b appears, which makes the agent b transition back into rs. At this point the queue contains two elements, c and a. The same repeats for c, until finally a can enter into cs. Summing up, what has to be searched within the transition sequence is the sub-sequence $\boxed{b\;|\;b\;|\;c\;|\;c\;|\;a}$, which amounts to (in CAFEOBJ notation):

```
find( double( droplast( find( A, Q ))) & A, T )
```

(The actual code is slightly different, but semantically the same.)

Here the doubling of the Q_A is achieved by first finding the sub-sequence within the queue up to A (which includes A), and then dropping the last element, which is A.

The functions mentioned above are exactly those needed to compute this waittime function. Due to the necessity of error handling, the definition becomes a bit lengthy.

4   Verification of Properties

Our aim is to verify the progress property and the entrance property. These properties are now expressed in the following ways:

- At any transition, the waiting time of agents not changing section decreases.
- If the waiting time of an agent reaches 0, then a section change happens.

Combining these two properties, and assuming fairness, we can show both, that every agent will eventually enter into the queue, and every agent in the queue will eventually gain access to the resource, i.e., enter into the critical section.

In the following let us assume that the following variable definitions are in effect:

```
vars S SS : MetaState . var Q : Queue . var AS : Aobs .
var A : Aid . var C : Label . var QQ : TransSeq .
```

Then the CAFEOBJ implementation of the first property is as follows:

```
pred wtd-allaid : Aobs MetaState MetaState .
eq[:m-and wtd-allaid]:
 wtd-allaid( ( ( lb[A]: C ) AS ) , S, SS ) =
    ( ( sec( ( lb[A]: C ) ,S) == sec( ( lb[A]: C ) , SS) ) implies
         ( waittime( A, S ) > waittime ( A, SS ) ) ) .
```

And the one of the second property:

```
pred wtzerochange-allaid : Aobs MetaState MetaState .
eq[:m-and wtzerochange-allaid]:
 wtzerochange-allaid( ( ( lb[ A ]: C ) AS ) , S , SS ) =
    ( ( waittime( A, S ) == 0 ) implies
         ( sec( ( lb[ A ]: C ) , S ) =/= sec( ( lb[ A ]: C ) , SS) ) ) .
```

Here we have to note that the sec operator computes the actual section SS and not the one given by C.

These two properties alone do not function as inductive invariant, so several more have to be included. In addition, we are lifting also the properties used in the original specification to the meta level by making the new operators simply operate on the projections. Again, the interested reader is referred to [17] for the full source.

4.1   Proof Score with Patterns

The method described in Sect. 2.4 is used here in combination with a covering set of patterns. We mention only the definition of cover set here, but details on this methodology will be presented at [18] and a forthcoming article:

Definition 2 (cover set). *Assume a set $S \subseteq$ State of states (ground terms of sort State) is given. A finite set of state patterns $C = \{C_1, \ldots, C_n\}$ is called cover set for S if for every $s \in S$ there is a substitution δ from the variables X occurring in C to the set of all ground terms, and a state pattern $C \in C$ such that $\delta(C) = s$.*

Practically this means, that we give a set of state terms that need to cover all possible ground instances of state terms. For the base case, a set of 13 state patterns has been given. We list only the cases for rs, the cases for the other sections are parallel.

```
eq s1 = (q $ empty) .
eq s2 = (empQ $ ((lb[b1]: rs) as)) . ...
eq s8 = ((b1 & q) $ ((lb[b1]: rs) as)) . ...
eq s11 = ((b1 & q) $ ((lb[b2]: rs) as)) . ...
```

For the meta-level we combine these patterns with patterns for the transition sequence, where once b1 is at the head of the transition sequence, and once another identifier b2, amounting to 26 different meta state patterns:

```
eq n1 = ( s1 ^ ( b1 & t ) ) . eq n2 = ( s2 ^ ( b1 & t ) ) . ...
eq l1 = ( s1 ^ ( b2 & t ) ) . eq l2 = ( s2 ^ ( b2 & t ) ) . ...
```

We conclude this section with a discussion of the search predicate, actually family of search predicates, in CAFEOBJ.

4.2 The CafeOBJ Search Predicate

During a proof over reachable states by induction on the transitions, we need a method that provides *all* possible successors of a certain state. The CAFEOBJ search predicate we use is $S = (*, 1) \Rightarrow + S'$ suchThat $prop(S, S')$, where S and S' are states, and $prop(S, S')$ is a Boolean value. This is a tricky predicate and full discussion is available in an upcoming reference manual for CAFEOBJ, but in this case it does the following:

- It searches all successor states S' that are reachable in exactly one step from the left side state S (here 1 stands for maximal one step, and + for at least one step). Call the set of successors $Succ(S)$.
- It checks all those states determined by the first step whether the property given in Bool holds. If there is at least one successor state where it holds, the whole predicate returns true, i.e., what is returned is $\exists S' \in Succ(S) : prop(S, S')$.

This can be used with a double negation to implement an exhaustive search in all successor states by using the equality:

$$\forall S' \subset Succ(S) : prop(S, S') \quad \Leftrightarrow \quad \neg \exists S' \in Succ(S) . prop(S, S')$$

This leads to the following, admittedly not very beautiful, definition of the inductive invariant condition, where we use SS for the S' in the above equality:

```
pred inv-condition : MetaState MetaState .
eq inv-condition(S:MetaState,SS:MetaState) =
  ( not ( S =(*,1)=>+ SS suchThat
      ( not ( inv-prop(S, SS) == true)))) .
```

Here inv-prop is the set of inductive invariant properties we have mentioned above.

Note that the operator used here has access not only to the original or the successor state, but to both states. This peculiar feature allows us to prove properties like decrease of waiting time, which is impossible if there is no access to both values in the same predicate. As pointed out above, CAFEOBJ actually includes a whole set of search predicates which allows searching for arbitrary, but given depth, but verifications making use of these predicates are still to come.

The final step in the proof score is to verify that both the initial condition and the inductive invariant condition do actually hold on all the state patterns by reducing the expressions to true:

```
red init-condition(n1) . red init-condition(n2) .  ...
red inv-condition(n1,SS:MetaState) .
red inv-condition(n2,SS:MetaState) . ...
```

This concludes the discussion of the specification methodology and proof score code.

5   Discussion and Conclusion

Using the method laid out above, we have formally verified the three properties given in the beginning, *mutual exclusion property* (only at most one agent at a time is in the critical section), *progress property* (an agent in the queue will eventually gain access to the resource), and *entrance property* (every agent will eventually enter into the queue). While the original base specification and proof score verified the first two properties, it also required the assumption that the number of agents is finite. In our case, this assumption is superseded by the assumption of fairness of the transition sequence, which is by itself necessary to verify the third property.

This methodology also opens up further options in the specification. By requiring acquisition of the resource within a reasonable time, and providing requirements on the transition sequence that the reasonable-time condition is fulfilled, we believe it is possible to specify and verify time-critical systems.

We have to note that what we call here *progress property* has already been shown in different settings [1–3,5,13]. The key contribution is the change of focus onto behaviour algebras as specification methodology, and as a consequence the extension to the *entrance property*, meaning that an agent always gets a chance to enter the queue. In addition, we extended the proof of the progress property to infinitely many agents. Of course, every actual instance of the protocol will encompass only finitely many agents, but the proof provided here is uniform for any number of agents. The current work also serves as an example of reflecting meta-properties into specifications, allowing for the verification of additional properties.

Assuming a meta-level fairness property to prove liveness properties of the specification might be considered a circular argument, but without regress to meta-level fairness, *no* proof of the entrance property can be achieved. Keeping this in mind, our goal is to provide a reasonably simple and intuitive definition of fairness on the meta-level, that can be used for verification of the necessary properties, similar to any axiomatic approach where trust is based on simple axioms.

Future work we are foreseeing centers around the following points:

– Adaption of the methodology to other protocols: One probable candidate is the already mentioned Alternating Bit Protocol, where proofs for liveness properties are hard to obtain in other systems. We believe that a meta-specification similar to the one given here can be employed for this protocol, as well as others.
– Automatization of the method: Most of the steps done during lifting the specification to the meta-level are semi-automatic. It might be interesting to provide built-in functionality to extend a given specification based on states with transition sequences.
– Relation to other methods in the field: As mentioned in the section on related work, targeting liveness properties is an active research area. While our approach seems to be unique in using behavioral algebras, we will compare the methodologies taken by other researchers.

– Characterization of strength of properties: We have seen that the mutual exclusion property can be proven by only looking at states, that the progress property only needs access to a state and its successor, but the entrance property needs access to all future states. We are working on a formal description of this concept called n-visibility.

We conclude with recapitulating the major contributions of this work: First of all, it provides an example for the inclusion of meta-concepts in a formal specification. Reflecting these meta-properties allows for the verification of additional properties, in particular liveness properties.

Furthermore, it is an example of a specification that spans all the corners of the CAFEOBJ cube, in particular mixing co-algebraic methods, infinite stream representation via hidden sorts, with transition sequence style modeling.

References

1. Bae, K., Meseguer, J.: Predicate abstraction of rewrite theories. In: Dowek, G. (ed.) RTA-TLCA 2014. LNCS, vol. 8560, pp. 61–76. Springer, Heidelberg (2014)
2. Bae, K., Meseguer, J.: Infinite-state model checking of LTLR formulas unsing narrowing. In: WRLA 2014, 10th International Workshop on Rewriting Logic and its Applications, to appear
3. Bjørner, N., Browne, A., Colón, M., Finkbeiner, B., Manna, Z., Sipma, H., Uribe, T.E.: Verifying temporal properties of reactive systems: a step tutorial. Form. Methods Syst. Des. 16(3), 227–270 (2000)
4. Chandy, K.M., Misra, J.: Parallel Program Design—A Foundation. Addison-Wesley, Boston (1989)
5. Chetali, B.: Formal verification of concurrent programs using the Larch prover. IEEE Trans. Softw. Eng. 24(1), 46–62 (1998)
6. Futatsugi, K.: Generate and check methods for invariant verification in CafeOBJ. In: JAIST Research Report IS-RR-2013-006, http://hdl.handle.net/10119/11536 (2013)
7. Futatsugi, K.: Generate and check method for verifying transition systems in CafeOBJ. Submitted for publication (2014)
8. Futatsugi, K., Găină, D., Ogata, K.: Principles of proof scores in CafeOBJ. Theor. Comput. Sci. 464, 90–112 (2012)
9. Goguen, J.A., Lin., K.: Behavioral verification of distributed concurrent systems with BOBJ. In: QSIC, pp. 216–235. IEEE Computer Society (2003)
10. Iida, S., Meseguer, J., Ogata, K. (eds.): Specification, Algebra, and Software. LNCS, vol. 8373, pp. 520–540. Springer, Heidelberg (2014)
11. Meseguer, J.: Twenty years of rewriting logic. J. Log. Algebr. Program. 81(7–8), 721–781 (2012)
12. Ogata, K., Futatsugi, K.: State machines as inductive types. IEICE Trans. Fundam. Electron. Commun. Comput. Sci. E90–A(12), 2985–2988 (2007)
13. Ogata, K., Futatsugi, K.: Proof score approach to verification of liveness properties. IEICE Trans. 91–D(12), 2804–2817 (2008)
14. Ogata, K., Futatsugi, K.: A combination of forward and backward reachability analysis methods. In: Dong, J.S., Zhu, H. (eds.) ICFEM 2010. LNCS, vol. 6447, pp. 501–517. Springer, Heidelberg (2010)

15. Paulson, L.C.: Mechanizing UNITY in Isabelle. ACM Trans. Comput. Log. **1**(1), 3–32 (2000)
16. Podelski, A., Rybalchenko, A.: Transition invariants. In: LICS, pp. 32–41. IEEE Computer Society (2004)
17. Preining, N.: Specifications in CafeOBJ http://www.preining.info/blog/cafeobj/
18. Preining, N., Futatsugi, K., Ogata, K.: Proving liveness properties using abstract state machines and n-visibility. In: Talk at the 22nd International Workshop on Algebraic Development Techniques WADT 2014, Sinaia, Romania, September 2014
19. Rybakov, V.: Linear temporal logic with until and next, logical consecutions. Ann. Pure Appl. Log. **155**(1), 32–45 (2008)
20. Stiliadis, D., Varma, A.: Latency-rate servers: a general model for analysis of traffic scheduling algorithms. IEEE/ACM Netw. **6**(5), 611–624 (1998)
21. Wierman, A.: Fairness and scheduling in single server queues. Surv. Oper. Res. Manag. Sci. **16**(1), 39–48 (2011)

Program Synthesis

A Hybrid Method for the Verification and Synthesis of Parameterized Self-Stabilizing Protocols

Amer Tahat and Ali Ebnenasir[✉]

Department of Computer Science, Michigan Technological University,
Houghton, MI 49931, USA
{atahat,aebnenas}@mtu.edu

Abstract. This paper presents a hybrid method for verification and synthesis of parameterized self-stabilizing protocols where algorithmic design and mechanical verification techniques/tools are used hand-in-hand. The core idea behind the proposed method includes the automated synthesis of self-stabilizing protocols in a limited scope (i.e., fixed number of processes) and the use of theorem proving methods for the generalization of the solutions produced by the synthesizer. Specifically, we use the Prototype Verification System (PVS) to mechanically verify an algorithm for the synthesis of weakly self-stabilizing protocols. Then, we reuse the proof of correctness of the synthesis algorithm to establish the correctness of the generalized versions of synthesized protocols for an arbitrary number of processes. We demonstrate the proposed approach in the context of an agreement and a coloring protocol on the ring topology.

Keywords: Mechanical verification · Program synthesis · Self-stabilization · Parameterized systems

1 Introduction

Self-stabilization is an important property of dependable distributed systems as it guarantees *convergence* in the presence of transient faults. That is, from *any* state/configuration, a Self-Stabilizing (SS) system recovers to a set of legitimate states (a.k.a. *invariant*) in a finite number of steps. Moreover, from its invariant, the executions of an SS system satisfy its specifications and remain in the invariant; i.e., *closure*. Nonetheless, design and verification of convergence are difficult tasks [10,19] in part due to the requirements of (i) recovery from arbitrary states; (ii) recovery under distribution constraints, where processes can read/write only the state of their neighboring processes (a.k.a. their *locality*), and (iii) the non-interference of convergence with closure. Methods for algorithmic design of convergence [3,4,13,16] can generate only the protocols that

---

This work was partially supported by the NSF grant CCF-1116546.

M. Proietti and H. Seki (Eds.): LOPSTR 2014, LNCS 8981, pp. 201–218, 2015.
DOI: 10.1007/978-3-319-17822-6_12

202 A. Tahat and A. Ebnenasir

are correct up to a limited number of processes and small domains for variables. Thus, it is desirable to devise methods that enable automated design of parameterized SS systems, where a *parameterized* system includes several sets of symmetric processes that have a similar code up to variable re-naming. The proposed method in this paper has important applications in both hardware [11] and software [12] networked systems, be it a network-on-chip system or the Internet.

Numerous approaches exist for mechanical verification of self-stabilizing systems most of which focus on synthesis and verification of specific protocols. For example, Qadeer and Shankar [32] present a mechanical proof of Dijkstra's token ring protocol [10] in the Prototype Verification System (PVS) [34]. Kulkarni *et al.* [29] use PVS to mechanically prove the correctness of Dijkstra's token ring protocol in a component-based fashion. Prasetya [31] mechanically proves the correctness of a self-stabilizing routing protocol in the HOL theorem prover [18]. Tsuchiya *et al.* [36] use symbolic model checking to verify several protocols such as mutual exclusion and leader election. Kulkarni *et al.* [8,28] mechanically prove (in PVS) the correctness of algorithms for automated addition of fault tolerance; nonetheless, such algorithms are not tuned for the design of convergence. Most existing automated techniques [6,14,16,27] for the design of fault tolerance enable the synthesis of non-parametric fault-tolerant systems. For example, Kulkarni and Arora [27] present a family of algorithms for automated design of fault tolerance in non-parametric systems, but they do not explicitly address self-stabilization. Abujarad and Kulkarni [4] present a method for algorithmic design of self-stabilization in locally-correctable protocols, where the local recovery of all processes ensures the global recovery of the entire distributed system. Farahat and Ebnenasir [13,16] present algorithms for the design of self-stabilization in non-locally correctable systems. Jacobs and Bloem [22] show that, in general, synthesis of parameterized systems from temporal logic specifications is undecidable. They also present a semi-decision procedure for the synthesis of a specific class of parameterized systems in the absence of faults.

The **contributions** of this paper are two-fold: a hybrid method (Fig. 1) for the synthesis of parameterized self-stabilizing systems and a reusable PVS theory for mechanical verification of self-stabilization. The proposed method includes a synthesis step and a theorem proving step. Our previous work [13,16] enables the synthesis step where we take a non-stabilizing protocol and generate a self-stabilizing version thereof that is correct by construction up to a certain number of processes. This paper investigates the second step where we use the theorem prover PVS to prove (or disprove) the correctness of the synthesized protocol for an arbitrary number of processes; i.e., *generalize* the synthesized protocol. The synthesis algorithms in [13,16] incorporate weak and strong convergence in existing network protocols; i.e., adding convergence. *Weak* (respectively, *Strong*) convergence requires that from every state there exists an execution that (respectively, every execution) reaches an invariant state in finite number of steps. To enable the second step, we first mechanically prove the correctness of the Add_Weak algorithm from [16] that adds weak convergence. As a result, any protocol generated by Add_Weak will be correct by construction.

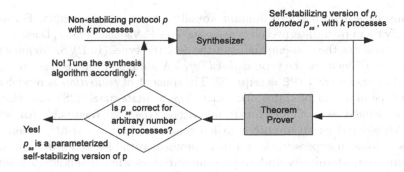

Fig. 1. A Hybrid method for the synthesis of parameterized self-stabilizing protocols.

Moreover, the mechanical verification of Add_Weak provides a reusable theory in PVS that enables us to verify the generalizability of small instances of different protocols generated by an implementation of Add_Weak. If the mechanical verification succeeds, then it follows that the synthesized protocol is in fact correct for an arbitrary number of processes. Otherwise, we use the feedback of PVS to determine why the synthesized protocol cannot be generalized and re-generate a protocol that addresses the concerns reported by PVS. We continue this cycle of *synthesize and generalize* until we have a parameterized protocol. Notice that the theory developed in mechanical proof of Add_Weak can also be reused for the mechanical verification of self-stabilizing protocols designed by means other than our synthesis algorithms. We demonstrate this reusability in the context of a coloring protocol (Sect. 6) and a binary agreement protocol (in [35]).

Organization. Section 2 introduces basic concepts and presents their formal specifications in PVS. Then, Sect. 3 formally presents the problem of adding convergence to protocols. Sections 4 and 5 respectively present the specification and verification of Add_Weak in PVS. Section 6 demonstrates the reusability and generalizability properties in the context of a graph coloring protocol. Section 7 discusses related work and Sect. 8 makes concluding remarks and presents future extensions of this work.

2   Formal Specifications of Basic Concepts

In this section, we define protocols, state predicates, computations and convergence, and present their formal specifications in PVS. The definitions of protocols and convergence are adapted respectively from [28] and [10].

2.1   Protocols

A protocol includes a set of processes, a set of variables and a set of transitions. Since we would like the specification of a protocol to be as general as possible, we impose little constraints on the notions of state, transitions, etc. Thus, the notations *state, variable*, and *domain* are all abstract and nonempty. Formally, we specify them by uninterpreted types state: Type+, Variable: Type+, Dom: Type+, where '+' denotes the non-emptiness of the declared type. (In PVS, "*definedType* : TYPE+ = []"declares the type *definedType*.) A *state predicate* is a set of states specified as StatePred: TYPE = set[state]. The concept of *transition* is modeled as a tuple type of a pair of states, Transition: Type = [state,state] [28]. Likewise, an *action* is defined as a set of transitions, Action:Type =set[Transition]. An action can be considered as an atomic guarded command "*grd* → *stmt*", where *grd* denotes a Boolean expression in terms of protocol variables and *stmt* is a set of statements that atomically update program variables when *grd* holds. An action is *enabled iff* (if and only if) its guard *grd* evaluates to true. We assume that Dom, Variable, and Action are finite types in our PVS specifications. A *process* is a tuple of a subset of variables that are readable by that process, a subset of variables that are writable by that process, and its set of transitions. A *protocol prt* is a tuple of a finite set of processes, variables, and finite set of transitions.

$$\text{p\_process} : \text{TYPE+} = [\text{set}[\text{Variable}], \text{set}[\text{Variable}], \text{set}[\text{Transition}]]$$

$$\text{nd\_Protocol} : \text{TYPE+} = [\text{set}[\text{p\_process}], \text{set}[\text{Variable}], \text{set}[\text{Transition}]]$$

The *projection of a protocol prt on a state predicate I*, denoted $PrjOnS(prt, I)$, includes the set of transitions of *prt* that start in I and end in I. One can think of the projection of *prt* as a protocol that has the same set of processes and variables as those of *prt*, but its transition set is a subset of *prt*'s transitions confined in I. We model this concept by defining the following function, where Z is instantiated by transitions of *prt*. (proj_k is a built-in function in PVS that returns the k-th element of a tuple.)

$$\text{PrjOnS}(Z: \text{Action}, I: \text{StatePred}): \text{Action} =$$
$$\{t:\text{Transition} \mid t \in Z \wedge \text{proj\_1}(t) \in I \wedge \text{proj\_2}(t) \in I\}$$

Example: Coloring on a ring of n processes with $m > 2$ colors. The coloring protocol, denoted $TR(m,n)$, includes $n > 3$ processes located along a bidirectional ring. Each process P_j has a local variable c_j with a domain of $m > 2$ values representing m colors. Thus, the set of variables of $TR(m,n)$ is $V_{TR(m,n)} = \{c_0, c_1, ..., c_{n-1}\}$. As an example of a state predicate, consider the states where no two neighboring processes have the same color. Formally, $I_{coloring} = \forall j : 0 \leq j < n : c_j \neq c_{j\oplus 1}$. Each process P_j $(0 \leq j < n)$ has the following action:

$$A_j : (c_j = c_{j\ominus 1}) \vee (c_j = c_{j\oplus 1}) \rightarrow c_j := other(c_{j\ominus 1}, c_{j\oplus 1}) \tag{1}$$

If P_j has the same color as that of one of its neighbors, then P_j uses the function "$other(c_{j\ominus 1}, c_{j\oplus 1})$" to non-deterministically set c_j to a color different from $c_{j\ominus 1}$ and $c_{j\oplus 1}$. The projection of the actions A_j (for $0 \leq j < n$) on the predicate $I_{coloring}$ is empty because no action is enabled in $I_{coloring}$. The coloring protocol has applications in several domains such as scheduling, bandwidth allocation, register allocation, etc. It is known that if $m > d$, where d is the max degree in the topology graph of the protocol, then the coloring problem is solvable. For this reason, we have $m > 2$ for the ring.

2.2 Distribution and Atomicity Models

We model the impact of distribution in a shared memory model by considering read and write restrictions for processes with respect to variables. Due to inability of a process P_j in reading some variables, each transition of P_j belongs to a *group* of transitions. For example, consider two processes P_0 and P_1 each having a Boolean variable that is not readable for the other process. That is, P_0 (respectively, P_1) can read and write x_0 (respectively, x_1), but cannot read x_1 (respectively, x_0). Let $\langle x_0, x_1 \rangle$ denote a state of this program. Now, if P_0 writes x_0 in a transition $(\langle 0,0 \rangle, \langle 1,0 \rangle)$, then P_0 has to consider the possibility of x_1 being 1 when it updates x_0 from 0 to 1. As such, executing an action in which the value of x_0 is changed from 0 to 1 is captured by the fact that a group of two transitions $(\langle 0,0 \rangle, \langle 1,0 \rangle)$ and $(\langle 0,1 \rangle, \langle 1,1 \rangle)$ is included in P_0. In general, a transition is included in the set of transitions of a process iff its associated group of transitions is included. Formally, any two transitions (s_0, s_1) and (s_0', s_1') in a group of transitions formed due to the read restrictions of a process P_j meet the following constraints, where r_j denotes the set of variables P_j can read: $\forall v : v \in r_j : (v(s_0) = v(s_0')) \wedge (v(s_1) = v(s_1'))$ and $\forall v , v \notin r_j . (v(s_0) - v(s_1)) \wedge (v(s_0') - v(s_1'))$, where $v(s)$ denotes the value of a variable v in a state s that we represent by the Val(v, s) function in PVS. To enable the reusability of our PVS specifications, we specify our distribution model as a set of axioms so one can mechanically prove convergence under different distribution and atomicity models.

In the following formal specifications, v is of type Variable, p is of type p_process, t and t' are of type Transition, and non_read and transition_group are functions that respectively return the set of unreadable variables of the process p and the set of transitions that meet $\forall v : v \in r_j : (v(s_0) = v(s_0')) \wedge (v(s_1) = v(s_1'))$ for a transition $t = (s_0, s_1)$ and its groupmate $t' = (s_0', s_1')$.

> AXIOM subset?(proj_2(p),proj_1(p)) // *Writable variables are a subset of readable variables.*

> AXIOM member(t',transition_group(p, t, prt)) AND member(v,Non_read(p, prt))
> IMPLIES Val(v,proj_1(t)) = Val(v,proj_2(t)) AND Val(v,proj_1(t')) = Val(v,proj_2(t'))

member(x,X) and subset?(X,Y) respectively represent the membership and subset predicates in a set-theoretic context.

Example: Read/Write restrictions in $TR(3, n)$. In the coloring protocol, each process P_j can read $\{c_{j \ominus 1}, c_j, c_{j \oplus 1}\}$, and is allowed to write only c_j, where $\oplus$ and $\ominus$ denote addition and subtraction modulo n respectively. For a process P_j, each transition group includes 3^{n-3} transitions because P_j can read only the state of itself and its left and right neighbors; there is one transition in the group corresponding to each valuation of unreadable variables.

2.3 Computation

A *computation* of a protocol *prt* is a sequence A of states, where $(A(i), A(i+1))$ represents a transition of *prt* executed by some action of *prt*. In a more general term, a computation of any set of transitions Z, is a sequence of states in which every state can be reached from its predecessor by a transition in Z. Thus, we define the following function to return the set of computations generated by the set of transitions Z.

COMPUTATION(Z: Action): set[sequence[state]]={ A: sequence[state]
$|\forall(n : nat) : ((A(n), A(n+1)) \in Z)\}$

A *computation prefix* of a protocol *prt* is a *finite* sequence of states where each state is reached from its predecessor by a transition of *prt*. Kulkarni *et al.* [28] specify a prefix as an infinite sequence in which only a finite number of states are used. By contrast, we specify a computation prefix as a finite sequence type. We believe that it is more natural and more accurate to model the concept of prefix by finite sequences. Our experience also shows that modeling computation prefixes as finite sequences simplifies formal specification and verification of reachability and convergence while saving us several definitions that were required in [28] to capture the length of the prefix. We first define a type for finite sequences of states; i.e., Pos_F_S. Then, we use the predicate Condi_prefix?(A,Z) that holds when all transitions $(A(i), A(i+1))$ of a sequence A belong to a set of transitions Z. The notation A'length denotes the length of the sequence A, A'seq(i) returns the i-th element of sequence A and below[k] represents natural values less than k. The function PREFIX returns the set of computation prefixes generated by transitions of Z.

Pos_F_S: TYPE = {c:finite_sequence[state] | c'length > 0}

Condi_Prefix?(A:Pos_F_S,Z:Action):bool= FORALL(i: below[A'length-1]):
member((A'seq(i), A'seq(i+1)), Z)

PREFIX(Z: Action): set[Pos_F_S]= {A:Pos_F_S | Condi_Prefix?(A,Z) }

Example: A computation of $TR(3,5)$. Consider an instance of $TR(m,n)$ where $n = 5$ and $m = 3$. Thus, $c_j \in \{0,1,2\}$ for $0 \le j < 5$. Let $\langle c_0, c_1, c_2, c_3, c_4 \rangle$ denote a state of the protocol. Starting from a state $\langle 0,1,2,2,0 \rangle$, the following sequence of transitions could be taken: P_2 executes $(\langle 0,1,2,2,0 \rangle, \langle 0,1,0,2,0 \rangle)$ and P_0 executes $(\langle 0,1,0,2,0 \rangle, \langle 2,1,0,2,0 \rangle)$.

2.4 Closure and Convergence

A state predicate I is *closed* in a protocol *prt* iff every transition of *prt* that starts in I also terminates in I [5,19]. The closed predicate checks whether a set of transitions Z is actually closed in a state predicate I.

closed?(I: StatePred, Z: Action): bool = FORALL (t:Transition | (member(t,Z)) AND
member(proj_1(t), I)) : member(proj_2(t), I)

A protocol *prt* *weakly converges* to a non-empty state predicate I iff from every state s, there exists at least one computation prefix that reaches some state in I [5,19]. A *strongly converging* protocol guarantees that every computation from s will reach some state in I. Notice that any strongly converging protocol is also weakly converging, but the reverse is not true in general. A protocol *prt* is weakly (respectively, strongly) self-stabilizing to a state predicate I iff (1) I is closed in *prt*, and (2) *prt* weakly (respectively, strongly) converges to I.

Example: Closure and convergence of $TR(3,5)$. Notice that the actions A_j, where $0 \leq j < 5$, are closed in $I_{coloring}$ since no action is enabled in $I_{coloring}$. Moreover, starting from any state (in the 3^5 states of the state space of $TR(3,5)$), every computation will reach a state in $I_{coloring}$. The computation of $TR(3,5)$ presented in this section is an example of a converging computation.

3 Problem Statement

The problem of adding convergence (from [16]) is a transformation problem that takes as its input a protocol *prt* and a state predicate I that is closed in *prt*. The output of Problem 1 is a revised version of *prt*, denoted *prt*$_{ss}$, that converges to I from any state. Starting from a state in I, *prt*$_{ss}$ generates the same computations as those of *prt*; i.e., *prt*$_{ss}$ behaves similar to *prt* in I.

Problem 1. **Add Convergence**

- **Input:** (1) A protocol *prt*; (2) A state predicate I such that I is closed in *prt*; and (3) A property of L_s converging, where $L_s \in \{$weakly, strongly$\}$.
- **Output:** A protocol *prt*$_{ss}$ such that : (1) I is unchanged; (2) the projection of *prt*$_{ss}$ on I is equal to the projection of *prt* on I, and (3) *prt*$_{ss}$ is L_s converging to I. Since I is closed in *prt*$_{ss}$, it follows that *prt*$_{ss}$ is L_s self-stabilizing to I.

Previous work [16,19] shows that weak convergence can be added in polynomial time (in the size of the state space), whereas adding strong convergence is known to be an NP-complete problem [26]. Farahat and Ebnenasir [13,16] present a sound and complete algorithm for the addition of weak convergence and a set of heuristics for efficient addition of strong convergence. While one of our objectives is to develop a reusable proof library (in PVS) for mechanical verification of both weak and strong convergence, the focus of this paper is mainly on enabling the mechanical verification of weak convergence for parameterized systems. Algorithm 1 provides an informal and self-explanatory representation of the Add_Weak algorithm presented in [16].

Mechanical verification of the soundness of Add_Weak ensures that any protocol synthesized by Add_Weak is correct by construction. Moreover, the lemmas and theorems developed in mechanical verification of Add_Weak provide a

---

Algorithm 1. Add_Weak

Input: *prt*:nd_Protocol, *I*: statePred;
Output: set[Transition]; // Set of transitions of a weakly self-stabilizing version of *prt*.
1: Let Δ_{prt} be the set of transition groups of *prt*.
2: Let $\Delta_{converge}$ be the set of transition groups that adhere to read/write restrictions of processes of *prt*, but exclude any transition starting in *I*;
3: $\Delta_{ws} = \Delta_{prt} \cup \Delta_{converge}$;
4: $no\_Prefix := \{s : state \mid (s \notin I) \wedge (\text{there is no computation prefix using transitions of } \Delta_{ws} \text{ that can reach a state in } I)\}$
5: If ($no\_Prefix \neq \emptyset$) then weak convergence cannot be added to *prt*; return;
6: return Δ_{ws};

---

reusable framework for mechanical verification of different protocols that we generate using our synthesis tools [16,25]. The verification of synthesized protocols increases our confidence in the correctness of the *implementation* of Add_Weak and helps us to generalize small instances of weakly converging protocols to their parameterized versions.

4   Specification of Add_Weak

This section presents the highlights of the formal specification of Add_Weak in PVS. (The complete PVS specifications are available at http://asd.cs.mtu.edu/projects/mechVerif/ss.html.) We start by specifying the basic components used in the Add_Weak algorithm, namely the transition predicates $\Delta_{prt}, \Delta_{converge}$ and Δ_{ws}, and the state predicate no_Prefix.

Notation. In the subsequent formal specifications, we use the identifiers Delta_prt, Delta_Converge and Delta_ws corresponding to the variables $\Delta_{prt}, \Delta_{converge}$ and Δ_{ws} in Add_Weak. The function transition_groups_proc(p,prt) returns the set of transition groups of a process *p* of a protocol *prt*.

Delta_Converge(prt:nd_Protocol,I:StatePred):set[set[Transition]]= {gg:set[Transition] |
Exists (p:p_process | member(p,proj_1(prt))):member(gg,transition_groups_proc(p,prt))
AND FORALL (t:Transition | member(t,gg)): NOT member(proj_1(t),I) }

We find it useful to define a dependent type of all prefixes *A* of a set of transitions *Z* and we call it PREFIX(Z). Furthermore, we formally specify the concept of reachability as follows:

Reach_from?(Z:Action,A:PREFIX(Z),s0:state,
I: StatePred):bool = Exists (j:below[A'length]): A'seq(0)= s0 AND member(A'seq(j),I)

The predicate Reach_from returns true iff a state predicate *I* is reachable from a state *s0* using computation prefixes of *Z*. To specify the set of states no_Prefix, we first specify a predicate noPrefixExists that determines if for a protocol *prt*, a state predicate *I* and a state s_0, no state in *I* can be reached from s_0 by computation prefixes of *prt*.

noPrefixExists?(prt:nd_Protocol,I:StatePred,s0:state):bool= FORALL (g:Action,
A:PREFIX(g)| member(g,Delta_ws(prt,I))AND member(A,PREFIX(g)) AND A'seq(0)=
s0):NOT (Reach_from?(g,A,s0,I))

We then specify the state predicate no_Prefix in Add_Weak as follows:

no_Prefix(prt:nd_Protocol,I:StatePred):set[state]= {s0:state | NOT(member(s0,I)) AND
noPrefixExists?(prt,I,s0)}

We also specify Add_Weak as a function that returns a set of transitions.

Add_weak(prt:nd_Protocol,
I:{X:StatePred | closed?(X,proj_3(prt))}):set[Transition] = COND
empty?(no_Prefix(prt,I)) − > Delta_ws(prt,I), ELSE − > proj_3(prt) ENDCOND

5   Verification of Add_Weak

In order to prove the soundness of Add_Weak, we check if (1) I is unchanged;
(2) the projection of Δ_{ws} on I is equal to the projection of Δ_{prt} on I, and (3)
Δ_{ws} is weakly converging to I. The first constraint holds trivially since no step
of Add_Weak adds/removes a state to/from I. Next, we present a set of lemmas
and theorems that prove the other two constraints of Problem 1.

5.1   Verifying the Equality of Projections on Invariant

In this section, we prove that Constraint 2 of Problem 1 holds for the output of
Add_Weak, denoted by a protocol whose set of transitions is Δ_{ws}. Our proof obligation is to show that the projection of Δ_{ws} on I is equal to the projection of Δ_{prt}
on I. We decompose this into two set inclusion obligations of PrjOnS(Delta_prt,I)
⊆ PrjOnS(Delta_ws,I) and PrjOnS(Delta_ws,I) ⊆ PrjOnS(Delta_prt,I). Notice that, by
assumption, closed?(I,Delta_prt) is true.

Lemma 1. PrjOnS(Delta_prt,I) *is a subset of* PrjOnS(Delta_ws,I).

Proof. The proof is straightforward since by construction we have $\Delta_{ws} = \Delta_{prt} \cup \Delta_{converge}$.

Lemma 2. PrjOnS(Delta_ws,I) *is a subset of* PrjOnS(Delta_prt,I).

Proof. If a transition $t = (s_0, s_1)$ is in PrjOnS(Delta_ws,I) then $s_0 \in I$. Since
$\Delta_{ws} = \Delta_{prt} \cup \Delta_{converge}$, either $t \in \Delta_{prt}$ or $t \in \Delta_{converge}$. By construction,
$\Delta_{converge}$ excludes any transition starting in I including t. Thus, t must be in
Δ_{prt}. Since $s_0 \in I$, it follows that $t \in$ PrjOnS(Delta_prt,I).

Theorem 1. PrjOnS(Delta_ws,I) = PrjOnS(Delta_prt,I).

5.2   Verifying Weak Convergence

In this section, we prove the weak convergence property (i.e., Constraint 3 of Problem 1) of the output of Add_Weak. Specifically, we show that from any state $s_0 \in \neg I$, there is a prefix A in PREFIX(Delta_ws) such that A reaches some state in I. Again, we observe that, an underlying assumption in this section is that closed?(I,Delta_prt) holds. For a protocol prt and a predicate I that is closed in prt and a state $s \notin I$, we have:

Lemma 3. *If* empty?(no_Prefix(prt,I)) *holds then* noPrefixExists?(prt,I,s) *returns false for any* $s \notin I$.

Lemma 4. *If from every state there is a prefix reaching I (i.e.,* noPrefixEx- ists?(prt,I,s) *returns false for any* $s \notin I$), *then there exists a sequence of states A and a set of transitions Z such that* Z $\in$ Delta_ws, A $\in$ PREFIX(Z), A(0)=s *holds, and* Reach_from?(Z,A,s,I) *returns true for any* $s \notin I$.

Lemma 4 implies that when Add_Weak returns, the revised version of prt guarantees that there exists a computation prefix to I from any state outside I; hence weak convergence. This is due to the fact that A is a prefix of Δ_{ws}.

Theorem 2. *If* empty?(no_Prefix(prt,I)) *holds and* $s \notin I$ *then there exists a sequence of states A that starts from s and A* $\in$ PREFIX(Delta_ws) *and* Reach_from? (Delta_ws(prt,I),A,s,I) *returns true.*

6   Reusability and Generalizability

In this section, we demonstrate how the lemmas and theorems proved for the soundness of Add_Weak can be reused in proving the correctness of a graph coloring protocol and in generalizing it. *Reusability* enables us to instantiate the abstract concepts/types (e.g., state predicate I and actions of a protocol) for a concrete protocol and reuse the mechanical proof of Add_Weak to prove the weak convergence of that protocol. *Generalizability* determines whether a small instance of a protocol synthesized by our implementation of Add_Weak [16] can be proven to be correct for an arbitrary number of processes. For instance, we have used the Stabilization Synthesizer (STSyn) [16] tool to automatically generate the 3-coloring protocol presented in Sect. 2 for rings of up to 40 processes (i.e., $n < 41$). Nonetheless, due to scalability issues, STSyn cannot synthesize a self-stabilizing 3-coloring protocol for $n > 40$. In this section, we apply the proposed approach of *synthesize in small scale and generalize* to prove (or disprove) that the synthesized 3-coloring protocol is correct for rings of size greater than 40 and with more than 2 colors (i.e., $m > 2$).

6.1   PVS Specification of Coloring

This section presents the PVS specification of $TR(m, n)$. First, we instantiate the basic types in the PVS specification of Add_Weak for the coloring protocol.

Then, we present the specifications of some functions that we use to simplify the mechanical verification. Finally, we specify the processes and the protocol itself.

Defining a State of $TR(m,n)$**.** We first define the parameterized type COLORS: below[m] to capture the colors and the size of variable domains. Then, we define a *state* of the ring as a finite sequence of colors of length n; i.e., STC: NONEMPTY_TYPE {s:finseq | s'length=n}.

Position of Each Process in the Ring. Since each variable c_j holds two pieces of information namely the process position in the ring and its color, we declare the tuple type color_pos:TYPE+=[COLORS,below[n]] to represent a pair (color$_j$, position$_j$) for each process P_j.

Detecting Good/Bad Neighbors. The predicate is_nbr?(K:color_pos,L:color_pos) returns true iff K and L are two neighboring processes; i.e., mod(abs(K'2-L'2),n) $\leq$ 1, where K'2 denotes the second element of the pair K (which is the position of K in the ring). Likewise, we define the predicate is_bad_nbr?(K:color_pos,L:color_pos) that holds iff is_nbr?(K,L) holds and K'1 = L'1. To capture the *locality* of process P_j, we define the non-empty dependent type set_of_nbrs(K:color_pos):TYPE+ = {l:color_pos | is_nbr?(K,L) }. Likewise, we define the type set_of_bad_nbrs(K:color_pos): TYPE ={L:color_pos | is_bad_nbr?(K,L)} to capture the set of neighbors of a process that have the same color as that process. The function nbr_colors(K:color_pos):set [COLORS] returns the set of colors of the immediate neighbors of a process.

Functions. In order to simplify the verification of convergence (in Sect. 6.2), we associate the subsequent functions with a global state. The association of a global state to functions and types enables us to import the PVS theory of Add_Weak with the following types [STC,below[m], color_pos, [color_pos, STC $\rightarrow$ below[m]]] to reuse its already defined types and functions in specifying $TR(m,n)$ as follows. We define the following functions to simplify the verification process:

- The function ValPos(s:STC,j:below[n]):color_pos=(s'seq(j),j) that returns the color and the position of process j in a global state s as a tuple of type color_pos. An example use of this function is Val(s:STC,L:color_pos)= ValPos(s,L'2)'1.
- The predicate nbr_is_bad?(s:STC,j:below[n]):bool = nonempty?(set_of_bad_nbrs (ValPos(s,j))) returns true iff for an arbitrary state s and a process j the set of bad neighbors of process j is nonempty; we refer to such a case by saying s *is corrupted at process j*. Notice that an illegitimate state can be corrupted at more than one position.
- The predicate nbr_is_good?(s:STC,j:below[n]):bool returns the negation of the predicate nbr_is_bad?(s:STC,j:below[n]):bool.
- The predicate is_LEGT?(s:STC):bool= Forall (j:below[n]): nbr_is_good?(s,j) returns true iff s is a legitimate state. Thus, the set of illegitimate states is specified as S_ill:TYPE= { s:STC | not is_LEGT?(s)}.

Specification of a Process of TR(m,n). For an arbitrary global state s and a process j, we define the function READ_p which returns all readable variables of process j.

- READ_p(s:STC, j:below[n]): set[color_pos]= {L:set_of_nbrs(ValPos(s,j)) | TRUE } .

Similarly, we define the function WRITE_p which returns the variables that process j can write.

- WRITE_p(s:STC,j:below[n]): set[color_pos]= {L:color_pos | L = ValPos(s,j)}

We now define the function DELTA_p that returns the set of transitions belonging to process j if process j is corrupted in the global state s; i.e., j has a bad neighbor.

- DELTA_p(s:STC,j:below[n]):set[Transition] ={tr: Transition | Exists
  (c:set_of_bad_nbrs(ValPos(s,j))): tr = (s,action(s,j,ValPos(s,j),c)) }

The function action(s,j,ValPos(s,j),c) returns the state reached when process j acts to correct its corrupted state. Formally, we define action(s,j,ValPos(s,j),c) as follows:

- action(s:STC,j:below[n],K:color_pos, C:set_of_bad_nbrs(K)): STC = (# length := n, seq
  := (LAMBDA (i:below[n]):IF i= j THEN other(ValPos(s,j)) ELSE s(i) ENDIF) #)

(The LAMBDA abstractions in PVS enable us to specify binding expressions similar to quantified statements in predicate logic.) We specify the function *other* to randomly choose a new color other than the corrupted one. To this end, we use the *epsilon* function over the full set of colors minus the set of colors of the neighbors of the corrupted process. Formally, we have

- other(K:color_pos):COLORS = epsilon(difference(fullset_colors,nbr_colors(K))), where
  fullset_colors:set[COLORS]= {cl:COLORS | TRUE}.

Thus, the specification of a process of the protocol $TR(m,n)$ is as follows:

- Process_j(s:STC,j:below[n]): p_process = (READ_p(s,j), WRITE_p(s,j),DELTA_p(s,j))

The Parameterized Specification of the TR(m,n) Protocol. We define the $TR(m,n)$ protocol as the type TR_m_n(s:STC):nd_Protocol =(PROC_prt(s), VARB_prt(s),DELTA_prt(s)), where the parameters are defined as follows:

- PROC_prt(s:STC): set[p_process]={p:p_process | Exists (j:below[n]):p = Process_j(s,j)}
- VARB_prt(s:STC): set[color_pos]= {v:color_pos | Exists (j:below[n]):member(v,
  WRITE_p(s,j))}
- Delta_prt(s:STC): set[Transition]= {tr:Transition | Exists (j:below[n]):member(tr,
  DELTA_p(s,j))} .

6.2   Mechanical Verification of Parameterized Coloring

We now prove the weak convergence of $TR(m, n)$ for $m > 2$ and $n > 40$. To this end, we show that the set no_Prefix of $TR(m, n)$ is actually empty. The proof of emptiness of no_Prefix is based on a prefix constructor function that demonstrates the existence of a computation prefix σ of $TR(m, n)$ from any arbitrary illegitimate state s such that σ includes a state in I. Subsequently, we instantiate Theorem 2 for $TR(m, n)$.

Theorem 3. *IF (1) $TR(m, n)$ is closed in the state predicate $I_{coloring}$ (i.e.,* closed?$(I_{coloring}, (TR(m, n))`3))$; *(2) from every state there is a computation prefix to $I_{coloring}$ (i.e.,* empty?(no_Prefix$(TR(m, n), I_{coloring})$) *holds), and (3) $s \notin I_{coloring}$;*
   THEN there exists a sequence of states σ built from the transitions of $TR(m, n)$ (i.e., $\sigma \in$ PREFIX(Delta_ws$(TR(m, n), I_{coloring})$)*) such that σ starts in s and reaches a state in $I_{coloring}$ (i.e.,* Reach_from?$(\sigma, s, I_{coloring})$*).*

Proof. We show that Add_Weak will generate a weakly stabilizing version of protocol $TR(m, n)$ for any $n > 3$ and $m > 2$. To this end, we show that $TR(m, n)$ has an empty no_Prefix set and instantiate Theorem 2 for $TR(m, n)$. Let s be an arbitrary illegitimate state. We define a sequence of states σ that starts in s and terminates in I such that all transitions of σ belong to transitions of $TR(m, n)$ (i.e., proj_3(TR(m,n))). Before building a prefix from a particular illegitimate state s, we would like to identify a segment of the ring that is correct in the sense that the local predicate $((c_i \neq c_{i-1}) \wedge (c_i \neq c_{i+1}))$ is correct for processes from 1 to j; i.e., $\forall i : 1 \leq i \leq j : ((c_i \neq c_{i-1}) \wedge (c_i \neq c_{i+1}))$. For this purpose, we define three auxiliary functions.

- *Local corrector.* The first one is a corrector function that applies the *other* function on ValPos(s,j) if process j is corrupted; otherwise, it leaves c_j as is.
  - localCorrector(s:S_ill, j:below[n]):COLORS = COND nbr_is_good?(s,j) → s'seq(j), nbr_is_bad?(s,j) → other(ValPos(s,j)) ENDCOND
- *Segment corrector.* The function segCorrector(s,j) takes an illegitimate state s and an index j of the ring, and returns a global state where all processes from 1 to j have good neighbors. The rest of the processes have the same local state as in s. Since we model a global state of a ring of n processes as a sequence of n colors, we can represent the application of the segCorrector function on the j-th process in a global state s as ⟨ localCorrector(s,0),...,localCorrector(s,j),s(j+1),...,s(n-1)) ⟩, where the colors of processes from $j + 1$ to $n - 1$ remain unchanged by segCorrector.
  - segCorrector(s:S_ill,j:below[n]):STC= (# length := n, seq := (LAMBDA (i:below[n]):IF i ≤ j THEN localCorrector(s,i) ELSE s'seq(i) ENDIF) #)
- *Global corrector.* The function globalCorrector below is especially useful because constructing the appropriate computation prefix from s to some invariant state in $I_{coloring}$ directly is not straightforward. Since for all $j < n$ we do not know whether applying the local corrector function on a state s:S_ill at a process j will result in a legitimate state, we use globalCorrector to build

a sequence of states of length n formed by applying the segment corrector function consecutively at all processes regardless of being corrupted or not.
globalCorrector(s:S_ill): Pos_F_S = (# length := n, seq := (LAMBDA (i:below[n]): segCorrector(s,i)) #)

Each element j in the sequence that globalCorrector(s) returns is the image of the function segCorrector(s,j), which is a state that is correct up to process j. We define the function min_legit(globalCorrector(s)) over the sequence globalCorrector(s) to return the minimum index for which the corresponding state is legitimate.

min_legit(A:Pos_F_S):{i:integer | i≥-1} = COND not empty?(legitStatesIndices(A)) − > min(legitStatesIndices(A)), else − > -1 ENDCOND

The min_legit function first uses the function legitStatesIndices to compute the indices of the legitimate states in a sequence of global states A.

legitStatesIndices(A:Pos_F_S):set[below[A'length]]={i:below[A'length] | is_LEGT? (A'seq(i)) }

The Prefix-Constructor Function. Now, we are ready to define the constructor function of the required prefix from any illegitimate s:S_ill to I as follows:
    constPrefix(s:S_ill):Pos_F_S = (# length:= min_legit(globalCorrector(s))+2, seq:= (LAMBDA (i:below[min_legit(globalCorrector(s))+2]): if i=0 then s else segCorrector(s,i-1) endif) #)
The last element of constPrefix(s) is the first legitimate state of globalCorrector(s). Thus, all states before the last state in constPrefix(s) are illegitimate. Moreover, each application of the corrector function corresponds to a transition that starts in an illegitimate state. This is true until reaching a legitimate state. Thus, all involved transitions in the sequence globalCorrector(s) are in $TR(m, n)$, thereby making constPrefix(s) a computation prefix of $TR(m, n)$. To show the correctness of this argument, it is sufficient to show the existence of at least one index that is equal to min_legit(globalCorrector(s)). Thus, the sequence constPrefix(s) is well-defined for any state s outside I and reaches I as required. Thus, it is sufficient for weak convergence to show the correctness of the following two properties of the function segCorrector.

1. If a process j is corrupted then the color of process j in the state segCorrector (s,n-1) is the same as what the local corrector function selects for it; i.e., ValPos(segCorrector(s,n-1), j)'1= other(ValPos(s,j)).
2. If a process j is corrupted then in the state segCorrector(s,n-1) the process j is not corrupted (i.e., does not have a corrupted neighbor as well).

Notice that by the above two properties we show that the state generated by segCorrector(s,n-1) is legitimate. We prove these properties as two lemmas.

Lemma 5. *If* nonempty?(set_of_bad_nbrs(ValPos(s,j))) *holds then* ValPos(segCorrector (s,n-1), j)'1 = other(ValPos(s,j)).

Lemma 6. *If* nonempty?(set_of_bad_nbrs(ValPos(s,j))) *holds then* empty?(set_of_bad_nbrs((other(ValPos(s,j)), j))).

The mechanical proofs of these lemmas follow directly by expanding the required definitions and using the following axiom of choice.

AXIOM nonempty?(set_of_bad_nbrs(ValPos(sl,j))) IMPLIES empty?(C: color_pos| is_bad_nbr?((other(ValPos(sl, j)), j), C))

We need the axiom of choice because we use the epsilon function in the definition of the other() function. This way we show that there is always a different color that can correct the locality of a corrupted process.

7 Discussion and Related Work

This section discusses the impact of the proposed approach and the related work. Self-stabilization is an important property for networked systems, be it a network-on-chip system or the Internet. There are both hardware [11] and software systems [12] that benefit from the resilience provided by self-stabilization. Thus, it is important to have an abstract specification of self-stabilization that is independent from hardware or software. While several researchers [29,32] have utilized theorem proving to formally specify and verify the self-stabilization of specific protocols, this paper presents a problem-independent specification of weak convergence that enables potential reuse of efforts in the verification of convergence of different protocols.

One of the fundamental impediments before automated synthesis of self-stabilizing protocols from their non-stabilizing versions is the scalability problem. While there are techniques for parameterized synthesis [22,24] of concurrent systems, such methods are not directly useful for the synthesis of self-stabilization due to several factors. First, such methods are mostly geared towards synthesizing concurrent systems from formal specifications in some variant of temporal logic. Second, in the existing parameterized synthesis methods the formal specifications are often parameterized in terms of local liveness properties of individual components (e.g., progress for each process), whereas convergence is a global liveness property. Third, existing methods often consider the synthesis from a set of initial states that is a proper subset of the state space rather than the entire state space itself (which is the case for self-stabilization). With this motivation, our contributions in this paper enable a hybrid method based on synthesis and theorem proving that enables the generalization of small instances of self-stabilizing protocols generated by our tools [25].

Related Work. Kulkarni and Bonakdarpour's work [8,28] is the closest to the proposed approach in this paper. As such, we would like to highlight some differences between their contributions and ours. First, in [28], the authors focus on mechanical verification of algorithms for the addition of fault tolerance to concurrent systems in a high atomicity model where each process can read and write all system variables in one atomic step. One of the fault tolerance requirements they consider is nonmasking fault-tolerance, where a nonmasking system guarantees recovery to a set of legitimate states from states reachable by faults and not necessarily from the entire state space. Moreover, in [8], Kulkarni and Bonakdarpour investigate the mechanical verification of algorithms for the addition of multiple levels of fault tolerance in the high atomicity model. In this paper, our focus is on self-stabilization in distributed systems where recovery should be provided from any state and high atomicity actions are not feasible.

Methods for the verification of parameterized systems can be classified into the following major approaches, which do not directly address SS systems. Abstraction techniques [15,21,30] generate a finite-state model of a parameterized system and then reduce the verification of the parameterized system to the verification of its finite model. Network invariant approaches [20,23,37] find a process that satisfies the property of interest and is invariant to parallel composition. Logic program transformations and inductive verification methods [17,33] encode the verification of a parameterized system as a constraint logic program and reduce the verification of the parameterized system to the equivalence of goals in the logic program. In regular model checking [2,9], system states are represented by grammars over strings of arbitrary length, and a protocol is represented by a transducer. Abdulla *et al.* [1] also investigate reachability of unsafe states in symmetric timed networks and prove that it is undecidable to detect livelocks in such networks. Bertrand and Fournier [7] also focus on the verification of safety properties for parameterized systems with probabilistic timed processes.

8   Conclusion and Future Work

This paper focuses on exploiting theorem proving for the generalization of synthesized self-stabilizing protocols that are correct in a finite scope (i.e., up to a small number of processes). We are particularly interested in weak stabilization where reachability to legitimate states is guaranteed from any state. The contributions of this paper comprise a component of a hybrid method for verification and synthesis of parameterized self-stabilizing network protocols (see Fig. 1). This paper specifically presents a mechanical proof for the correctness of the Add_Weak algorithm from [16] that synthesizes weak convergence. This mechanical proof provides a reusable theory in PVS for the proof of weakly stabilizing systems in general (irrespective of how they have been designed). The success of mechanical proof for a small synthesized protocol shows the generality of the synthesized solution for arbitrary number of processes. We have demonstrated the proposed approach in the context of a binary agreement protocol (in [35]) and a graph coloring protocol (Sect. 6).

We will extend this work by reusing the existing PVS theory for mechanical proof of algorithms (in [16]) that design strong convergence. Moreover, we are currently investigating the generalization of more complicated protocols (e.g., leader election, maximal matching, consensus) using the proposed approach.

References

1. Abdulla, P.A., Jonsson, B.: Model checking of systems with many identical timed processes. Theor. Comput. Sci. **290**(1), 241–264 (2003)
2. Abdulla, P.A., Jonsson, B., Nilsson, M., Saksena, M.: A survey of regular model checking. In: Gardner, P., Yoshida, N. (eds.) CONCUR 2004. LNCS, vol. 3170, pp. 35–48. Springer, Heidelberg (2004)
3. Abujarad, F., Kulkarni, S.S.: Multicore constraint-based automated stabilization. In: Guerraoui, R., Petit, F. (eds.) SSS 2009. LNCS, vol. 5873, pp. 47–61. Springer, Heidelberg (2009)
4. Abujarad, F., Kulkarni, S.S.: Automated constraint-based addition of nonmasking and stabilizing fault-tolerance. Theor. Comput. Sci. **412**(33), 4228–4246 (2011)
5. Arora, A., Gouda, M.G.: Closure and convergence: a foundation of fault-tolerant computing. IEEE Trans. Softw. Eng. **19**(11), 1015–1027 (1993)
6. Attie, P.C., Arora, A., Emerson, E.A.: Synthesis of fault-tolerant concurrent programs. ACM Trans. Program. Lang. Syst (TOPLAS) **26**(1), 125–185 (2004)
7. Bertrand, N., Fournier, P.: Parameterized verification of many identical probabilistic timed processes. In: LIPIcs-Leibniz International Proceedings in Informatics, vol. 24 (2013)
8. Bonakdarpour, B., Kulkarni, S.S.: Towards reusing formal proofs for verification of fault-tolerance. In Workshop in Automated Formal Methods (2006)
9. Bouajjani, A., Jonsson, B., Nilsson, M., Touili, T.: Regular model checking. In: Emerson, E.A., Sistla, A.P. (eds.) CAV 2000. LNCS, vol. 1855, pp. 403–418. Springer, Heidelberg (2000)
10. Dijkstra, E.W.: Self-stabilizing systems in spite of distributed control. Commun. ACM **17**(11), 643–644 (1974)
11. Dolev, S., Haviv, Y.A.: Self-stabilizing microprocessor: analyzing and overcoming soft errors. IEEE Trans. Comput. **55**(4), 385–399 (2006)
12. Dolev, S., Yagel, R.: Self-stabilizing operating systems. In: Proceedings of the twentieth ACM symposium on Operating systems principles, pages 1–2. ACM (2005)
13. Ebnenasir, A., Farahat, A.: Swarm synthesis of convergence for symmetric protocols. In: European Dependable Computing Conference, pp. 13–24 (2012)
14. Ebnenasir, A., Kulkarni, S.S., Arora, A.: FTSyn: a framework for automatic synthesis of fault-tolerance. Int. J. Softw. Tools Technol. Transf. **10**(5), 455–471 (2008)
15. Emerson, E.A., Namjoshi, K.S.: On reasoning about rings. Int. J. Found. Comput. Sci. **14**(4), 527–550 (2003)
16. Farahat, A., Ebnenasir, A.: A lightweight method for automated design of convergence in network protocols. ACM Trans. Auton. Adapt. Syst. (TAAS) **7**(4), 38:1–38:36 (2012)
17. Fioravanti, F., Pettorossi, A., Proietti, M., Senni, V.: Generalization strategies for the verification of infinite state systems. TPLP **13**(2), 175–199 (2013)
18. Gordon, M.J.C., Melham, T.F.: Introduction to HOL: A Theorem proving Environment for Higher Order Logic. Cambridge University Press, Cambridge (1993)
19. Gouda, M.G.: The theory of weak stabilization. In: Datta, A.K., Herman, T. (eds.) WSS 2001. LNCS, vol. 2194, p. 114. Springer, Heidelberg (2001)
20. Grinchtein, O., Leucker, M., Piterman, N.: Inferring network invariants automatically. In: Furbach, U., Shankar, N. (eds.) IJCAR 2006. LNCS (LNAI), vol. 4130, pp. 483–497. Springer, Heidelberg (2006)
21. Ip, C.N., Dill, D.L.: Verifying systems with replicated components in murphi. Form. Methods Syst. Design **14**(3), 273–310 (1999)

22. Jacobs, S., Bloem, R.: Parameterized synthesis. In: Flanagan, C., König, B. (eds.) TACAS 2012. LNCS, vol. 7214, pp. 362–376. Springer, Heidelberg (2012)

23. Kesten, Y., Pnueli, A., Shahar, E., Zuck, L.D.: Network invariants in action. In: Brim, L., Jančar, P., Křetínský, M., Kučera, A. (eds.) CONCUR 2002. LNCS, vol. 2421, pp. 101–115. Springer, Heidelberg (2002)

24. Khalimov, A., Jacobs, S., Bloem, R.: PARTY Parameterized Synthesis of Token Rings. In: Sharygina, N., Veith, H. (eds.) CAV 2013. LNCS, vol. 8044, pp. 928–933. Springer, Heidelberg (2013)

25. Klinkhamer, A., Ebnenasir, A.: A parallel tool for automated synthesis of self-stabilization. http://asd.cs.mtu.edu/projects/protocon/

26. Klinkhamer, A., Ebnenasir, A.: On the complexity of adding convergence. In: Arbab, F., Sirjani, M. (eds.) FSEN 2013. LNCS, vol. 8161, pp. 17–33. Springer, Heidelberg (2013)

27. Kulkarni, S.S., Arora, A.: Large automating the addition of fault-tolerance. In: Joseph, M. (ed.) FTRTFT 2000. LNCS, vol. 1926, p. 82. Springer, Heidelberg (2000)

28. Kulkarni, S.S., Bonakdarpour, B., Ebnenasir, A.: Mechanical verification of automatic synthesis of fault-tolerance. Int. Symp. Log.-based Program Synth. Transform. **3573**, 36–52 (2004)

29. Kulkarni, S.S., Rushby, J., Shankar, N.: A case-study in component-based mechanical verification of fault-tolerant programs. In: 19th IEEE International Conference on Distributed Computing Systems - Workshop on Self-Stabilizing Systems, pp. 33–40 (1999)

30. Pnueli, A., Xu, J., Zuck, L.D.: Liveness with $(0, 1, \infty)$-Counter Abstraction. In: Brinksma, Ed, Larsen, Kim Guldstrand (eds.) CAV 2002. LNCS, vol. 2404, pp. 107–. Springer, Heidelberg (2002)

31. Prasetya, I.S.W.B.: Mechanically verified self-stabilizing hierarchical algorithms. Tools and Algorithms for the Construction and Analysis of Systems (TACAS'97), volume 1217 of Lecture Notes in Computer Science, pages 399–415 (1997)

32. Qadeer, S., Shankar, N.: Verifying a self-stabilizing mutual exclusion algorithm. In: Gries, D., de Roever, W.-P. (eds.) IFIP International Conference on Programming Concepts and Methods (PROCOMET 1998), pp. 424–443. Chapman & Hall, Shelter Island, NY (1998)

33. Roychoudhury, A., Narayan Kumar, K., Ramakrishnan, C.R., Ramakrishnan, I.V., Smolka, S.A.: Verification of Parameterized Systems Using Logic Program Transformations. In: Graf, S. (ed.) TACAS 2000. LNCS, vol. 1785, p. 172. Springer, Heidelberg (2000)

34. Shankar, N., Owre, S., Rushby, J.M.: The PVS Proof Checker: A Reference Manual. Computer Science Laboratory, SRI International, Menlo Park, CA, Feb. 1993. A new edition for PVS Version 2 is released in 1998

35. Tahat, A., Ebnenasir, A.: A hybrid method for the verification and synthesis of parameterized self-stabilizing protocols. Technical Report CS-TR-14-02, Michigan Technological University, May 2014. http://www.mtu.edu/cs/research/papers/pdfs/Technical%20Report%2014-02.pdf

36. Tsuchiya, T., Nagano, S., Paidi, R.B., Kikuno, T.: Symbolic model checking for self-stabilizing algorithms. IEEE Trans. Parallel Distrib. Syst. **12**(1), 81–95 (2001)

37. Wolper, P., Lovinfosse, V.: Verifying properties of large sets of processes with network invariants. In: International Workshop on Automatic Verification Methods for Finite State Systems, pp. 68–80 (1989)

Drill and Join: A Method for Exact Inductive Program Synthesis

Remis Balaniuk[(✉)]

Universidade Católica de Brasília, Brasília, Brazil
remis@robotics.stanford.edu

Abstract. In this paper we propose a novel semi-supervised active machine learning method, based on two recursive higher-order functions that can inductively synthesize a functional computer program. Based on properties formulated using abstract algebra terms, the method uses two combined strategies: to reduce the dimensionality of the Boolean algebra where a target function lies and to combine known operations belonging to the algebra, using them as a basis to build a program that emulates the target function. The method queries for data on specific points of the problem input space and build a program that exactly fits the data. Applications of this method include all sorts of systems based on bitwise operations. Any functional computer program can be emulated using this approach. Combinatorial circuit design, model acquisition from sensor data, reverse engineering of existing computer programs are all fields where the proposed method can be useful.

1 Introduction

Induction means reasoning from specific to general. In the case of inductive learning from examples, the general rules are derived from input/output (I/O) examples or answers from questions. Inductive machine learning has been successfully applied to a variety of classification and prediction problems [1,3].

Inductive program synthesis (IPS) builds from examples the computation required to solve a problem. The problem must be formulated as a task of learning a concept from examples, referred to as *inductive concept learning*. A computer program is automatically created from an incomplete specification of the concept to be implemented, also referred as the *target function* [3].

Research on inductive program synthesis started in the seventies. Since then it has been studied in several different research fields and communities such as artificial intelligence (AI), machine learning, inductive logic programming (ILP), genetic programming, and functional programming [3,8].

One basic approach to IPS is to simply enumerate programs of a defined set until one is found which is consistent with the examples. Due to combinatorial explosion, this general enumerative approach is too expensive for practical use. Summers [9] proposed an analytical approach to induce functional Lisp programs without search. However, due to strong constraints imposed on the forms of I/O-examples and inducible programs in order to avoid search, only relatively simple

© Springer International Publishing Switzerland 2015
M. Proietti and H. Seki (Eds.): LOPSTR 2014, LNCS 8981, pp. 219–237, 2015.
DOI: 10.1007/978-3-319-17822-6_13

functions can be induced. Several variants and extensions of Summers' method have been proposed, like in [7]. An overview is given in [10]. Kitzelmann [6] proposed a combined analytical and search-based approach. Albarghouthi [5] proposed ESCHER, a generic algorithm that interacts with the user via I/O examples, and synthesizes recursive programs implementing intended behavior.

Hybrid methods propose the integration of inductive inference and deductive reasoning. Deductive reasoning usually requires high-level specifications as a formula in a suitable logic, a background theory for semantic correctness specification, constraints or a set of existing components as candidate implementations. Some examples of hybrid methods are the syntax-guided synthesis [12], the sciduction methodology [17], the component-based synthesis [2] and the oracle-guided component-based program synthesis approach [16].

Inductive program synthesis is usually associated to functional programming [8]. In functional code the output value of a function depends only on the arguments that are input to the function. It is a declarative programming paradigm. Side effects that cause change in state that do not depend on the function inputs are not allowed. Programming in a functional style can usually be accomplished in languages that aren't specifically designed for functional programming. In early debates around programming paradigms, conventional imperative programming and functional programming were compared and discussed. J. Backus [14], in his work on programs as mathematical objects, supported the functional style of programming as an alternative to the "ever growing, fat and weak" conventional programming languages. Backus identified as inherent defects of imperative programming languages their inability to effectively use powerful combining forms for building new programs from existing ones, and their lack of useful mathematical properties for reasoning about programs. Functional programming, and more particularly function-level programming, is founded on the use of combining forms for creating programs that allow an algebra of programs.[1]

Our work is distinguishable from previous works on IPS in a number of ways:

- Our method is based on function-level programming.
- Our method is based on active learning. Active learning is a special case of machine learning in which the learning algorithm can control the selection of examples that it generalizes from and can query one or more oracles to obtain examples. The oracles could be implemented by evaluation/execution of a model on a concrete input or they could be human users. Most existing IPS methods supply the set of examples at the beginning of the learning process, without reference to the learning algorithm. Some hybrid synthesis methods use active learning, as in [12,16,17], to generate examples to a deductive procedure. Our method defines a learning protocol that queries the oracle to obtain the desired outputs at new data points.

---

[1] Function-level programming, as proposed by Backus [14], is a particular, constrained type of functional programming where a program is built directly from programs that are given at the outset, by combining them with program-forming operations or functionals.

- Our method generates programs on a very low-level declarative language, compatible with most high-level programming languages. Most existing methods are conceived considering and restricted to specific high-level source languages. Our whole method is based on Boolean algebra. Inputs and outputs are bit vectors and the generated programs are Boolean expressions. The synthesized program can also be used for a combinatorial circuit design describing the sequences of gates required to emulate the target function. Research on reconfigurable supercomputing is very interested in providing compilers that translate algorithms directly into circuit design expressed in an hardware description language. They want to avoid the high cost of having to hand-code custom circuit designs [13].
- The use of Boolean algebra and abstract algebra concepts at the basis of the method defines a rich formalism. The space where a program is to be searched, or synthesized, corresponds to a well-defined finite family of operations that can be reused, combined and ordered.
- Our method can be applied to general purpose computing. A program generated using our method, computing an output bit vector from an input bit vector, is equivalent to a system of Boolean equations or a set of truth tables. However, bit vectors can also be used to represent any kind of complex data types, like floating point numbers and text strings. A functionally complete set of operations performed on arbitrary bits is enough to compute any computable value. In principle, any Boolean function can be built-up from a functionally complete set of logic operators. In logic, a functionally complete set of logical connectives or Boolean operators is one which can be used to express all possible truth tables by combining members of the set into a Boolean expression. Our method synthesizes Boolean expressions based on the logic operators set $\{XOR, AND\}$ which is functionally complete.
- If the problem has a total functional behavior and enough data is supplied during the learning process our method can synthesize the exact solution.

This paper is organized as follows. Sections 2 and 3 review some relevant mathematical concepts. Sections 4 and 5 describe the mathematics of the method. Section 6 presents the method itself and how the programs are synthesized. Section 7 presents the main algorithms. Section 8 shows a Common Lisp implementation of the simplest version of the method. Sections 9, 10 and 11 close the document discussing the method, possible applications and future work.

2   Boolean Ring, F2 Field, Boolean Polynomials and Boolean Functions

A Boolean ring is essentially equivalent to a Boolean algebra, with ring multiplication corresponding to conjunction ($\wedge$) and ring addition to exclusive disjunction or symmetric difference ($\oplus$ or XOR). In Logic, the combination of operators $\oplus$ (XOR or exclusive OR) and $\wedge$ (AND) over elements *true*, *false* produce the Galois field F2 which is extensively used in digital logic and circuitry [11].

This field is functionally complete and can represent any boolean function obtainable with the system $(\wedge\vee)$ and can also be used as a standard algebra over the set of the integers modulo 2 (binary numbers 0 and 1)[2]. Addition has an identity element (*false*) and an inverse for every element. Multiplication has an identity element (*true*) and an inverse for every element but *false*.

Let B be a Boolean algebra and consider the associated Boolean ring. A Boolean polynomial in B is a string that results from a finite number of Boolean operations on a finite number of elements in B. A multivariate polynomial over a ring has a unique representation as a xor-sum of monomials. This gives a normal form for Boolean polynomials:

$$\bigoplus_{J \subset \{1,2,\ldots,n\}} a_J \prod_{j \in J} x_j \tag{1}$$

where $a_J \in B$ are uniquely determined. This representation is called algebraic normal form, Zhegalkin polynomials or Reed–Muller expansion [18]. A Boolean function of n variables $f : \mathbb{Z}_2^n \to \mathbb{Z}_2$ can be associated with a Boolean polynomial by deriving an algebraic normal form.

3   Abstract Algebra and Higher Order Functions

Let us consider a generic functional setting having as domain the set of bit strings of a finite, defined, length and as range the set $\{true, false\}$ or the binary numbers 0 and 1 represented by one bit. This setting can represent the inputs and output of a logic proposition, a Boolean function, a truth table or a fraction of a functional program corresponding to one of its output bits[3].

In abstract algebra terms, this setting will define a finitary Boolean algebra consisting of a finite family of operations on $\{0,1\}$ having the input bit string as their arguments. The length of the bit string will be the arity of the operations in the family. An n-ary operation can be applied to any of 2^n possible values of its n arguments. For each choice of arguments an operation may return 0 or 1, whence there are 2^{2^n} n-ary possible operations in the family. In this functional setting, IPS could be seen as the synthesis of an operation that fits a set of I/O examples inside its family. The Boolean algebra defines our program space.

Let V^n be the set of all binary words of length n, $|V^n| = 2^n$. The Boolean algebra B on V^n is a vector space over $\mathbb{Z}_2$. Because it has 2^{2^n} elements, it is of dimension 2^n over $\mathbb{Z}_2$. This correspondence between an algebra and our program space defines some useful properties. The operations in a family need not be all explicitly stated. A basis is any set of operators from which the remaining operations can be obtained by composition. A Boolean algebra may be defined from any of several different bases. To be a basis is to yield all other operations by composition, whence any two bases must be intertranslatable.

---

[2] Throughout this paper we will use indistinctively 0, F or *false* for the binary number 0 and 1, T or *true* for the binary number 1.

[3] Bit strings can represent any complex data type. Consequently, our functional setting includes any functional computer program having fixed length input and output.

A basis is a linearly independent spanning set. Let $v_1, \ldots, v_m \in B$ be a basis of B. $Span(v_1, \ldots, v_m) = \{\lambda_1 \wedge v_1 \oplus \ldots \oplus \lambda_m \wedge v_m \mid \lambda_1, \ldots, \lambda_m \in \mathbb{Z}_2\}$. The dimension $dim(B)$ of the Boolean algebra is the minimum m such that $B = span(v_1, \ldots, v_m)$[4].

The method proposed in this paper consists of two combined strategies: reducing the dimensionality of the Boolean algebra where a target function lies and combining known operations in the algebra, using them as a basis to synthesize a program that emulates the target function.

Both strategies are implemented using two recursive higher order functions that we created and that we named the *drill* and the *join*.

In computer science higher-order functions are functions that take one or more functions as input and output a function. They correspond to linear mappings in mathematics.

4 The *Drill* Function

We define the set F_m of functions $f : \mathbb{Z}_2^p \times \mathbb{Z}_2^q \to \mathbb{Z}_2$ containing Boolean functions belonging to a Boolean algebra of dimension m, described in polynomial form:

$$f(X, Y) = \bigoplus_{i-1}^{m} g_i(X) \wedge h_i(Y) \tag{2}$$

where $g_i : \mathbb{Z}_2^p \to \mathbb{Z}_2$ and $h_i : \mathbb{Z}_2^q \to \mathbb{Z}_2$ are also Boolean functions. Note that the polynomial representations used in 1 and 2 are equivalent and interchangeable. Equation 2 simply splits its input space in two disjoint subsets: $p + q = n$. In 1 lower case is used to indicate atomic Boolean variables while in 2 upper case is used to indicate Boolean vectors.

Considering a function $f \in F_m$, a chosen $X_0 \in \mathbb{Z}_2^p$ and a chosen $Y_0 \in \mathbb{Z}_2^q$ such that $f(X_0, Y_0) \neq 0$, we define the *drill* higher-order function:

$$\mathbb{IF}_{X_0 Y_0} = \mathbb{IF}(f(X, Y), X_0, Y_0) = f(X, Y) \oplus (f(X_0, Y) \wedge f(X, Y_0)) \tag{3}$$

Note that the function $\mathbb{IF}$ outputs a new function and has as inputs the function f and instances of X and Y, defining a position on f input space.

Theorem: If $f \in F_m$ and $f(X_0, Y_0) \neq 0$ then $\overline{f} = \mathbb{IF}_{X_0 Y_0} \in F_r$ and $r \leq m - 1$[5,6].

---

[4] The dimension of the Boolean algebra will also determine the minimum number of monomials required to define each of its Boolean polynomials in algebraic normal form.

[5] We will use the overline notation to distinguish between modules (Boolean functions and Boolean spaces) belonging to the target function subspace (without overline) and modules belonging to a lower dimension linear subspace generated by the *drill* linear mapping (with overline).

[6] F_r is a set containing Boolean functions belonging to a Boolean algebra of dimension r.

Proof: Consider $W = span(h_1, \ldots, h_m)$. Consequently $dim(W) \leq m$. The linear operator $h \in W \rightarrow h(Y_0)$ is not the zero map because the hypothesis forbids $h_i(Y_0) = 0$ for all $i = 1, \ldots, n$. Consequently, the vector subspace $\overline{W} = \{h \in W | h(Y_0) = 0\}$ has $dim(\overline{W}) \leq m - 1$. Notice that for all $X \in \mathbf{Z}_2^p$ we have $\overline{f}(X, \cdot) \in \overline{W}$. In fact:

$$\overline{f}(X, Y_0) = f(X, Y_0) \oplus (f(X_0, Y_0) \wedge f(X, Y_0)) = 0 \qquad (4)$$

Let $r = dim(\overline{W})$ and $\overline{h_i}, i = 1, \ldots, r$ be a spanning set such that $\overline{W} = span(\overline{h_1}, \ldots, \overline{h_r})$. For all $X \in \mathbf{Z}_2^p$, $\overline{f}(X, \cdot)$ can be represented as a linear combination of the $\overline{h_i}$, the coefficients depending on X. In other words, there exist coefficients $\overline{g_i}(X)$ such that:

$$\overline{f}(X, \cdot) = \bigoplus_{i=1}^{r} \overline{g_i}(X) \wedge \overline{h_i} \qquad (5)$$

or written differently and remaining that $r = dim(\overline{W})$ and consequently $r \leq m - 1$:

$$\overline{f}(X, Y) = \bigoplus_{i=1}^{r} \overline{g_i}(X) \wedge \overline{h_i}(Y) \qquad (6)$$

$\square$

As an illustration let us consider a Boolean function $f : \mathbf{Z}_2 \times \mathbf{Z}_2 \rightarrow \mathbf{Z}_2$ whose behavior can be described by Table 1.

Table 1. Truth table for $f(x, y)$

x	y	$f(x, y)$
F	F	T
T	F	F
F	T	T
T	T	T

One possible representation for this function would be $f(x, y) = y \vee (\neg x \wedge \neg y)$. f is of dimension 2 (no shorter representation is possible). Note that this representation is given here for the illustration's sake. The method does not require high-level definitions, only I/O examples.

Respecting the stated hypothesis we can pick $x_0 = F$ and $y_0 = F$ once $f(F, F) = T$. The partial functions obtained will be: $f(x_0, y) = y \vee (T \wedge \neg y)$ and $f(x, y_0) = F \vee (\neg x \wedge T)$. Applying the *drill* function we obtain:

$$\overline{f}(x, y) = \mathbb{F}(f(x, y), x_0, y_0) = (y \vee (\neg x \wedge \neg y)) \oplus ((y \vee (T \wedge \neg y)) \wedge (F \vee (\neg x \wedge T))) = (x \wedge y) \qquad (7)$$

We can see that $\overline{f}(x, y)$ is of dimension 1, confirming the stated theorem.

5   The *Join* Function

Consider now the set F_m of Boolean functions $f : \mathbb{Z}_2^n \to \mathbb{Z}_2$ and $v_1, \ldots, v_m \in F_m$ a basis. The functions in this set can be described in polynomial form as:

$$f(X) = \bigoplus_{i=1}^{m} \lambda_i \wedge v_i(X) \tag{8}$$

where $\lambda_i \in \mathbb{Z}_2$ are the coefficients[7].

Considering a function $f \in F_m$, a chosen $X_j \in \mathbb{Z}_2^n$ such that $f(X_j) \neq 0$ and a chosen function v_j belonging to the basis such that $v_j(X_j) \neq 0$, we define the *join* function:

$$\mathbb{H}_{X_j v_j} = \mathbb{H}(f(X), X_j, v_j) = f(X) \oplus v_j(X) \tag{9}$$

Theorem: If $f \in F_m$, $f(X_j) \neq 0$ and $v_j(X_j) \neq 0$, then $\overline{\overline{f}} = \mathbb{H}_{X_j v_j} \in F_r$ and $r \leq m - 1$[8].

Proof: Consider $W = span(v_1, \ldots, v_m)$. Consequently $dim(W) \leq m$. The linear operator $v \in W \to v(X_j)$ is not the zero map otherwise $v_j(X_j) = 0$. Consequently, the vector subspace $\overline{\overline{W}} = \{f \in W | f(X_j) = 0\}$ has $dim(\overline{\overline{W}}) \leq m - 1$. We can see that $\overline{\overline{f}} \in \overline{\overline{W}}$. In fact:

$$\overline{\overline{f}}(X_j) - f(X_j) \oplus v_j(X_j) = 0 \tag{10}$$

Let $r = dim(\overline{\overline{W}})$ and $\overline{\overline{v_i}}, i = 1, \ldots, r$ be a spanning set such that $\overline{\overline{W}} = span(\overline{\overline{v_i}}, \ldots, \overline{\overline{v_r}})$. The function $\overline{\overline{f}}$ can be represented as a linear combination of the $\overline{\overline{v_i}}$. In other words, and remaining that $r = dim(\overline{\overline{W}})$ so $r \leq m - 1$, there exist coefficients $\overline{\overline{\lambda_i}}$ such that:

$$\overline{\overline{f(X)}} = \bigoplus_{i-1}^{r} \overline{\overline{\lambda_i}} \wedge \overline{\overline{v_i(X)}} \tag{11}$$

$\square$

We can use the same function $f : \mathbb{Z}_2 \times \mathbb{Z}_2 \to \mathbb{Z}_2$ described by Table 1 to illustrate the behavior of the *join* higher-order function. f belongs to a Boolean

---

[7] Note that in Eq. 8 we are considering the target function as having one single vector input X while in 2 the same target function has two vector inputs X and Y. These different notations for the same target function can be understood as X in Eq. 8 being the concatenation of X and Y from Eq. 2 or X and Y in Eq. 2 being a split of X from Eq. 8 in two vector subspaces. It follows that *drill* can be applied only to target Boolean functions having at least two input variables while *join* can be applied to target Boolean functions of any arity.

[8] We will use the double overline notation to distinguish between modules (Boolean functions and Boolean spaces) belonging to the target function subspace (without overline) and modules belonging to a lower dimension linear subspace generated by the *join* linear mapping (with double overline).

algebra of dimension 2^2 which can be defined, for instance, by the following spanning set: $v_1(x,y) = x, v_2(x,y) = y, v_3(x,y) = x \wedge y, v_4(x,y) = T$. Respecting the stated hypothesis we can pick $X_j = (T,T)$ and $v_j = v_1$ once $f(T,T) = T$ and $v_1(T,T) = T$. Applying the *join* function we obtain:

$$\overline{\overline{f}}(x,y) = \mathbb{IH}(f(X), X_j, v_j) = (y \vee (\neg x \wedge \neg y)) \oplus x = (x \wedge y) \tag{12}$$

We can see that $\overline{\overline{f}}(x,y)$ is of dimension 1, confirming the stated theorem.

6   The *Drill & Join* Program Synthesis Method

Drill and *join* are used to define a program synthesis method. Considering an active learning framework, the input function $f(X,Y)$ on $\mathbb{IF}$, defined in Eq. 3, and the input function $f(X)$ on $\mathbb{IH}$, defined in Eq. 9, represent an external unknown concept from which it is possible to obtain data by means of queries (input-output examples).

This unknown concept could be, for instance, some physical phenomenon that a machine with sensors and actuators can actively experiment, an algorithm to be translated in a hardware description language, a computer program that one would like to emulate or optimize or a decision process to be implemented on a computer for which one or more experts are able to answer required questions.

In order to understand the method it is important to notice two important properties of both higher-order functions:

- $\mathbb{IF}$ and $\mathbb{IH}$ can be applied recursively: if $f(X,Y) \in F_m$ then $\overline{f_1}(X,Y) = \mathbb{IF}(f(X,Y), X_0, Y_0) \in F_{m-1}$ and $\overline{f_2}(X,Y) = \mathbb{IF}(\overline{f_1}(X,Y), X_1, Y_1) \in F_{m-2}$[9]. Similarly, if $f(X) \in F_m$ then $\overline{\overline{f_1}}(X) = \mathbb{IH}(f(X), X_0, v_0) \in F_{m-1}$ and $\overline{\overline{f_2}}(X) = \mathbb{IH}(\overline{\overline{f_1}}(X), X_1, v_1) \in F_{m-2}$[10]. Each recursion generates a new function belonging to an algebra of a lower dimension.
- The recursion ends when the higher-order functions become the zero map: $\mathbb{IF}(f(X,Y), X_i, Y_i) = 0 \Leftrightarrow f(X,Y) = (f(X_i,Y) \wedge f(X,Y_i))$ and similarly, $\mathbb{IH}(f(X), X_i, v_i) = 0 \Leftrightarrow f(X) = v_i(X)$.

The first property enables us to apply the same higher order function recursively in order to gradually reduce the dimensionality of the initial problem. The second defines a stop condition. As a result, the output program is obtained.
$$\forall X, Y \in \mathbb{Z}_2^p \times \mathbb{Z}_2^q : \overline{f_{m+1}}(X,Y) = \mathbb{IF}(\overline{f_m}(X,Y), X_m, Y_m) = 0 \Leftrightarrow$$
$$\overline{f_m}(X,Y) = (\overline{f_m}(X_m,Y) \wedge \overline{f_m}(X,Y_m))$$
Replacing $\overline{f_m}(X,Y) = \mathbb{IF}(\overline{f_{m-1}}(X,Y), X_{m-1}, Y_{m-1})$ gives us:

$$\overline{f_{m-1}}(X,Y) \oplus (\overline{f_{m-1}}(X_{m-1},Y) \wedge \overline{f_{m-1}}(X,Y_{m-1})) = (\overline{f_m}(X_m,Y) \wedge \overline{f_m}(X,Y_m)) \tag{13}$$

---

[9] The notation $\overline{f_i}$ indicates a function resulting from the i-th recursive *drill* linear mapping.

[10] The notation $\overline{\overline{f_i}}$ indicates a function resulting from the i-th recursive *join* linear mapping.

and consequently:

$$\overline{f_{m-1}}(X,Y) = (\overline{f_{m-1}}(X_{m-1},Y) \wedge \overline{f_{m-1}}((X,Y_{m-1})) \oplus (\overline{f_m}(X_m,Y) \wedge \overline{f_m}(X,Y_m)) \tag{14}$$

Tracking the recursion back to the beginning we obtain:

$$f(X,Y) = \bigoplus_{i=1}^{m} \overline{f_i}(X_i,Y) \wedge \overline{f_i}(X,Y_i) \tag{15}$$

Equation 15 tells us that the original target function f can be recreated using the partial functions $\overline{f}$ obtained using the *drill* function. The partial functions $\overline{f}$ are simpler problems, defined on subspaces of the original target function.

Similarly:

$$\forall X \in \mathbb{Z}_2^n : \overline{\overline{f_{m+1}}}(X) = \mathbb{IH}(\overline{\overline{f_m}}(X), X_m, v_m) = 0 \Leftrightarrow \overline{\overline{f_m}}(X) = v_m(X) \tag{16}$$

Replacing $\overline{\overline{f_m}}(X) = \mathbb{IH}(\overline{\overline{f_{m-1}}}(X), X_{m-1}, v_{m-1})$ gives us:

$$\overline{\overline{f_{m-1}}}(X) \oplus v_{m-1}(X) = v_m(X) \tag{17}$$

and:

$$\overline{\overline{f_{m-1}}}(X) = v_{m-1}(X) \oplus v_m(X) \tag{18}$$

Tracking the recursion back to the beginning we obtain:

$$f(X) = \bigoplus_{i=1}^{m} v_i(X) \tag{19}$$

Equation 19 tells us that the original target function f can be recreated using the partial functions $\overline{\overline{f}}$ obtained using the *join* function and the basis v.

Note that if the *drill* initial condition cannot be established, i.e., no $(X_0, Y_0) :$ $f(X_0, Y_0) \neq 0$ can be found, the target function is necessary $f(X, Y) = F$. On the same way if no $X_0 : f(X_0) \neq 0$ can be found to initiate *join* the target function is the zero map $f(X) = F$. If it exists a $X_j : f(X_j) \neq 0$ but no $v_j : v_j(X_j) \neq 0$ the basis was not chosen appropriately.

The $\mathbb{IF}$ higher order function defines a double recursion. For each step of the dimensionality reduction recursion two subspace synthesis problems are defined: $\overline{f_i}(X_i, Y)$ and $\overline{f_i}(X, Y_i)$. Each of these problems can be treated as a new target function in a Boolean algebra of reduced arity once part of the arguments is fixed. They can be recursively solved using $\mathbb{IF}$ or $\mathbb{IH}$ again.

The combination of $\mathbb{IF}$ and $\mathbb{IH}$ defines a very powerful inductive method. The $\mathbb{IF}$ higher-order function alone requires the solution of an exponential number of subspace synthesis problems in order to inductively synthesize a target function. The $\mathbb{IH}$ higher-order function requires a basis of the Boolean algebra. Considering the 2^n cardinality of a basis, it can be impractical to require its prior existence in large arity algebras. Nevertheless, both functions combined can drastically reduce the number of queries and the prior bases definition.

The method, detailed on the next sections, uses the following strategies:

Table 2. Truth table for the *drill* steps

x	y	$f(x,y)$	$f(x_0,y)$	$f(x,y_0)$	$\overline{f_1}(x,y)$	$\overline{f_1}(x_1,y)$	$\overline{f_1}(x,y_1)$	$\overline{f_2}(x,y)$
F	F	T	T	T	F	F	F	F
T	F	F	T	F	F	F	T	F
F	T	T	T	T	F	T	F	F
T	T	T	T	F	T	T	T	F

- Bases are predefined for low arity input spaces, enabling the *join* function.
- Synthesis on large arity input spaces begin using the **IF** function.
- Previously synthesized programs on a subspace can be memorized in order to compose a basis on that subspace.
- At each new synthesis, if a basis exists use **IH**, otherwise use **IF**.

To illustrate the functioning of the whole method let us use again the same function $f : \mathbb{Z}_2 \times \mathbb{Z}_2 \to \mathbb{Z}_2$ described by Table 1.

Table 2 shows the same target function and the transformation steps performed using the *drill* function. We apply the *drill* function one first time $\overline{f_1}(X,Y) = \mathbf{IF}(f(X,Y),X_0,Y_0)$ with $x_0 = F$, $y_0 = F$, defining two subspace problems: $f(x_0,y)$ and $f(x,y_0)$. Both problems can be solved using the *join* function and the basis $v_0(x) = x$, $v_1(x) = T$. $\overline{f_1}$ is not the zero map, requiring a second recursion of *drill*: $\overline{f_2} = \mathbf{IF}(\overline{f_1}(X,Y),X_1,Y_1)$ with $x_1 = T$, $y_1 = T$. Two new subspace problems are defined: $\overline{f_1}(x_1,y)$ and $\overline{f_1}(x,y_1)$ and they can be solved using *join* and the same basis again. $\overline{f_2}(X,Y) = F$ is finally the zero map, stopping the recursion.

To illustrate the use of the *join* function let us consider the reconstruction of $\overline{\overline{f}}(X) = f(x,y_0)$ detailed in the fifth column of Table 2. One first application $\overline{f_1}(X) = \mathbf{IH}(f(X),X_0,v_0)$ with $x_0 = F$ and $v_0(x) = x$ will not result in the zero map, as shown on the fifth column of Table 3, requiring a recursive call $\overline{f_2}(X) = \mathbf{IH}(\overline{f_1}(X),X_1,v_1)$ with $x_1 = T$ and $v_1(x) = T$ which will result the zero map.

Using Eq. 19 we can find a representation for the partial target function: $f(X) = f(x,y_0) = v_0(x) \oplus v_1(x) = x \oplus T$. The same process can be used to find representations for all one-dimensional problems: $f(x_0,y) = T$, $\overline{f_1}(x_1,y) = y$ and $\overline{f_1}(x,y_1) = x$.

Having solved the partial problems we can use Eq. 15 to build the full target function: $f(x,y) = (f(x_0,y) \wedge f(x,y_0)) \oplus (\overline{f_1}(x_1,y) \wedge \overline{f_1}(x,y_1)) = (T \wedge (x \oplus T)) \oplus (y \wedge x)$ which is equivalent to our initial representation.

Table 3. Truth table for the *join* steps

x	$f(x)$	$v_0(x)$	$v_1(x)$	$\overline{\overline{f_1}}(x)$	$\overline{\overline{f_2}}(x)$
F	T	F	T	T	F
T	F	T	T	T	F

Each higher order function underlies a query protocol. At each recursion of IF one or more queries for data are made in order to find $X_i, Y_i \in \mathbb{Z}_2^p \times \mathbb{Z}_2^q$: $\overline{f_i}(X_i, Y_i) \neq 0$. At each recursion of IH a position $X_i \in \mathbb{Z}_2^p : f(X_j) \neq 0$ and a function $v_i : v_i(X_i) \neq 0$ from the basis are chose. The queries require data from the target function and must be correctly answered by some kind of oracle, expert, database or system. Wrong answers make the algorithms diverge. Data is also necessary to verify the recursion stop condition. A full test of $\overline{f_i}(X, Y) = 0$ or $\overline{\overline{f_i}}(X) = 0$, scanning the whole input space, can generate proof that the induced result exactly meets the target function. Partial tests can be enough to define candidate solutions subject to further inspection.

7   The Main Algorithms of the *Drill & Join* method

To explain how the *drill* and *join* higher-order functions can be used as program-forming functionals we propose the following two algorithms.

Drill takes as initial inputs the target function: *fn* and the dimension of its input space: *inputs*. Deeper inside the recursion *fn* corresponds to $\overline{f_m}(X, Y)$, *inputs* defines the dimension of its subspace and *initial* indicates its first free dimension. *Drill* returns a synthesized functional program that emulates *fn*.

```
1: procedure DRILL(fn, inputs, optional: initial = 0)
2:     if have a basis for this subspace then
3:         return JOIN(fn,inputs,initial);
4:     end if
5:     pos ←find a position inside this subspace where fn is not null;    ▷ Use a protocol to
   query fn (active learning)
6:     if pos = null then
7:         return FALSE;                                    ▷ Stop condition: fn is the zero map.
8:     end if
9:     fa(args) = fn(concatenate(args, secondhalf(pos)));              ▷ f(X, Y₀)
10:    fb(args) = fn(concatenate(firsthalf(pos), args));              ▷ f(X₀, Y)
11:    fc(args) = fn(args) ⊕ (fa(firsthalf(args)) ∧ fb(secondhalf(args))) ;
12:                                                        ▷ f̄ₘ₊₁(X, Y)
13:    pa = DRILLl(fa, inputs/2, initial);           ▷ Recursive call to synthesize f(X, Y₀)
14:    pb = DRILL(fb, inputs/2, initial + inputs/2);         ▷ Recurs. call for f(X₀, Y)
15:    pc = DRILL(fc, inputs, initial);                ▷ Recursive call for f̄ₘ₊₁(X, Y)
16:    return '((' pa 'AND' pb ')' XOR pc ')';               ▷ Returns the program
17: end procedure
```

Note that *fa*, *fb* and *fc* are new target functions based on *fn* and *pa*, *pb* and *pc* are programs obtained by recursively calling *drill*. *firsthalf* and *secondhalf* split a vector in two halves.

Join takes as input a function *fn*, belonging to a Boolean algebra. The algorithm requires a basis for this Boolean algebra, materialized as an array of functions: *basis[]* and an array of programs emulating the basis: *basisp[]*. The algorithm creates a program to emulate *fn* combining the program basis.

```
1: procedure JOIN() fn, optional: initial = 0
2:     pos ←find a position inside this subspace where fn is not null;    ▷ Use a protocol to
   query fn (active learning)
```

```
 3:      if pos = null then
 4:          return FALSE;                              ▷ Stop condition: fn is the zero map.
 5:      end if
 6:      v ←select a function v from basis[] such that v(pos) is not 0;
 7:      vp ←get the program from basisp[] that emulates v;
 8:      fa(args) = fn(args) ⊕ v(args);
 9:      pa = JOIN(fa, initial);                        ▷ Recursive call for $\overline{\overline{f_{m+1}}}(X)$
10:      return '(' pa 'XOR' vp ')' ;                   ▷ Returns the program.
11: end procedure
```

8   A Common Lisp Version of the *Drill & Join* Method

Common Lisp is a natural choice of programming language to implement the *drill & join* method. Lisp functions can take other functions as arguments to build and return new functions. Lisp lists can be interpreted and executed as programs.

The following code implements the simplest version of the method. It synthesizes a Lisp program that emulates a function *fn* by just querying it. The function *fn* must accept bit strings as inputs and must return one bit as the answer. In this simple illustration *fn* is another Lisp function but in real use it would be an external source of data queried throughout an adequate experimental protocol.

Drill takes the unknown function *fn* and its number of binary arguments *nargs* as input. It returns a list composed of logical operators, logical symbols *nil* and *true* and references to an input list of arguments. The output list can be executed as a Lisp program that emulates *fn*.

```
(defun drill (fn nargs &optional (ipos 0) (slice 0))
(let ((base   (list nil #'(lambda(x) (first x)) #'(lambda(x) t) ) )
      (basep (list nil #'(lambda(x) (list 'nth x 'args)) #'(lambda(x) t))))
 (if (= nargs 1)
  (join fn base basep ipos)
  (let ((pos (findpos fn nargs slice)))
   (if (null pos)
    nil
    (labels ((fa (args) (funcall fn (append args (cdr pos))))
             (fb (args) (funcall fn (append (list (car pos)) args)))
             (fc (args) (xor (funcall fn args)
                   (and (fa (list (car args))) (fb (cdr args)))))))
     (let ((sa (drill #'(lambda(args) (fa args)) 1 ipos))
           (sb (drill #'(lambda(args) (fb args)) (1- nargs) (1+ ipos)))
           (sc (drill #'(lambda(args) (fc args)) nargs ipos (1+ slice))))
      (if (and (atom sa) sa)
          (setq r1 sb)
          (setq r1 (list 'and sa sb)))
      (if (null sc)
          r1
          (list 'xor r1 sc)))))))))
```

In this implementation the split of the input space is done by choosing the first argument to be X and the rest to be Y. *Drill* will call *Join* when the recursion is down to just one input function ($nargs = 1$). A basis for one input bit functions $\{f(x) = x, f(x) = T\}$ is defined directly inside *drill* as the functions list *base* and the programs list *basep* which are passed as arguments to the *join* function.

```
(defun join (fn base basep &optional (ipos 0))
 (let ((pos (findpos fn 1)))
  (if (null pos)
   nil
   (let ((fb (findbase base basep pos)))
    (labels ((fa (args) (xor (funcall fn args) (funcall (nth 0 fb) args))))
     (let ((r (join #'(lambda(args) (fa args)) base basep ipos)))
      (return-from join  (list 'xor (funcall (nth 1 fb) ipos)  r)))))))))
```

The *findpos* function is used to test if *fn* is the zero map performing a full search on the *fn* input space if necessary. It stops when a non-zero answer is found and returns its position. The full search means that no inductive bias was used.

```
(defun findpos (fn nargs)
 (loop for i from 0 to (1- (expt 2 nargs)) do
  (let ((l (make-list nargs)) (j i) (k 0))
   (loop do  (if (= (mod j 2) 1) (setf (nth k l) t))
             (incf k) (setq j (floor j 2)))
   while (> j 0))
  (if (funcall fn l)  (return-from findpos l)))))
```

The *findbase* function is used inside *join* to find a function from the basis respecting the constraint $v_j(X_j) \neq 0$.

```
(defun findbase (base basep pos)
    (loop for i from 1 to  (1- (list-length base)) do
       (if (funcall (nth i base)pos)
          (let ((ba (nth i base)) (bp (nth i basep)))
           (setq fb (list ba bp))
           (delete (nth 0 fb) base)  (delete (nth 1 fb) basep)
           (return-from findbase fb)))))
```

The method queries the target function (*funcall fn*) only inside *findpos*, in order to find a non-zero position.

Using the Lisp code provided above it is possible to check the consistency of the method. Applied to our illustration described by Table 1 the generated program would be:

```
(XOR (AND (XOR T (NTH 0 ARGS)) T) (AND (NTH 0 ARGS) (NTH 1 ARGS)))
```

As a more advanced illustration, let us consider the "unknown" target function to be the Fibonacci sequence. To avoid bulky outputs we will limit the illustration to have as input an unsigned integer between 0 and 63. Consequently, the input of the *fn* function can be a six bit long bit string. The range of the output (between 1 and 6557470319842) requires a 64 bits unsigned integer. The target function $fibonacci(n)$ computes a long integer corresponding to the $n - th$

position of the Fibonacci sequence. To translate integers to lists of $\{NIL, T\}$ handled by the *drill* and *join* lisp code we use their binary representation. The translation is done by the routines longint2bitlist, bitlist2longint, 6bitlist2int and 6int2bitlist, included in the appendices, which are called from the function $myfibonacci(n)$

```
(defun myfibonacci(n)(let ((r (6bitlist2int n)))(longint2bitlist(fibonacci r))))
```

In order to synthesize a program able to emulate the whole target function we need to call *Drill* for each output bit and generate a list of boolean expressions.

```
(defun synthesis (fn nargs nouts filename)
    (with-open-file (outfile filename :direction :output)
        (let ((l (make-list nouts)))
            (loop for i from 0 to (1- nouts) do
              (labels ((fa (args) (nth i (funcall fn args))))
                (let ((x (drill #'(lambda(args) (fa args)) nargs))) (setf (nth i l) x))))
            (print l outfile)))))
```

To run the generated program we need to compute each output bit and translate the bit string into an integer:

```
(defun runprogram(filename v)
 (with-open-file (infile filename)
   (setq s (read infile))  (setq args (6int2bitlist v))
   (let ((r (make-list (list-length s))))
     (loop for i from 0 to (1- (list-length s)) do
        (setf (nth i r) (eval (nth i s))))
     (print (bitlist2longint r)))))
```

A full check on the whole target function input space shows that the synthesized program exactly emulates the target function. To illustrate how the generated programs looks like we show below the expression that computes the first output bit of the Fibonacci sequence:

```
(XOR (AND (XOR (AND (XOR (AND (NTH 0 ARGS) (XOR T (NTH 1 ARGS))) (AND (XOR T
(NTH 0 ARGS)) (NTH 1 ARGS))) T) (AND (XOR (AND (XOR T (NTH 0 ARGS)) T) (AND
(NTH 0 ARGS) (NTH 1 ARGS))) (NTH 2 ARGS))) (XOR (AND (XOR (AND (XOR T (NTH 3 ARGS))
T) (AND (NTH 3 ARGS) (NTH 4 ARGS))) (XOR T (NTH 5 ARGS))) (AND (XOR (AND (NTH 3 ARGS)
(XOR T (NTH 4 ARGS))) (AND (XOR T (NTH 3 ARGS)) (NTH 4 ARGS))) (NTH 5 ARGS)))) (AND
(XOR (AND (XOR (AND (XOR T (NTH 0 ARGS)) T) (AND (NTH 0 ARGS) (NTH 1 ARGS))) (XOR T
(NTH 2 ARGS))) (AND (XOR (AND (NTH 0 ARGS) (XOR T (NTH 1 ARGS))) (AND (XOR T
(NTH 0 ARGS)) (NTH 1 ARGS))) (NTH 2 ARGS))) (XOR (AND (XOR (AND (NTH 3 ARGS)
(XOR T (NTH 4 ARGS))) (AND (XOR T (NTH 3 ARGS)) (NTH 4 ARGS))) T) (AND (XOR (AND
(XOR T (NTH 3 ARGS)) T) (AND (NTH 3 ARGS) (NTH 4 ARGS))) (NTH 5 ARGS)))))
```

The generated program is basically a single Boolean expression that explicitly references an input list called *args*. The full program consists of a list of those Boolean expressions.

9 Discussion

The implementation presented on Sect. 8 has didactic purposes only. A number of enhancements are possible. Bases can be dynamically built on subspaces by memorizing programs previously created avoiding drilling down to smaller subspaces and reducing the number of queries to the target function.

An interesting aspect of the generated programs is the fact that the computation of each output bit is completely independent of the others. There is a specific program for each output bit, enabling parallel processing.

Empirical comparisons between the proposed method and existing ones are difficult because of the conceptual differences between them. Existing benchmark frameworks, as CHStone [15] and SyGuS [12] tend to be specific to a certain synthesis approach. CHStone was conceived for C-based high-level synthesis and SyGuS for syntax-guided hybrid synthesis (inductive and deductive). Conceptual comparisons can be done but without objective results. The proposed method can handle target functions that others probably cannot, but because it works at the bit level of inputs and output, the number of examples required for learning and testing tend to be larger than in other methods. Most existing inductive methods use static example databases while our method is based on active learning and requires an experimental protocol in order to query the target concept during the learning process. Our method requires a predefined input space, with fixed length, and does not handle dynamic input lists like in Lisp programs generated by most variants and extensions of Summers [9] analytical approach. But on the other side, the simplicity of the code generated by our method, based on only two logical operators, enables its compilation in almost any conventional programming language, even on hardware description languages. The loop-free declarative nature of the generated programs brings predictability at runtime in terms of execution time and use of machine resources. The example proposed in Sect. 8 showed how to synthesize a declarative program to emulate the Fibonacci sequence. The generated program requires, for any six bits input, exactly 1804 low-level, bitwise logic operations to compute the corresponding Fibonacci number. An equivalent imperative program will require local variables, loop or recursion controls and a number of variable assignments and arithmetic operations proportional to the input value. The method does not require any prior knowledge about the target function, as a background theory, types of variables or operation on types, like in hybrid methods [12].

Nevertheless, the goal of this paper is not to prove that we created a better method but to present a new concept on inductive program synthesis. Future work will be necessary in order to assess the advantages and disadvantages of the proposed method in each possible field of application when other methods are also available.

For practical use of our method it is important to be able to estimate the effort required to synthesize a program. In typical active learning, there is usually a cost element associated with every query. This cost depends on the characteristics of the target concept and the associated experimental protocol used to query it. The synthesis effort will depend on the number of queries to be required and the cost of each query. If the implementation of the method is based on a full verification of the zero map, like in our illustration presented on Sect. 8, a full scan of the target function input space will be necessary and the number of queries will depend only on the size of the input space in bits. As a future work an inductive bias can be proposed in order to avoid the full scan and then reduce the number of queries.

10   Applications

Applications of this method include all sorts of systems based on bitwise operations, given that the learning problem can be described in functional form. We successfully applied the method to target functions handling different complex data types, as floating-point numbers and texts.

Any functional computer program can be emulated using our approach. There can be a number of reasons to perform reverse engineering of existing computer programs: the lack of a source code, the need to translate an executable program to run on a different computational platform, the intent to optimize an inefficient implementation.

The method can also be used to translate algorithms from a high-level language directly into combinational circuit design expressed in an hardware description language. The code generated by the proposed method can be easily mapped into a netlist (sequence of circuit gates).

Machine learning is another field of application. Machines having sensors and actuators, like robots, can acquire direct models using the method. The machine can actively experiment a physical phenomena and synthesize a program to predict the result of possible actions.

11   Conclusion

We have presented *Drill & Join*, a generic method that actively interacts with an external concept (system, function, oracle or database) via I/O examples, and synthesizes programs. Generated programs are based on Boolean expressions. The method is not restricted to any specific form of functional learning problem or target function and does not require any background knowledge to be applied. The only requirement is that the external source of data is consistent. Our work presents a number of interesting questions for future consideration. The combination of the *drill* and the *join* higher-order functions and the dynamic construction of bases on subspaces via memorization of generated programs can drastically reduce the number of recursive calls and queries. Further investigation is necessary on how to explore these dynamic bases on large input spaces. The stop condition of the algorithms, based on a full verification of the zero map, requires a complete scan of the learning input space. Partial verifications can compromise the convergence of the algorithms. Investigations on inductive biases adequate to the method are necessary.

Acknowledgments. The author would like to thank Pierre-Jean Laurent from the Laboratoire de Modelisation et Calcul- LMC-IMAG at the Universite Joseph Fourier, Grenoble, France for his contributions concerning the mathematical proofs of the proposed method and Emmanuel Mazer from the Institut National De Recherche en Informatique et en Automatique- INRIA- Rhne Alpes, France for his assistance and helpful contributions to this research.

12   Appendices

12.1   Details of the Fibonacci Sequence Program Synthesis

The translation between bit lists and integers is made using the following routines:

```
(defun 6bitlist2int(l)
  (let ((r 0))
    (loop for i from 0 to 5 do
        (let ((x (nth i l)))
            (if x (setq r (+ r (expt 2 (- 5 i)))))))
      (return-from 6bitlist2int r)))

(defun 6int2bitlist(n)
  (let ((l (make-list 6)))
    (loop for j from 5 downto 0 do
        (if (= (mod n 2) 1) (setf (nth j l) t))
        (setq n (floor n 2)))
      (return-from 6int2bitlist l)))

(defun longint2bitlist(n)
   (let ((l (make-list 64)))
      (loop for j from 63 downto 0 do
          (if (= (mod n 2) 1) (setf (nth j l) t)) (setq n (floor n 2)))
       (return-from longint2bitlist l)))

(defun bitlist2longint(l)
   (let ((r 0))
     (loop for i from 0 to 63 do
         (let ((x (nth i l)))
           (if x (setq r (+ r (expt 2 (- 63 i))))))))
    (return-from bitlist2longint r)))
```

The Fibonacci sequence can be implemented in Lisp as:

```
(defun fibonacci (n)
   (let ((a 0) (b 1))
      (loop do
         (setq b (+ a b)) (setq a (- b a)) (setq n (1- n))
        while (> n 1))
     (return-from fibonacci b)))
```

Calling the synthesis procedure and then testing the generated program:

```
(synthesis (function myfibonacci) 6 64 "fibo6bits.txt")
(defun verificaInt664(fn filename)
    (print "Input a number between 0 e 63 (6 bits):")
    (with-open-file (infile filename)
      (setq s (read infile))
      (loop do
          (setf v (read))
          (if (< v 0) (return-from verificaInt664))
          (print "Function result:")  (print (funcall fn v))
          (setq args (6int2bitlist v))
          (let ((r (make-list (list-length s))))
              (loop for i from 0 to (1- (list-length s)) do
                  (setf (nth i r) (eval (nth i s))))
              (print "Synthesized program result:") (print (bitlist2longint r)))
        while (> v 0))))
(verificaInt664 (function fibonacci) "fibo6bits.txt")
```

References

1. Kotsiantis, S.B.: Supervised machine learning: a review of classification techniques. Informatica **31**, 249–268 (2007)
2. Gulwani, S., Jha, S., Tiwari, A., Venkatesan, R.: Synthesis of Loop-free Programs. SIGPLAN Not. **46**(6), 62–73 (2011). doi:10.1145/1993316.1993506
3. Kitzelmann, E.: Inductive programming: a survey of program synthesis techniques. In: Schmid, U., Kitzelmann, E., Plasmeijer, R. (eds.) AAIP 2009. LNCS, vol. 5812, pp. 50–73. Springer, Heidelberg (2010)
4. Stone, M.H.: The theory of representations of Boolean Algebras. Trans. Am. Math. Soc. **40**, 37–111 (1936)
5. Albarghouthi, A., Gulwani, S., Kincaid, Z.: Recursive program synthesis. In: Sharygina, N., Veith, H. (eds.) CAV 2013. LNCS, vol. 8044, pp. 934–950. Springer, Heidelberg (2013)
6. Kitzelmann, E.: A combined analytical and search-based approach for the inductive synthesis of functional programs. Kunstliche Intelligenz **25**(2), 179–182 (2011)
7. Kitzelmann, E., Schmid, U.: Inductive synthesis of functional programs: an explanation based generalization approach. J. Mach. Learn. Res. **7**, 429–454 (2006)
8. Kitzelmann, E.: Analytical inductive functional programming. In: Hanus, M. (ed.) LOPSTR 2008. LNCS, vol. 5438, pp. 87–102. Springer, Heidelberg (2009)
9. Summers, P.D.: A methodology for LISP program construction from examples. J. ACM **24**(1), 162–175 (1977)
10. Smith, D.R.: The synthesis of LISP programs from examples. A survey. In: Biermann, A.W., Guiho, G., Kodratoff, Y. (eds.) Automatic Program Construction Techniques, pp. 307–324. Macmillan, New York (1984)
11. Sasao, T.: Switching Theory for Logic Synthesis. Springer, Boston (1999). ISBN: 0-7923-8456-3
12. Alur, R., Bodik, R., Juniwal G. et al.: Syntax-guided synthesis, FMCAD, pp. 1–17. IEEE (2013)
13. Tripp, J.L., Gokhal, M.B., Peterson, K.D.: Trident: from high-level language to hardware circuitry. IEEE - Comput. **40**(3), 28–37 (2007). 0018–9162/07
14. Backus, J.: Can programming be liberated from the von Neumann style? a functional style and Its algebra of programs. Commun. ACM **21**(8), 613–641 (1978)

15. Hara, Y., Tomiyama, H.I., Honda, S., Takada, H.: Proposal and quantitative analysis of the CHStone Benchmark program suite for practical C-based High-level synthesis. J. Inf. Process. **17**, 242–254 (2009)
16. Jha, S., Gulwani, S., Seshia, S.A., Tiwari, A.: Oracle-guided Component-based Program Synthesis. In: ICSE (2010)
17. Seshia, S.A.: Sciduction: combining induction, deduction, and structure for verification and synthesis. In: DAC, pp. 356–365 (2012)
18. McCluskey, E.J.: Introduction to the Theory of Switching Circuits. McGrawHill Book Company, New York (1965). Library of Congress Catalog Card Number 65–17394

Program Derivation

Functional Kleene Closures

Nikita Danilenko[✉]

Institut Für Informatik, Christian-Albrechts-Universität Kiel,
Olshausenstraße 40, 24098 Kiel, Germany
nda@informatik.uni-kiel.de

Abstract. We present a derivation of a purely functional version of
Kleene's closure algorithm for Kleene algebras (with tests) that contain
a subset where the closure is already known. In particular, our result is
applicable to the Kleene algebra of square matrices over a given Kleene
algebra. Our approach is based solely on laws imposed on Kleene algebras
and Boolean algebras. We implement our results in the functional programming language Haskell for the case of square matrices and discuss a
general implementation. In this process we incorporate purely algebraic
improvements like the use of commutativity to obtain a concise and opti-
mised functional program. Our overall focus is on a functional program
and the computational structures from which it is composed. Finally, we
discuss our result particularly in light of alternative approaches.

1 Introduction

The Kleene closure is a well established computational paradigm with numerous
applications, e.g. in regular expressions. Also, it is possible to define the Kleene
closure for matrices of Kleene algebras, which provides a unified approach to
additional problems like the all-pair-shortest-path problem in graph theory or
matrix multiplication or even inversion of matrices. Matrix algebras come with
the additional benefit of being able to represent problems from other branches
like reachability in graphs (Boolean matrices), the Dijkstra algorithm (matrices
over the so-called tropical Kleene algebra, cf. Sect. 6) or the CYK algorithm
(matrices over rules). This is to say that using matrices is not a restriction, but
simply a general view for many different problems, while allowing all of the usual
algebraic means associated with matrices.

The vast amount of problems captured by the computational scheme of this
closure has led to a lot of research in this area, which includes several implementations in different programming languages. These are usually given as imperative (pseudo-)code, but there has been little development of a functional program.
While there are several more or less canonical implementations in a functional
programming language, they are usually based upon a translation of a given
algorithm, but not on a purely functional approach. Clearly, there is a difference
between a program in a functional language and a functional program, which
is particularly important when dealing with algorithms. By definition an algorithm has a sequential look-and-feel that allows following a set of instructions

M. Proietti and H. Seki (Eds.): LOPSTR 2014, LNCS 8981, pp. 241–258, 2015.
DOI: 10.1007/978-3-319-17822-6_14

step by step. Such a construct fits well in the context of imperative languages, but may not be suited for a functional definition, since functional programs are usually inherently compositional and not necessarily computed in sequence.

In this paper we generalise an approach taken to compute a specific instance of the Kleene closure in [1] to the general case and present a purely functional program that can be used to compute the said closure. Our functions are prototypic by design, but the modularity of their components can be easily used to improve upon the implementation. This article is structured as follows.

- We recall the necessary preliminaries of a Kleene algebra and provide a definition of the Kleene closure that employs an auxiliary function.
- Using the algebraic reasoning we derive a recursive variant of this function.
- We implement the obtained recursion in Haskell, where we additionally employ Kleene algebra laws to improve performance.
- For comparison we implement the Kleene algorithm in three additional ways and perform tests.

To the best of our knowledge such a derivation has not been done so far. All of the presented code is given in Haskell [9]. In the course of the text we will refer to certain Haskell functions and modules all of which can be found using Hoogle (http://haskell.org/hoogle). A polished version of the code presented in this article is available at https://github.com/nikitaDanilenko/functionalKleene.

2   Algebraic Preliminaries

In the following we will deal with Kleene algebras according to the definition given in [7]. All definitions and consequences in this section are mentioned in the above source and the non-elementary results are proved as well; we include all of these for the sake of completeness only. We begin by the definition of a Kleene algebra, which can be split into two parts – the notion of an idempotent semiring and the concept of the star closure.

Definition 1 (Idempotent Semiring and its Order). *A structure* $(S, +, \cdot, 0, 1)$ *where* $+, \cdot$ *are binary functions is called an* **idempotent semiring** *iff all of the following conditions hold:*

(ISR1) $+$ *is associative and commutative with* 0 *as its neutral element*
(ISR2) $\cdot$ *is associative with* 1 *as its neutral element and distributes over* $+$
(ISR3) 0 *is annihilating (i.e.* $\forall s \in S : 0 \cdot s = 0 = s \cdot 0$*)*
(ISR4) $+$ *is idempotent (i.e.* $\forall s \in S : s + s = s$*)*

The axioms **(ISR1)** *and the first part of* **(ISR2)** *can be read as:* $(S, +, 0)$ *is a commutative monoid and* $(S, \cdot, 1)$ *is a monoid. The explicit multiplication symbol* $\cdot$ *is omitted when there is no risk of ambiguity. We assume that multiplication binds tighter than addition. In an idempotent semiring we define for all* $s, t \in S$:

$$s \leq t :\Longleftrightarrow s + t = t.$$

Then $\leq$ *is an order and* $+, \cdot$ *are monotonic (w.r.t.* $\leq$*) in both components.*

With these preliminaries we are ready to define a Kleene algebra.

Definition 2 (Kleene Algebra). *A structure* $(K, +, \cdot, {}^*, 0, 1)$, *where* $+, \cdot$ *are binary functions and* * *is a unary function, is called* **Kleene algebra** *(KA for short) iff all of the following hold:*

(KA1) $(K, +, \cdot, 0, 1)$ *is an idempotent semiring.*
(KA2) $\forall a \in K : 1 + a \cdot a^* \leq a^* \wedge 1 + a^* \cdot a \leq a^*$.
(KA3) $\forall a, b \in K : (b \cdot a \leq b \rightarrow b \cdot a^* \leq b) \wedge (a \cdot b \leq b \rightarrow a^* \cdot b \leq b)$.

Of the very numerous known properties of Kleene algebras we will use those that we summarise in the following lemma. All of these properties are proved in [7].

Theorem 1 (Properties of * and the Kleene Closure). *Let* $(K, +, \cdot, {}^*, 0, 1)$ *be a Kleene algebra. We define* ${}^+ : K \rightarrow K$, $a \mapsto a \cdot a^*$, *called* **Kleene closure**. *Then the following hold:*

(1) $\forall a \in K : 1 + a \cdot a^* = a^*$.   *(fixpoint)*
(2) $\forall a, b \in K : 1 + a \cdot b = b \rightarrow a^* \leq b$.   *(least fixpoint)*
(3) $\forall a, b \in K : (a + b)^* = a^*(b \cdot a^*)^*$.   *(decomposition)*
(4) $\forall a, b, x \in K : a \cdot x = x \cdot b \rightarrow a^* \cdot x = x \cdot b^*$.   *(star commutativity)*
(5) $\forall a \in K : 1 + a^+ = a^* = (1 + a)^+$.

For the remainder of this section we assume that $(K, +, \cdot, {}^*, 0, 1)$ is a Kleene algebra and $n \in \mathbb{N}_{>0}$. Additionally, we will use the convention that * and ${}^+$ bind tighter than multiplication. For many algebraic structures it is possible to lift the structure to the level of square matrices over the structure, which we denote by $K^{n \times n}$. Kleene algebras are such a structure due to the following result. We write 0_n and 1_n to denote the zero and identity $n \times n$-matrices respectively. In the context of matrices we use "+" and "$\cdot$" to denote matrix addition and multiplication, respectively.

Theorem 2 ($K^{n \times n}$ as a Kleene Algebra). $(K^{n \times n}, +, \cdot, 0_n, 1_n)$ *is an idempotent semiring. Additionally, there is a function* ${}^* : K^{n \times n} \rightarrow K^{n \times n}$ *such that* $(K^{n \times n}, +, \cdot, {}^*, 0_n, 1_n)$ *is a Kleene algebra.*

The first statement of this theorem is simple, but tedious to prove. The second is usually proved by defining an actual mapping ${}^* : K^{n \times n} \rightarrow K^{n \times n}$. Throughout the literature one usually finds two particular definitions, which we mention here for the purpose of comparison. Let $a \in K^{n \times n}$. For simplicity of notation we will index matrices starting from 0 instead of 1.

The first definition states that it is possible to compute matrices $a^{(0)}, \ldots, a^{(n)}$ such that $a^{(0)} := a$ and $a^{(n)} = a^+$, where for all $i, j, k \in \mathbb{N}_{<n}$ one has

$$a_{i,j}^{(k+1)} := a_{i,j}^{(k)} + a_{i,k}^{(k)} \cdot \left(a_{k,k}^{(k)}\right)^* \cdot a_{k,j}^{(k)}. \tag{i}$$

This definition provides two ways of computing a^* – either as $a^* = 1_n + a^+$ or as $a^* = (1_n + a)^+$, by Theorem 1.(5).

The second approach is based upon choosing $l, m \in \mathbb{N}_{>0}$ such that $n = l + m$ and splitting A into submatrices $a = \left(\begin{smallmatrix} p & q \\ r & s \end{smallmatrix} \right)$ where $p \in K^{l \times l}$, $s \in K^{m \times m}$, and q, r have corresponding dimensions. Then one computes $x := (p + q \cdot s^\star \cdot r)^\star$ and sets

$$a^\star := \begin{pmatrix} x & x \cdot q \cdot s^\star \\ s^\star \cdot r \cdot x & s^\star + s^\star \cdot r \cdot x \cdot q \cdot s^\star \end{pmatrix}. \tag{ii}$$

Since all of the matrices used in this definition have strictly smaller dimensions than a these computations can be used recursively for the computation of $a^\star$.

Both definitions are elegant in different ways – the first one is easy to translate in graph-theoretic terms and easy to implement in an imperative language, while the second one has a foundation in automata theory. Still, both definitions are rather algorithms (the second one can be even considered a non-deterministic one, since the actual decomposition does not matter), since they describe a sequence of steps that lead to $a^\star$. From a complexity point of view the second definition provides an additional challenge, since it contains matrix multiplication. It is not apparent at the first glance, what the exact complexity is, while the first one is clearly cubic in n.

3   A Functional Approach

In this section we will develop a functional definition of the Kleene closure. Our approach is a direct generalisation of the methods used in [1]. For the sake of simplicity, but also for that of generality we need a slightly more sophisticated structure than a Kleene algebra, namely a so-called Kleene algebra with tests. Tests are elements of the Kleene algebra that "behave" like elements of a Boolean algebra. We use a similar notation and definition as in [8].

Definition 3 (Kleene algebra with tests). *A Kleene algebra with tests (KAT for short) is a structure $(K, B, +, \cdot, {}^\star, 0, 1)$ such that all of the following conditions hold:*

(KAT1) $(K, +, \cdot, {}^\star, 0, 1)$ *is a Kleene algebra.*
(KAT2) $B \subseteq K$.
(KAT3) $(B, +_B, \cdot_B, 0, 1)$ *is a Boolean algebra (i.e. a distributive and complementary lattice), where $+_B, \cdot_B$ are the operations of K restricted to B.*

We suppress the explicit negation operation for simplicity, since it is easily recovered from the complementarity that provides complements and distributivity, which can be used to show that complements are unique. Note that for any Kleene algebra there is a trivial set of tests namely $B_0 = \{0, 1\}$. Abstract tests provide an elegant means to express Boolean conditions without leaving the Kleene algebra. This allows logic reasoning about Kleene algebra elements in terms of Kleene algebra elements themselves. For example, an *if-then-else* assignment can be written as

$$ite(b, x, y) := b \cdot x + \bar{b} \cdot y,$$

where $b \in \{0,1\}$, $x, y \in K$ and $\bar{b}$ denotes the negation of b. This construct is similar to the Shannon decomposition of Boolean functions, but is more general since the values x, y can be arbitrary Kleene algebra elements.

All tests $b \in B$ satisfy $b \leq 1$, because in Boolean algebras 1 is the largest element and thus $b + 1 = 1$, which yields $b \leq 1$ in K.

From now on we assume that $(K, B, +, \cdot, {}^*, 0, 1)$ is a Kleene algebra with tests. We then consider the following function.

$$\tau : K \times B \to K, \quad (a, b) \mapsto a \cdot (b \cdot a)^*.$$

This function is a translation of the function in relational terms from [1]. Just as its relational version, τ has the following properties for all $a \in K$:

$$\begin{aligned}
\tau(a, 0) &= a \cdot (0 \cdot a)^* = a \cdot (0^*) = a \cdot 1 = a, \\
\tau(a, 1) &= a \cdot (1 \cdot a)^* = aa^* = a^+.
\end{aligned} \tag{iii}$$

To deal with tests between 0 and 1 we take a similar approach as in [1] and study the recursion properties of τ. We observe that for all $a \in K$ and all $b_1, b_2 \in B$ we get the following chain of equations:

$$\begin{aligned}
\tau(a, b_1 + b_2) &= a \cdot ((b_1 + b_2) \cdot a)^* && \text{definition of } \tau \\
&= a\,(b_1 a + b_2 a)^* && \text{distributivity} \\
&= a\left((b_1 a)^* \left((b_2 a)\,(b_1 a)^*\right)^*\right) && \text{by Theorem 1.(3)} \\
&= \left(a\,(b_1 a)^*\right)\left(b_2\left(a\,(b_1 a)^*\right)\right)^* && \text{associativity} \\
&= \tau(a, b_1) \cdot (b_2 \cdot \tau(a, b_1))^* && \text{definition of } \tau \\
&= \tau(\tau(a, b_1), b_2) && \text{definition of } \tau.
\end{aligned}$$

In summary we get for all $a \in K$ and all $b_1, b_2 \in B$

$$\tau(a, b_1 + b_2) = \tau(a, b_1)\,(b_2 \tau(a, b_1))^* = \tau(\tau(a, b_1), b_2). \tag{iv}$$

This property is a generalisation of the one derived in [1], both in terms of the decomposition of the underlying set as well as the algebraic structure at hand, but the steps on the way are exactly those used in the above source.

Now consider the application of the above recursion to a finite subset $B' \subseteq B$, such that[1] $B' = \{b_0, \ldots, b_{m-1}\}$ for some $m \in \mathbb{N}$ and $\sum_{i=0}^{m-1} b_i = 1$. Then we can use the above formula and compute $a_0 := a$ and $a_{i+1} := \tau(a_i, b_i)$ for all $i \in \{1, \ldots, m\}$ to obtain $a_m = a^+$. Note that this is very similar to the construction given in the introduction only that it does not depend on matrices.

This computational paradigm is captured by a left-fold. Note that it does not depend on a particular order of traversal of the partition elements, but is intrinsic to the actual computation. In the above example we split the sum into its first summand and the rest, while we could have taken the last element and the

---

[1] Such a set where $b_i < 1$ for all $i \in \mathbb{N}_{<m}$ does not necessarily exist. However, if B is finite, we can simply choose the atoms of B (i.e. the upper neighbours of 0) as B'.

corresponding rest as well without changing the result. Left-folds are favoured over right-folds in strict languages since they are usually more efficient (tail-call). In a lazy setting tail-calls can become more complex, because the accumulation parameter is not evaluated until needed, while its construction grows increasingly more complex. Since we are looking for a solution in a lazy functional language, we may need to transform the above recursion into a (non-generic) right-fold.

Our aim is to use the function τ for the computation of the Kleene closure of a given algebra K. To do that we will determine the Kleene closure for a specific subset of K, namely $\{ba \mid b \in B, a \in K\}$ and then apply τ in a recursive fashion as described above, assuming that there is a decomposition of 1 as a sum of finitely many tests. By Eq. (iv) the knowledge of * on the above set is enough to compute the Kleene closure of any element of the Kleene algebra.

Let us summarise the result of this section in the following theorem.

Theorem 3 (Recursive Computation of τ). *Let $(K, B, +, \cdot, ^*, 0, 1)$ be a KAT and $\tau : K \times B \to K$, $(a, b) \mapsto a(ba)^*$. Then the following hold:*

(1) For all $a \in K$ and $b, c \in B$ we get $\tau(a, b + c) = \tau(a, b)(c\tau(a, b))^*$.

(2) For all $m \in \mathbb{N}$ and $b \in B^m$ such that $\sum_{i=0}^{m-1} b_i = 1$, all $a \in K$, all $i \in \mathbb{N}_{<m}$ setting $n_i := \sum_{j=0, j \neq i}^{m-1} b_j$ we find that $a^+ = \tau(a, 1) = \tau(a, n_i)(b_i \tau(a, n_i))^*$.

Proof. The first claim is just a rephrasal of Eq. (iv). For the second let $m \in \mathbb{N}$, $b \in B^m$, $a \in K$, $i \in \mathbb{N}_{<m}$ and n_i as required. Then $1 = n_i + b_i$ and thus

$$a^+ \overset{\text{Eq. (iii)}}{=} \tau(a, 1) = \tau(a, n_i + b_i) \overset{(1)}{=} \tau(a, n_i)(b_i \tau(a, n_i))^* . \qquad \square$$

In statement (2) of the previous theorem the b_i is removed from the sum for clarity, but the theorem obviously holds without this removal as well, because in idempotent semirings (e.g. the Boolean algebra B) addition is idempotent. As for applications of the last equality, the motivation is that a good choice of b allows a simple computation of $(b_i \tau(a, n_i))^*$, which can then in turn be applied to compute a^+ or a^* by iteration.

4   Application to Square Matrices

In this section we apply the technique from the previous section to the Kleene algebra of square matrices with entries from a Kleene algebra. We use the Boolean algebra of partial identities as test, which we elaborate shortly. For the remainder of this section let K be a Kleene algebra and $n \in \mathbb{N}$. Additionally, we do not differentiate between $K^{1 \times 1}$ and K, just as we consider the sets K^n and $K^{1 \times n}$ to be the same. To be perfectly accurate we should use isomorphisms between these sets, but we omit these to avoid unnecessary clutter.

To avoid confusion, we will index constants by their algebra throughout this section and make ample use of brackets to avoid indexing the star operation. Recall that 1_n is simply the identity matrix and 0_n is the zero matrix. For every $a \in K^{n \times n}$ we use a_j to address the j-th row of a and $a_{j,k}$ to address the component at the j-th row in the k-th column.

We define partial identities by using standard unit vectors.

Definition 4 (Standard Unit Vectors and Matrices, Partial Identity).
*We define for all $i \in \mathbb{N}_{<n}$ the **i-th standard unit row vector** $e_i \in K^n$ as*

$$e_i : \mathbb{N}_{<n} \to K, \ p \mapsto \delta_{i,p},$$

*where δ is the Kronecker-delta. Clearly, $e_i \cdot e_j^\top = \delta_{i,j}$ for all $i,j \in \mathbb{N}_{<n}$. Additionally for all $i,j \in \mathbb{N}_{<n}$ we define the **(i,j)-th standard unit matrix** as $e(i,j) := e_i^\top \cdot e_j$. For every $S \subseteq \mathbb{N}_{<n}$ the **partial identity on S** is defined as*

$$I_S := \sum_{i \in S} e(i,i)$$

and $\mathsf{PI}(K, n) := \{ I_T | T \subseteq \mathbb{N}_{<n} \}$.

As mentioned above, we use the set of all partial identities $\mathsf{PI}(K, n)$ as tests, which requires this set with the restricted operations of $K^{n \times n}$ to be a Boolean algebra. This is indeed true.

Lemma 1 (Partial Identities are a Boolean Algebra). *Let $B := \mathsf{PI}(K, n)$. Then the structure $(B, +_B, \cdot_B, 0_n, 1_n)$ is a Boolean algebra, where $+_B, \cdot_B$ are the operations of $K^{n \times n}$ restricted to B.*

Proof. We provide only an outline of the proof. Consider the function

$$f : (2^{\mathbb{N}_{<n}}, \cup, \cap, \emptyset, \mathbb{N}_{<n}) \to (B, +_B, \cdot_B, 0_n, 1_n), \ S \mapsto I_S.$$

Clearly, f is bijective, $f(\emptyset) = 0_n$ and $f(\mathbb{N}_{<n}) = 1_n$. A simple, but slightly lengthy, computation yields that f distributes over addition and multiplication as well. Thus f is a constant preserving lattice isomorphism. Since its domain is a Boolean algebra, so is its image, which is its range, because f is bijective. $\square$

To apply the function τ from the previous section, we need a finite decomposition of 1_n into tests. We choose a decomposition in standard unit matrices by setting $b_i := e(i,i)$ for all $i \in \mathbb{N}_{<n}$. As stated just after Theorem 3 this choice should provide a simple means to compute $(b_i a)^*$ for all $i \in \mathbb{N}_{<n}$ and all $a \in K^{n \times n}$. To see that this is the case, we need two auxiliary lemmas.

Lemma 2 (Homothetic Injection). *Let $\varphi : K \to K^{n \times n}$, $c \mapsto c \bullet 1_n$, where $\bullet : K \times K^{n \times n} \to K^{n \times n}$ is the scalar multiplication of a matrix. Then φ is a Kleene algebra homomorphism.*

Proof. Clearly, $\varphi(0_K) = 0_n$ and $\varphi(1_K) = 1_n$ by the very definition of scalar multiplication. The additivity and multiplicativity of φ are immediate consequences of this definition as well. The only difficulty is the star operation. Let $a \in K$. Then we have $1_n + \varphi(a) \cdot \varphi(a^*) = \varphi(1_K) + \varphi(a \cdot a^*) = \varphi(1_K + aa^*) = \varphi(a^*)$ by

Theorem 1.(1). By Theorem 1.(2) this provides $\varphi(a)^* \leq \varphi(a^*)$. Thus $\varphi(a)^*$ has non-zero entries at most along its diagonal. Let $i \in \mathbb{N}_{<n}$. Then we have:

$$(\varphi(a)^*)_{i,i} = (1_n + \varphi(a)\varphi(a)^*)_{i,i} = 1_K + ((a \bullet 1_n)\,\varphi(a)^*)_{i,i}$$
$$= 1_K + (a \bullet \varphi(a)^*)_{i,i} = 1_K + a\,(\varphi(a)^*)_{i,i}\,,$$

which again by Theorem 1.(2) yields $\varphi(a^*)_{i,i} = a^* \leq (\varphi(a)^*)_{i,i}$. This results in $\varphi(a^*) \leq \varphi(a)^*$ and since $\leq$ is an order, we finally conclude $\varphi(a^*) = \varphi(a)^*$.   □

Lemma 3. *Let* $c, v \in K$ *such that* $v^2 = cv$. *Then* $v^* = 1 + c^*v$.

Proof. Set $a := c$, $b := v$ and $x := v$. Then $ax = cv = vv = xb$, hence by Theorem 1.(4) we get $c^*v = a^*x = xb^* = vv^*$, which yields $v^* = 1 + vv^* = 1 + c^*v$.   □

With this lemma we can compute $(e(i,i)a)^*$ without using the star in $K^{n \times n}$.

Theorem 4 (Kleene Closure of Matrices). *Let* $a \in K^{n \times n}$, $b \in \mathsf{PI}(K, n)$ *and* $i \in \mathbb{N}_{<n}$. *Then we have:*

(1) For all $x \in K^{n \times n}$ *and* $k \in \mathbb{N}_{<n}$ *we have* $(e(k,k)x)^* = 1_n + ((x_{k,k})^*) \bullet e(k,k)x$.
(2) $\tau(a, b + e(i,i)) = \tau(a, b) + \tau(a, b)\big(((\tau(a,b)_{i,i})^*) \bullet e(i,i)\big)\tau(a,b)$.
(3) For every $j \in \mathbb{N}_{<n}$ *we additionally get*

$$\tau(a, b + e(i,i))_j = \tau(a,b)_j +' \big(\tau(a,b)_{j,i} \cdot (\tau(a,b)_{i,i})^*\big) * \tau(a,b)_i\,,$$

where $+'$ is the addition and $*$ is the scalar multiplication of vectors.

Proof. (1) Let $x \in K^{n \times n}$, $k \in \mathbb{N}_{<n}$ and φ as in Lemma 2. Then we get:

$$e(k,k)x \cdot e(k,k)x = e_k^\top e_k x e_k^\top e_k x = e_k^\top (x_{k,k}) e_k x = x_{k,k} \bullet \big(e_k^\top e_k x\big)$$
$$= \varphi(x_{k,k}) \cdot e(k,k)x\,.$$

By Lemmas 3 and 2 this yields that

$$(e(k,k)x)^* = 1_n + \varphi(x_{k,k})^* e(k,k)x = 1_n + \varphi((x_{k,k})^*)e(k,k)x$$
$$= 1_n + (x_{k,k})^* \bullet e(k,k)x\,.$$

(2) We calculate as follows:

$$\tau(a, b + e(i,i)) = \tau(a,b)\,(e(i,i)\tau(a,b))^* \qquad\qquad \text{by Theorem 3}$$
$$= \tau(a,b)\,\big(1_n + (\tau(a,b)_{i,i})^* \bullet e(i,i)\tau(a,b)\big) \qquad\quad \text{by (1)}$$
$$= \tau(a,b) + \tau(a,b)\Big((\tau(a,b)_{i,i})^* \bullet e(i,i)\Big)\tau(a,b)\,.$$

(3) First of all for every $d \in K^{n \times n}$, $c \in K$ and $j \in \mathbb{N}_{<n}$ the following holds

$$(d\,(c \bullet e(i,i)d))_j = e_j d\,(c \bullet e_i^\top e_i d) = e_j d e_i^\top\,(c \bullet e_i d) = (d_{j,i}c) * d_i\,. \qquad (!)$$

Let $j \in \mathbb{N}_{<n}$. Then the following holds:

$$\tau(a, b + e(i,i))_j = (\tau(a,b) + \tau(a,b)\,((\tau(a,b)_{i,i})^* \bullet e(i,i))\,\tau(a,b))_j \qquad \text{by (2)}$$
$$= \tau(a,b)_j +' (\tau(a,b)\,((\tau(a,b)_{i,i})^* \bullet e(i,i))\,\tau(a,b))_j$$
$$= \tau(a,b)_j +' (\tau(a,b)_{j,i}(\tau(a,b)_{i,i})^*) * \tau(a,b)_i\,. \qquad\qquad \text{by (!)}$$

□

5   A Functional Implementation

In this section we use the approach of the previous sections to obtain a functional implementation for Kleene closure of a square matrix. We discuss this restriction in a moment. The first step in the implementation is to encode the required algebraic structures. One particularly simple way to do that is to use type classes.

class *IdempotentSemiring* σ **where**
$(\oplus), (\odot)$ $\quad :: \sigma \to \sigma \to \sigma$
zero, one $\quad :: \sigma$
isZero, isOne $:: \sigma \to Bool$
class *IdempotentSemiring* $\kappa \Rightarrow$ *KleeneAlgebra* κ **where**
star $:: \kappa \to \kappa$

Additionally, we require instances of these type classes to satisfy the corresponding algebraic laws. Such an approach requires a user to check the necessary conditions, which may or may not be neglected in an application[2]. The predicates *isZero,isOne* are not explicit parts of the algebraic definition. In a theoretical context we can always compare values for equality, while in practice the equality of two objects may be undecidable. We require equality checks for constants only, since these are sufficient for a simple optimisation (otherwise one can require *KleeneAlgebra* to be a subclass of *Eq* thus allowing arbitrary equality checks).

For the most general case we define a type class *KAT* for KAT's as follows:

class *IdempotentSemiring* $\kappa \Rightarrow$ *KAT* κ **where**
isSimple $\quad :: \kappa \to Bool$
compute $\quad :: \kappa \to \kappa$
sparseTests $:: [\kappa]$

The intended semantics are that *sparseTests* is a list of tests such that their sum is *one* and *isSimple* is a predicate such that the Kleene closure of those $a :: \kappa$ with *isSimple* $a \equiv True$ can be computed with the function *compute*. Assuming *isSimple* $(t \odot a) \equiv True$ for all t in *sparseTests* we can use Theorem 3 as follows:

$katStar :: KAT\ \kappa \Rightarrow \kappa \to \kappa$
$katStar\ a \mid isSimple\ a = compute\ a$
$\quad\quad\quad\quad \mid otherwise\ = one \oplus katPlus\ a$

$katPlus :: KAT\ \kappa \Rightarrow \kappa \to \kappa$
$katPlus\ a = tau\ a\ sparseTests$

$tau :: KAT\ \kappa \Rightarrow \kappa \to [\kappa] \to \kappa$
$tau\ a\ [\,] \quad\quad = a$
$tau\ a\ (t : ts) = x \odot katStar\ (t \odot x)\ \textbf{where}\ x = tau\ a\ ts$

---

[2] To ensure that the requirements are met, one can use tools like Coq (cf. [2]) to make certain functions applicable only once the preconditions are checked (i.e. proved). However, in Haskell one usually requires certain laws implicitly (e.g. for *Monad* or *Functor*), which is why we do not explore the mentioned approach further.

However, this approach comes with proper restrictions. Most importantly, a condition is required that guarantees that after a finite numer of multiplications with tests any Kleene algebra element is "simple" in the sense that *isSimple* applied to this element yields *True*. Without such a condition, the above computation may not terminate.

Also, to use the above implementation for square matrices one has to provide a semiring instance for said matrices, which requires taking the matrix sizes into account, causing additional overhead (e.g. checking sizes by hand or encoding the sizes in the matrix type). Without this restriction, matrices can have different sizes and it is non-trivial to define a multiplication $\odot$ on the set of all square matrices $\bigcup\{K^{n \times n} | n \in \mathbb{N}_{>0}\}$ and a corresponding unit $\mathbb{1}$. For instance, there is no multiplication satisfying both of the following conditions for all $n \in \mathbb{N}_{>0}$:

(1) The restriction $\odot|_{(K^{n \times n})^2}$ is the usual matrix multiplication.
(2) Every $a \in K^{n \times n}$ that is invertible with respect to the usual matrix multiplication, is invertible with respect to $\odot$.

This is simply due to the fact that for all $n, m \in \mathbb{N}_{>0}$ such that $n \neq m$ one would obtain $\mathbb{1}_n = \mathbb{1}_n \cdot \mathbb{1}_n = \mathbb{1}_n \odot \mathbb{1}_n = \mathbb{1} = \mathbb{1}_m \odot \mathbb{1}_m = \mathbb{1}_m \cdot \mathbb{1}_m = \mathbb{1}_m$, which is false. Since the set of invertible matrices is non-trivial (for $n > 1$), the above condition is not just a corner case. In the implementation in [4] the invertibility condition is omitted, which allows an elegant implementation in which only certain diagonal matrices invertible[3]. This is achieved by another representation of scalar matrices (i.e. $c \bullet \mathbb{1}$ for every $c \in K$), which is independent of the matrix size. However, we choose a simpler representation below to avoid matrices of no fixed size.

For the sake of demonstration we use the specialised Theorem 4 for a prototypical implementation without a semiring instance for matrices, where said theorem provides us with a possible implementation assuming certain existing functions. Both equations given in the theorem require access to single values of a matrix, i.e. $A_{i,j}$. Thus if we want to use these equations for an implementation we need some representation of matrices that allows such queries. Also, the first equation uses matrix multiplication while the second one does not, but instead relies on the concept of rows. There are numerous ways for a representation of a matrix, particularly if it is sparse (i.e. small percentage of non-zero values) and different representations provide different features and caveats. For now we choose the adjacency list model as a middle ground between efficiency and simplicity. To that end we use the following notations.

```
type Row α = [(Int, α)]
type Mat α = [Row α]

(!) :: IdempotentSemiring σ ⇒ Row σ → Int → σ
[]              ! _ = zero
((i, v) : ivs) ! k  | i ≡ k     = v
                    | i < k     = ivs ! k
                    | otherwise = zero
```

---

[3] $\left(\begin{smallmatrix} 0 & 1 \\ 1 & 0 \end{smallmatrix}\right)$ is invertible over every semiring, thus omitting the invertibility is a restriction.

Additionally we assume and maintain the conditions that *Rows* are sorted in ascending order of their first components, that the second components are non-zero and that an $n \times n$-matrix A is represented by a list of rows a such that $A_i = a \,!!\, i$ and $A_{i,j} = (a \,!!\, i) \,!\, j$ for all $i, j \in \mathbb{N}_{<n}$, where $(!!)$ is the built-in Haskell function for accessing indices of a list. This representation is similar to the one in [1,3] and is provided for the completeness of the implementation.

To apply Theorem 4 we need a scalar multiplication and addition of rows. A straightforward implementation of the former is obtained by multiplying every second component of a row with a given scalar and tidying up the result to remove possible zeroes[4]. There is room for some canonic improvement, since scalar multiplication in case of $0, 1$ is particularly simple. To that end we can simply define the actual multiplication as follows.

```
(*) :: IdempotentSemiring σ ⇒ σ → Row σ → Row σ
s * row | isZero s  = []
        | isOne s   = row
        | otherwise = filter (¬ ∘ isZero ∘ snd) (map (λ(i, v) → (i, s ⊙ v)) row)
```

For all $x, y \in K$ with $x+y = 0$ we get $x = x+0 = x+(x+y) = (x+x)+y = x+y = 0$, which results in $y = 0 + y = x + y = 0$ and thus non-zero values are closed under addition. That is to say that in an idempotent semiring we don't need the tidying step $filter\ (\neg \circ isZero \circ snd)$ when adding vectors. Our precondition that the first components of the rows are increasingly sorted allows implementing the addition of rows in terms of a straightforward merging strategy.

```
(+') :: IdempotentSemiring σ ⇒ Row σ → Row σ → Row σ
[]                        +' y                 = y
x                         +' []                = x
x@((i, v) : ivs) +' y@((j, w) : jws) | i ≡ j   = (i, v ⊎ w) : (ivs +' jws)
                                     | i < j   = (i, v)     : (ivs +' y)
                                     | otherwise = (j, w)   : (x   +' jws)
```

The function τ which we want to use for the computation of the Kleene closure depends on a finite decomposition of 1 into tests. In our approach in the previous section we used the tests $\{e(i,i) \mid i \in \mathbb{N}_{<n}\}$. Every such test and every sum of such tests is uniquely determined by a subset of $\mathbb{N}_{<n}$, which we have established in the proof of Lemma 1. Thus instead of successively computing b_i and n_i as in Theorem 3 we can use the decomposition of *sets* in $(\mathbb{N}_{<n} \setminus \{i\}) \cup \{i\}$ instead of the corresponding test. To gather the indices of a matrix we zip the matrix with the list of natural numbers and ignoring the matrix values.

```
spine :: [α] → [Int]
spine = zipWith const [0..]
```

The Kleene closure of a is simply $\tau(a, f(\mathbb{N}_{<n}))$ for every $a \in K^{n \times n}$, where f is the isomorphism between tests and sets from Lemma 1. Disregarding the f and using lists to represent sets we can implement this in Haskell as follows.

---

[4] In Kleene algebras there can be zero divisors, i.e. elements $x, y \in K$ such that $x \neq 0$ and $y \neq 0$, but $xy = 0$. This is why we need to filter the zero values.

$kleeneClosure :: KleeneAlgebra \; \kappa \Rightarrow Mat \; \kappa \rightarrow Mat \; \kappa$
$kleeneClosure \; a = tau \; a \; (spine \; a)$

The function τ can then be defined in an inductive way. First the base case.

$tauMatrices :: KleeneAlgebra \; \kappa \Rightarrow Mat \; \kappa \rightarrow [Int] \rightarrow Mat \; \kappa$
$tauMatrices \; a \; [\,] = a$

This is simply a quotation of a property of τ. Finally, $\tau(a, f(\{k\} \cup S))$ is a matrix whose rows are given by Theorem 4.(3). We can then write

$tauMatrices \; a \; (k : s) = newMat \; k \; (tauMatrices \; a \; s)$

and thus delay the actual computation in the auxiliary function $newMat$ that captures the computation scheme of the above proposition.

$newMat :: KleeneAlgebra \; \kappa \Rightarrow Int \rightarrow Mat \; \kappa \rightarrow Mat \; \kappa$
$newMat \; i \; a = map \; (\lambda a_j \rightarrow (a_j \,!\, i) \odot star \; (a_i \,!\, i) * a_i \,+' \, a_j) \; a$ **where** $a_i = a \,!!\, i$

In essence this definition is the implementation of the equation from Theorem 4.(3). The only difference is that we swapped the arguments of $(+')$, which is a valid transformation, since $+$ is commutative, which makes $(+')$ commutative as well. Experiments have shown that this simple algebraic rule improves the running times by a factor between 1.5 and 2. This concludes the implementation.

We observe that the full definition of tau is

$tauMatrices \; a \; [\,] \quad\;\; = a$
$tauMatrices \; a \; (k : s) = newMat \; k \; (tauMatrices \; a \; s)$

which yields $tauMatrices \; a \equiv foldr \; newMat \; a$ by the universal property of $foldr$ (cf. [5]) that in turn can be η-reduced to $tauMatrices \equiv foldr \; newMat$.

Note that both versions, the generic and the specialised one, require a list of tests they traverse (where in the latter case plain integers are interpreted as tests). In both cases this list is traversed left-to-right and in the specialised case the resulting recursion is a proper right-fold structure. However, the order of the tests themselves is actually an implicit parameter. In the specialised case we used the natural order $[e(0,0), \ldots, e(n-1, n-1)]$ for simplicity, but any other order still yields the same result. The freedom of choice in this parameter is similar to the order choice for the Gaussian elimination or the computation of determinants[5]. In both cases any fixed order may be not the best suited one for all matrices. For instance it is simpler to eliminate a variable x_i if its row contains more zeroes than the one of x_0 and since the variable order does not matter, one can simply choose an order where x_i is eliminated before x_0. In case of reachability-based applications the breadth-first or depth-first order of indices (when interpreted as vertices of a graph) represent viable alternatives, too.

---

[5] Gaussian elimination is indeed computationally similar to the Kleene closure.

Also, lists may not be the best suited data structure for storing tests. Instead, it may be more convenient to compute the necessary tests in a tree-like structure and then to pass this structure to the algorithm, instead of a list. This tree can then be either transformed in a list or traversed directly, depending on the concrete implementation. To allow this structure, type classes can be used to specify the necessary operations, so that the resulting function has the type $tau :: (KleeneAlgebra\ \kappa, Structure\ \sigma) \Rightarrow \kappa \to \sigma\ \kappa \to \kappa$ where $Structure$ is the type class abstraction of said data structure that contains the tests.

6   Alternative Implementations and Comparison

In this section we discuss implementations of the Kleene closure function and test their complexities in terms of running times and space consumptions.

6.1   Using Arrays

As we have stated before, Eq. (i) provides an out-of-the-box algorithm in imperative languages using arrays. Haskell provides a number of different arrays (mutable and immutable) with fast access to their indices. For the sake of simplicity we used the most basic arrays that allow simple and pure code. The downside of these arrays is that modification is very costly. Fortunately, in our application we don't need to modify arrays but only build new ones. Also, by Eq. (i) we don't require n different arrays, but only two different ones, since $a^{(k+1)}$ is constructed from $a^{(k)}$ alone. This is to say that while we compute n different arrays, only two are kept in memory at any given time, namely the k-th and $1 + k$-th ones. This is similar to in-situ updates, because a complete array is discarded and replaced with new one.

```
newtype ArrayMat α = ArrMat { unArrMat :: Array (Int, Int) α}
kleeneClosureArray :: KleeneAlgebra κ ⇒ ArrayMat κ → ArrayMat κ
kleeneClosureArray (ArrMat a₀) = ArrMat (foldl newArray a₀ [0 .. n]) where
    newArray arr k    = listArray bnds (map (newValue arr k) positions)
    bnds              = bounds a
    positions         = range bnds
    n                 = snd (snd bnds)
    newValue a k (i,j) = ((a ! (i, k)) ⊙ star (a ! (k, k)) ⊙ (a ! (k, j))) ⊕ a ! (i, j)
```

Here (!) is the array query function. The local function $newValue$ produces the value $a_{i,j}^{(k+1)}$, $newArray$ maps this producer over all index pairs (i, j) and $foldl$ repeats this procedure for all k and basically acts as a `for`-loop.

6.2   Another List Version

Another version comes to mind by simply observing that Eq. (i) contains the index j at the same relative position. This allows the following rephrasal:

$$\forall i, k \in \mathbb{N}_{<n} : a_i^{(k+1)} = a_i^{(k)} +' \left(a_{i,k}^{(k)} \cdot \left(a_{k,k}^{(k)} \right)^* \right) * a_k^{(k)}. \tag{v}$$

This is strikingly similar to our recursion for τ. The essential difference is that the values of k are traversed in another direction and it is not quite obvious why this is a valid transformation. The specification of Eq. (v) can be used to obtain a third implementation, where $newMat$ is the function from Sect. 5.

$$kleeneClosureLeft :: KleeneAlgebra\ \kappa \Rightarrow Mat\ \kappa \rightarrow Mat\ \kappa$$
$$kleeneClosureLeft\ a = foldl\ (flip\ newMat)\ a\ (shape\ a)$$

6.3   Blockwise Implementation

The more recent work of Dolan [4] uses (a flipped version of) Eq. (ii) for an implementation. The complete implementation is presented in the above paper and we used the code by the author with only very minor variations. Matrices are represented by the data type

data $Matrix\ a = Matrix\ [[a]]\ |\ Scalar\ a$

and are either a list of rows or, in case the matrix is $c \bullet 1_n$ for some $c \in K$ by c alone. Every row contains all values at all positions. Avoiding unnecessary zeroes is not as simple in this case, because splitting matrices into blocks either requires reindexing the remaining block or removing the last element instead of the first as in the implementation in [4]. Both operations come with additional (constant) complexity, while additional zeroes require more space. The closure function on $Matrix$ is

$$kleeneClosureBlockwise :: KleeneAlgebra\ \kappa \Rightarrow Matrix\ \kappa \rightarrow Matrix\ \kappa$$
$$kleeneClosureBlockwise\ a = a \odot closure\ a$$

where $closure$ is the function from [4], which for a matrix a computes $a^\star$ and $(\odot)$ is the matrix multiplication function provided by the Kleene algebra instance for matrices, which is also given in Dolan's paper.

6.4   A Note on Complexity

Assuming full evaluation, our closure function has a cubic complexity in the matrix dimension. Suppose that said dimension is n. To compute $kleeneClosure$ of an $n \times n$ matrix, the function tau is called $n+1$ times. Every call of tau results in a call of $newMat$, which is essentially a map function over a list of length n. The function that is passed to map is linear in the size of the list element – the query function $(!)$ is linear in the size of the list element, as is $(\ast)$. The addition $(+')$ is linear in the sum of the sizes of both its arguments, but since both sizes are at most n, it is also linear in n. Thus $newMat$ is quadratic in n, which results in the cubic complexity of tau. This estimate is based upon the assumption that $(\odot),(\oplus)$ and $star$ of the underlying Kleene algebra are constant time functions. However, the constants for these functions may differ significantly.

The alternative list implementation *kleeneClosureLeft* is also cubic, because it is simply based on another traversal, but all the employed functions are the same as for *kleeneClosure*. The array based function *kleeneClosureArray* is canonically cubic in the matrix dimension, because n different arrays of size n^2 are constructed. Finally, in [4] the author calculates the complexity of his closure function to be cubic in the dimension as well.

That said, complexity is a more delicate matter in Haskell than a simple assumption about "complete evaluation". More precisely, one needs to carefully consider those parts of the computation, which are necessary for a result. These parts, however, may depend on further function calls, i.e. f (*kleeneClosure a*) may behave differently than *kleeneClosure a*, depending on the actual f. A computational model for these considerations is presented in [12]. Still, in our case the greatly simplified reasoning above is sufficient for an superficial estimate, which essentially puts all implemented versions in the same complexity class.

6.5 Comparison

We have implemented all closure functions in essentially the presented way. Additionally, we implemented a random matrix generation based upon shuffling. Given a density $d \in [0, 1]$ and a size n we compute the number of non-zero positions as $p - \lfloor d \cdot n^2 \rfloor$. Then the first p positions in the matrix are filled and then the matrix is shuffled. This technique is known to be uniformly distributed, is already implemented in Haskell and it depends only on a single random generator, which can be created using a single *Int*. Additionally, random generators in Haskell generate pseudo-random numbers due to referential transparency, i.e. taking a random number from a fixed random generator will always produce the same result. This makes testing of randomly generated data easily repeatable.

Table 1. Evaluation in the tropical Kleene algebra[a].

d \ n		100					250					500					750					1000			
		a	b*	b	l	r	a	b*	b	l	r	a	b*	b	l	r	a	b*	b	l	r	a	b*	b	l	r
	avg	5	5	5	1	1	22	43	43	1	1	129	–	–	4	4	–	–	–	8	8	–	–	–	14	13
0.001	max	19	17	16	1	1	181	242	251	2	2	1460	–	–	9	6	–	–	–	15	13	–	–	–	34	23
	sec	0.2	0.3	0.2	0.1	0.1	3.9	7.0	5.2	0.1	0.1	39.3	–	–	0.6	0.6	–	–	–	1.6	1.6	–	–	–	3.3	3.3
	avg	5	5	5	1	1	22	29	39	5	4	139	–	–	4	5	–	–	–	2	2	–	–	–	2	1
0.01	max	19	17	16	1	1	178	220	225	10	10	1572	–	–	28	24	–	–	–	58	44	–	–	–	111	90
	sec	0.2	0.3	0.3	0.1	0.1	5.3	5.9	7.0	1.4	1.1	63.1	–	–	36.3	30.8	–	–	–	150.5	124.2	–	–	–	408.7	351.3
	avg	5	5	5	1	1	23	42	42	4	3	139	–	–	6	5	–	–	–	4	1	–	–	–	1	1
0.025	max	18	16	17	1	1	186	241	229	7	5	1574	–	–	30	25	–	–	–	70	63	–	–	–	131	117
	sec	0.2	0.3	0.3	0.1	0.1	6.7	6.4	7.4	4.5	3.5	70.4	–	–	50.3	44.2	–	–	–	192.7	171.8	–	–	–	490.3	438.5
	avg	5	5	5	1	1	22	40	42	4	3	–	–	–	1	6	–	–	–	2	4	–	–	–	1	2
0.05	max	20	16	16	1	1	181	226	241	8	7	–	–	–	33	30	–	–	–	77	71	–	–	–	135	127
	sec	0.3	0.3	0.3	0.1	0.1	7.2	6.5	8.1	5.7	4.8	–	–	–	56.5	50.7	–	–	–	207.6	187.4	–	–	–	517.2	470.9
	avg	4	5	5	1	1	29	44	46	4	4	–	–	–	2	7	–	–	–	4	5	–	–	–	3	3
0.1	max	16	16	16	2	1	222	255	272	9	8	–	–	–	34	33	–	–	–	78	74	–	–	–	135	133
	sec	0.3	0.3	0.4	0.2	0.1	7.6	7.2	8.7	6.4	5.6	–	–	–	60.0	54.3	–	–	–	215.6	196.6	–	–	–	531.1	486.1

[a] The tropical KA is $(\mathbb{N} \cup \{\infty\}, \min, +, \infty, 0, {}^*)$, where * is the constant 1-function.

We generated three random number generators from the random number generator generated from the number 42 and ran all functions on different sizes

n, densities[6] d and Kleene algebras. In all cases the generation of the random matrix counts towards the total time to simulate pre-processed input. The result of every function is fully evaluated. The values in Tables 1, 2 and 3 show the average (avg) and the maximum (max) space consumptions in megabyte and the running time (sec) in seconds and all values are arithmetic means over all generators[7]. The letters are abbreviations for a(rray), b(lock), l(eft) and r(ight) and refer to the respective closure function. The computation b uses the function *kleeneClosureBlockwise* mentioned in Sect. 6.3 that requires a closure operation and a matrix multiplication, while the value denoted by b* is the one for the star closure operation only. It is known (cf. [11]) that matrix multiplication is in general as complex as the star closure computation, so that there is no difference in the asymptotic complexity between the star and the Kleene closure. The value "−" denotes an out-of-memory exception occurring with the standard settings.

Table 2. Evaluation in the Boolean Kleene algebra.

$d \backslash n$		100					250					500					750					1000				
		a	b*	b	l	r	a	b*	b	l	r	a	b*	b	l	r	a	b*	b	l	r	a	b*	b	l	r
	avg	5	5	5	1	1	22	44	43	1	1	129	−	−	4	4	−	−	−	8	8	−	−	−	14	13
0.001	max	19	17	15	1	1	181	242	250	2	2	1461	−	−	10	6	−	−	−	15	13	−	−	−	34	23
	sec	0.2	0.3	0.2	0.1	0.1	4.0	7.1	5.4	0.1	0.1	40.5	−	−	0.6	0.6	−	−	−	1.7	1.7	−	−	−	3.3	3.3
	avg	5	5	5	1	1	22	31	38	8	8	135	−	−	95	90	−	−	−	3	5	−	−	−	−	2
0.01	max	18	17	15	1	1	177	220	201	40	44	1528	−	−	536	524	−	−	−	948	235	−	−	−	−	776
	sec	0.2	0.3	0.3	0.1	0.1	3.3	5.8	6.6	0.8	0.9	20.8	−	−	17.4	16.3	−	−	−	67.9	60.1	−	−	−	−	163.7
	avg	5	5	5	1	1	23	45	44	8	7	139	−	−	2	1	−	−	−	−	6	−	−	−	−	−
0.025	max	18	16	16	3	3	185	243	248	37	26	1574	−	−	336	110	−	−	−	−	429	−	−	−	−	−
	sec	0.1	0.3	0.3	0.1	0.1	2.1	5.8	6.5	2.3	1.8	16.3	−	−	23.5	21.7	−	−	−	−	76.9	−	−	−	−	−
	avg	5	5	5	2	1	22	47	43	4	3	154	−	−	3	2	−	−	−	−	4	−	−	−	−	−
0.05	max	20	16	16	4	2	181	236	253	44	15	1668	−	−	379	145	−	−	−	−	520	−	−	−	−	−
	sec	0.1	0.2	0.3	0.1	0.1	1.9	5.7	5.8	2.8	2.4	16.0	−	−	25.7	22.8	−	−	−	−	85.5	−	−	−	−	−
	avg	3	5	5	1	1	26	50	45	4	3	−	−	−	4	2	−	−	−	−	1	−	−	−	−	−
0.1	max	11	16	16	4	2	211	257	267	49	21	−	−	−	394	169	−	−	−	−	572	−	−	−	−	−
	sec	0.1	0.2	0.2	0.1	0.1	1.9	6.3	6.0	3.1	2.7	−	−	−	27.0	24.4	−	−	−	−	84.4	−	−	−	−	−

We observe that the developed right-fold version is almost always better than all other versions in terms of time consumption and never worse in terms of space usage. The reason for the early failure in the Boolean semiring is Haskell's non-strictness, which in this case needs to be tamed with non-algebraic means. Still, the right-fold function manages all matrices of size 750 in the Boolean case, while all other versions do not. One typical improvement of tail-recursive functions in Haskell is the use of *strictness annotations* which (partially) evaluate a parameter before usage. We have experimented with this technique in the above implementations and found that it yields little to no improvement in case of the left-fold variant, but considerable (yet constant) improvements in the right-fold version. This is an indicator that the computational paradigm of a right-fold is conceptually better suited for the Kleene closure than that of a left-fold.

---

[6] In a matrix with size 1000 a density of 0.1 means 100000 entries, which is likely to yield a fully filled transitive closure. This is why we use such seemingly small values.

[7] The measurements were taken on an machine with an Intel Core i5-2520M CPU (4 × 2.5 GHz) with 8 GB of DDR3 RAM, running Ubuntu 12.04 and using GHC 7.6.3.

Table 3. Evaluation in the product Kleene algebra "tropical × Boolean".

$d \backslash n$		100					250					500					750					1000				
		a	b*	b	l	r	a	b*	b	l	r	a	b*	b	l	r	a	b*	b	l	r	a	b*	b	l	r
0.001	avg	14	13	13	1	1	125	126	130	1	1	–	–	–	3	4	–	–	–	7	7	–	–	–	12	13
	max	65	60	60	1	1	1107	940	981	2	2	–	–	–	6	6	–	–	–	13	13	–	–	–	23	23
	sec	0.8	1.0	1.0	0.1	0.1	19.8	19.5	20.9	0.1	0.1	–	–	–	0.6	0.6	–	–	–	1.7	1.6	–	–	–	3.2	3.2
0.01	avg	13	13	14	1	1	95	115	119	8	4	–	–	–	13	22	–	–	–	–	–	–	–	–	–	–
	max	60	65	70	1	1	650	843	873	20	13	–	–	–	703	411	–	–	–	–	–	–	–	–	–	–
	sec	0.8	1.0	1.1	0.1	0.1	18.2	19.4	23.9	2.4	1.6	–	–	–	62.2	53.7	–	–	–	–	–	–	–	–	–	–
0.025	avg	12	12	13	1	1	122	131	134	20	15	–	–	–	–	8	–	–	–	–	–	–	–	–	–	–
	max	48	60	64	2	2	680	987	1018	110	65	–	–	–	–	798	–	–	–	–	–	–	–	–	–	–
	sec	0.8	1.0	1.2	0.1	0.1	21.9	20.8	26.5	8.3	7.0	–	–	–	–	86.6	–	–	–	–	–	–	–	–	–	–
0.05	avg	13	12	13	3	2	126	166	141	17	10	–	–	–	–	7	–	–	–	–	–	–	–	–	–	–
	max	40	60	60	7	4	663	1088	919	186	99	–	–	–	–	1037	–	–	–	–	–	–	–	–	–	–
	sec	0.9	1.0	1.3	0.4	0.2	22.5	24.8	24.5	10.8	9.4	–	–	–	–	93.5	–	–	–	–	–	–	–	–	–	–
0.1	avg	13	13	14	3	2	130	137	139	14	8	–	–	–	–	6	–	–	–	–	–	–	–	–	–	–
	max	41	61	61	12	6	805	1044	1063	247	128	–	–	–	–	1161	–	–	–	–	–	–	–	–	–	–
	sec	1.0	1.0	1.3	0.5	0.4	23.3	21.5	24.5	12.6	10.4	–	–	–	–	98.4	–	–	–	–	–	–	–	–	–	–

7 Related Work and Discussion

There exist different approaches to a functional version of Kleene's algorithm. In [10] an implementation is given that is based upon Eq. (i) and the use of arrays. We have not compared this implementation to ours because it is very similar to our array implementation. The special case of transitive closures (closures over the Boolean KA) is treated in [6] using a monadic abstraction of lazy arrays that is implemented efficiently internally. The work [1] presents another efficient implementation of the closure operation over the Boolean KA. The closure function in [1] is significantly faster and less space consuming than ours, because it uses only lists of (bounded) integers and not association lists with arbitrary values. However, this function works only on Boolean matrices and its generalisation to arbitrary Kleene algebras is essentially our implementation.

While the articles [4, 6, 10] are very well-written, they do not feature an actual derivation of the implementation and depend on the chosen data types. We made an implementation choice, but only to supply the complete code. It is simple to abstract our implementation to a general one – the only place where we explicitly use the fact that matrices are lists of rows is the implementation of $newMat$ that requires a (!!). This implementation can be replaced by a more parametric one, which is parametrised over the container type for matrices, and then also over the one for rows. Thus any representation of matrices that supports the notion of rows and rows that support addition, query and scalar multiplication is suited for our implementation. We used KATs to derive a general version of a Kleene closure. While less general than Kleene algebras, the Kleene algebra of square matrices over a Kleene algebra is a KAT without additional preconditions, because the set of partial identities is a Boolean algebra (cf. Lemma 1). Thus our approach for square matrices is applicable in the same settings as mentioned above and the one we took before the specialisation is applicable to all KATs where 1 is decomposable into a (finite) set of tests. Note that the right-fold structure can yield partial values in the infinite case as well, if some information about $x + y$ can be extracted from x alone. This is a proper improvement of the other versions from the Sect. 6, because left-folds diverge on infinite lists.

We have dealt with a single algorithm in this article, but the employed techniques can transferred to more problems, too. For instance, a graph representation is used in [3] to compute maximum matchings. More generally, it is natural to take a row-based approach, since the rows of a matrix represent the successor lists of a given graph and thus successor-based algorithms usually have elegant algebraic representations. Many algorithms that are expressed in terms of a Kleene closure in some specific algebra can be rewritten as a matrix closure. In fact *every* such closure can be expressed as a matrix closure by Lemma 2.

We are confident that many more functional implementations can be obtained from algebraic specifications, especially in the field of graph algorithms. Additionally, the combination of algebraic reasoning and functional programming can reveal complexity bottlenecks through parametric abstraction over the structures. We have shown that it is possible to find a specification that is based upon modifying a value that *will be computed* next instead of one using values that *have already been computed*. While the latter is tail-recursive and can yield a performance gain in a strict setting, the former is better suited for a non-strict approach allowing propagation of partial values before the recursive application.

Acknowledgements. I thank Rudolf Berghammer for encouraging this work, Insa Stucke for comments and the reviewers for their much appreciated feedback.

References

1. Berghammer, R.: A functional, successor list based version of warshall's algorithm with applications. In: de Swart, H. (ed.) RAMICS 2011. LNCS, vol. 6663, pp. 109–124. Springer, Heidelberg (2011)
2. Bertot, Y., Castéran, P.: Interactive Theorem Proving and Program Development. Coq'Art: The Calculus of Inductive Constructions. Springer, Berlin (2004)
3. Danilenko, N.: Using relations to develop a haskell program for computing maximum bipartite matchings. In: Kahl, W., Griffin, T.G. (eds.) RAMICS 2012. LNCS, vol. 7560, pp. 130–145. Springer, Heidelberg (2012)
4. Dolan, S.: Fun with semirings: a functional pearl on the abuse of linear algebra. In: Morrisett G., Uustalu T. (eds.) ICFP, pp. 101–110. ACM (2013)
5. Hutton, G.: A tutorial on the universality and expressiveness of fold. J. Funct. Program. **9**(4), 355–372 (1999)
6. Johnsson, T.: Efficient graph algorithms using lazy monolithic arrays. J. Funct. Program. **8**(4), 323–333 (1998)
7. Kozen, D.: A completeness theorem for kleene algebras and the algebra of regular events. Inf. Comput. **110**, 366–390 (1994)
8. Kozen, D., Smith, F.: Kleene algebra with tests: completeness and decidability. In: van Dalen, D., Bezem, Marc (eds.) CSL 1996. LNCS, vol. 1258, pp. 244–259. Springer, Heidelberg (1997)
9. Marlow, S. The Haskell Report (2010). www.haskell.org/onlinereport/haskell2010
10. O'Connor, R. (2011). http://r6.ca/blog/20110808T035622Z.html
11. Pettorossi, A.: Techniques for Searching, Parsing, and Matching. Aracne, Roma (2013)
12. Sands, D.: A nave time analysis and its theory of cost equivalence. J. Log. Comput. **5**, 495–541 (1995)

Semantic Issues in Logic Programming

On Completeness of Logic Programs

Włodzimierz Drabent[✉]

Institute of Computer Science, Polish Academy of Sciences and IDA,
Linköpings Universitet, Linköping, Sweden
drabent@ipipan.waw.pl

Abstract. Program correctness (in imperative and functional programming) splits in logic programming into correctness and completeness. Completeness means that a program produces all the answers required by its specification. Little work has been devoted to reasoning about completeness. This paper presents a few sufficient conditions for completeness of definite programs. We also study preserving completeness under some cases of pruning of SLD-trees (e.g. due to using the cut).

We treat logic programming as a declarative paradigm, abstracting from any operational semantics as far as possible. We argue that the proposed methods are simple enough to be applied, possibly at an informal level, in practical Prolog programming. We point out importance of approximate specifications.

Keywords: Logic programming · Program completeness · Declarative programming · Approximate specification

1 Introduction

The notion of partial program correctness splits in logic programming into correctness and completeness. Correctness means that all answers of the program are compatible with the specification, completeness – that the program produces all the answers required by the specification.

In this paper we consider definite clause programs, and present a few sufficient conditions for their completeness. We also discuss preserving completeness under pruning of SLD-trees (by e.g. using the cut). We are interested in declarative reasoning, i.e. abstracting from any operational semantics, and treating program clauses as logical formulae. Our goal is simple methods, which may be applied – possibly informally – in actual practical programming.

Related Work. Surprisingly little work was devoted to proving completeness of programs. Hogger [15] defines the notion of completeness, but does not provide any sufficient conditions. Completeness is not discussed in the important monograph [1]. Instead, a characterization is studied of the set of computed instances of an atomic query, in a special case when the set is finite and the answers are ground. In the paper [18] of Kowalski completeness is discussed, but the example proofs concern only correctness. As a sufficient condition for completeness

© Springer International Publishing Switzerland 2015
M. Proietti and H. Seki (Eds.): LOPSTR 2014, LNCS 8981, pp. 261–278, 2015.
DOI: 10.1007/978-3-319-17822-6_15

of a program P he suggests $P \vdash T_S$, where T_S is a specification in a form of a logical theory. The condition seems impractical as it fails when T_S contains auxiliary predicates, not occurring in P. It also requires that all the models of P (including the Herbrand base) are models of the specification. But it seems that such specifications often have a substantially restricted class of models, maybe a single Herbrand model, cf. [6].

Deville [6] provides an approach where correctness and completeness of programs should follow from construction. No direct sufficient criteria for completeness, applicable to arbitrary programs, are given. Also the approach is not declarative, as it is based on an operational semantics of SLDNF-resolution.

Stärk [22] presents an elegant method of reasoning about a broad class of properties of programs with negation, executed under LDNF-resolutions. A tool to verify proofs mechanically was provided. The approach involves a rather complicated induction scheme, so it seems impossible to apply the method informally by programmers. Also, the approach is not fully declarative, as the order of literals in clause bodies is important.

A declarative sufficient condition for program completeness was given by Deransart and Małuszyński [5]. The approach presented here stems from [13], the differences are discussed in [11]. The main contribution since the former version [9,10] is proving completeness of pruned SLD-trees. The author is not aware of any other work on this issue.

This paper, except Sect. 4.2, is an abbreviated version of some parts of [11]. A full version of Sect. 4.2 appeared in [12]. The reader is referred to [11,12] for missing proofs, more examples and further discussion.

Preliminaries. We use the standard notation and definitions [1]. An atom whose predicate symbol is p will be called a *p-atom* (or an *atom for p*). Similarly, a clause whose head is a p-atom is a *clause for p*. In a program P, by *procedure p* we mean the set of the clauses for p in P.

We assume a fixed alphabet with an infinite set of function symbols. The Herbrand universe will be denoted by $\mathcal{HU}$, the Herbrand base by $\mathcal{HB}$, and the sets of all terms, respectively atoms, by $\mathcal{TU}$ and $\mathcal{TB}$. For an expression (a program) E by $ground(E)$ we mean the set of ground instances of E (ground instances of the clauses of E). $\mathcal{M}_P$ denotes the least Herbrand model of a program P.

By "declarative" (property, reasoning, ...) we mean referring only to logical reading of programs, thus abstracting from any operational semantics. In particular, properties depending on the order of atoms in clauses will not be considered declarative (as they treat equivalent conjunctions differently).

By a computed (respectively correct) answer for a program P and a query Q we mean an instance $Q\theta$ of Q where θ is a computed (correct) answer substitution [1] for Q and P. We often say just *answer* as each computed answer is a correct one, and each correct answer (for Q) is a computed answer (for Q or for some its instance). Thus, by soundness and completeness of SLD-resolution, $Q\theta$ is an answer for P iff $P \models Q\theta$.

Names of variables begin with an upper-case letter. We use the list notation of Prolog. So $[t_1, \ldots, t_n]$ $(n \geq 0)$ stands for the list of elements $t_1, \ldots, t_n$. Only a term of this form is considered a list. (Thus terms like $[a, a|X]$, or $[a, a|a]$, where a is a constant, are not lists). The set of natural numbers will be denoted by $\mathbb{N}$; $f: A \hookrightarrow B$ states that f is a partial function from A to B.

The next section introduces the basic notions of specifications, correctness and completeness. Also, advantages of approximate specifications are discussed. The section is concluded with a brief overview on proving correctness. Section 3 discusses proving program completeness. Section 4 deals with proving that completeness is preserved under pruning. We finish with a discussion.

2   Correctness and Completeness

2.1   Specifications

The purpose of a logic program is to compute a relation, or a few relations. A specification should describe these relations. It is convenient to assume that the relations are over the Herbrand universe. To describe such relations, one relation corresponding to each procedure of the program (i.e. to a predicate symbol), it is convenient to use a Herbrand interpretation. Thus a (formal) **specification** is a Herbrand interpretation, i.e. a subset of $\mathcal{HB}$.

2.2   Correctness and Completeness

In imperative and functional programming, correctness usually means that the program results are as specified. In logic programming, due to its non-deterministic nature, we actually have two issues: *correctness* (all the results are compatible with the specification) and *completeness* (all the results required by the specification are produced). In other words, correctness means that the relations defined by the program are subsets of the specified ones, and completeness means inclusion in the opposite direction. In terms of specifications and the least Herbrand models we define:

Definition 1. *Let P be a program and $S \subseteq \mathcal{HB}$ a specification. P is* **correct** *w.r.t. S when $\mathcal{M}_P \subseteq S$; it is* **complete** *w.r.t. S when $\mathcal{M}_P \supseteq S$.*

We will sometimes skip the specification when it is clear from the context. We propose to call a program **fully correct** when it is both correct and complete. If a program P is fully correct w.r.t. a specification S then, obviously, $\mathcal{M}_P = S$.

A program P is correct w.r.t. a specification S iff Q being an answer of P implies $S \models Q$. (Remember that Q is an answer of P iff $P \models Q$.) The program is complete w.r.t. S iff $S \models Q$ implies that Q is an answer of P. (Here our assumption on an infinite set of function symbols is needed [11].)

It is sometimes useful to consider local versions of these notions:

Definition 2. *A* **predicate** *p in P is* **correct** *w.r.t. S when each p-atom of $\mathcal{M}_P$ is in S, and* **complete** *w.r.t. S when each p-atom of S is in $\mathcal{M}_P$.*

An **answer** *Q is* **correct** *w.r.t. S when $S \models Q$.*

P is **complete for a query** *Q w.r.t. S when $S \models Q\theta$ implies that $Q\theta$ is an answer for P, for any ground instance $Q\theta$ of Q.*

Informally, P is complete for Q when all the answers for Q required by the specification S are answers of P. Note that a program is complete w.r.t. S iff it is complete w.r.t. S for any query iff it is complete w.r.t. S for any query $A \in S$.

2.3 Approximate Specifications

Often it is difficult, and not necessary, to specify the relations defined by a program exactly; more formally, to require that $\mathcal{M}_P$ is equal to a given specification. Often the relations defined by programs are not exactly those intended by programmers. For instance this concerns the programs in Chap. 3.2 of the textbook [23] defining predicates member/2, append/3, sublist/2, and some others. The defined relations are not those of list membership, concatenation, etc. However this is not an error, as for all intended queries the answers are as for a program defining the intended relations. The exact semantics of the programs is not explained in the textbook; such explanation is not needed. Let us look more closely at append/3.

Example 3. 1. The program APPEND

$$app(\,[H|K], L, [H|M]\,) \leftarrow app(\,K, L, M\,). \qquad\qquad app(\,[\,], L, L\,).$$

does not define the relation of list concatenation. For instance, APPEND $\models$ $app([\,], 1, 1)$. In other words, APPEND is not correct w.r.t.

$$S^0_{\text{APPEND}} = \{\, app(k, l, m) \in \mathcal{HB} \mid k, l, m \text{ are lists}, k * l = m \,\},$$

where $k * l$ stands for the concatenation of lists k, l. It is however complete w.r.t. S^0_{APPEND}, and correct w.r.t.

$$S_{\text{APPEND}} = \{\, app(k, l, m) \in \mathcal{HB} \mid \text{if } l \text{ or } m \text{ is a list then } app(k, l, m) \in S^0_{\text{APPEND}} \,\}.$$

Correctness w.r.t. S_{APPEND} and completeness w.r.t. S^0_{APPEND} are sufficient to show that APPEND will produce the required results when used to concatenate or split lists. More precisely, the answers for a query $Q = app(s, t, u)$, where t is a list or u is a list, are $app(s\theta, t\theta, u\theta)$, where $s\theta, t\theta, u\theta$ are lists and $s\theta * t\theta = u\theta$. (The lists may be non-ground.)

2. Similarly, the procedures member/2 and sublist/2 are complete w.r.t specifications describing the relation of list membership, and the sublist relation. It is easy to provide specifications, w.r.t. which the procedures are correct. For instance, member/2 is correct w.r.t. $S_{\text{MEMBER}} = \{\, member(t, u) \in \mathcal{HB} \mid$ if $u = [t_1, \ldots, t_n]$ for some $n \geq 0$ then $t = t_i$, for some $0 < i \leq n \,\}$.

3. The exact relations defined by programs are often misunderstood. For instance, in [7, Ex. 15] it is claimed that a program $Prog_1$ defines the relation of list inclusion. In our terms, this means that predicate *included* of $Prog_1$ is correct and complete w.r.t.

$$\left\{\, included(l_1, l_2) \in \mathcal{HB} \;\middle|\; \begin{array}{l} l_1, l_2 \text{ are lists,} \\ \text{every element of } l_1 \text{ belongs to } l_2 \end{array} \,\right\}.$$

However the correctness does not hold: The program contains a unary clause $included([\,], L)$, so $Prog_1 \models included([\,], t)$ for any term t.

The examples show that in many cases it is unnecessary to know the semantics of a program exactly. Instead it is sufficient to describe it approximately. An **approximate specification** is a pair of specifications S_{compl}, S_{corr}, for completeness and correctness. The intention is that the program is complete w.r.t. the former, and correct w.r.t. the latter: $S_{compl} \subseteq \mathcal{M}_P \subseteq S_{corr}$. In other words, the specifications S_{compl}, S_{corr} describe, respectively, which atoms have to be computed, and which are allowed to be computed. For the atoms from $S_{corr} \setminus S_{compl}$ the semantics of the program is irrelevant. By abuse of terminology, S_{corr} or S_{compl} will sometimes also be called approximate specifications.

2.4   Proving Correctness

We briefly discuss proving correctness, as it is complementary to the main subject of this paper. The approach is due to Clark [4].

Theorem 4 (Correctness). *A sufficient condition for a program P to be correct w.r.t. a specification S is*

> *for each ground instance $H \leftarrow B_1, \ldots, B_n$ of a clause of the program,
> if $B_1, \ldots, B_n \in S$ then $H \in S$.*

Example 5. Consider a program SPLIT and a specification describing how the sizes of the last two arguments of s are related ($|l|$ denotes the length of a list l):

$$s([], [], []). \tag{1}$$

$$s([X|Xs], [X|Ys], Zs) \leftarrow s(Xs, Zs, Ys). \tag{2}$$

$$S = \{\, s(l, l_1, l_2) \mid l, l_1, l_2 \text{ are lists}, 0 \leq |l_1| - |l_2| \leq 1 \,\}.$$

SPLIT is correct w.r.t. S, by Theorem 4 (the details are left for the reader, or see [11]). A stronger specification for which SPLIT is correct is shown in Example 11.

The sufficient condition is equivalent to $S \models P$, and to $T_P(S) \subseteq S$.

Notice that the proof method is declarative. The method should be well known, but is often neglected. For instance it is not mentioned in [1], where a more complicated method, moreover not declarative, is advocated. That method is not more powerful than the one of Theorem 4 [13]. See [11,13] for further examples, explanations, references and discussion.

3   Proving Completeness

We first introduce a notion of semi-completeness, and sufficient conditions under which semi-completeness of a program implies its completeness. Then a sufficient condition follows for semi-completeness. We conclude the section with a way of showing completeness directly without employing semi-completeness.

Definition 6. *A* level mapping *is a function* $| \ |: \mathcal{HB} \to \mathbb{N}$ *assigning natural numbers to atoms.*

A program P is **recurrent** *w.r.t. a level mapping* $| \ |$ *[1,3] if, in every ground instance* $H \leftarrow B_1, \ldots, B_n \in ground(P)$ *of its clause* $(n \geq 0)$, $|H| > |B_i|$ *for all* $i = 1, \ldots, n$. *A program is* recurrent *if it is recurrent w.r.t. some level mapping.*

A program P is **acceptable** *w.r.t. a specification S and a level mapping* $| \ |$ *if P is correct w.r.t. S, and for every* $H \leftarrow B_1, \ldots, B_n \in ground(P)$ *we have* $|H| > |B_i|$ *whenever* $S \models B_1, \ldots, B_{i-1}$. *A program is* acceptable *if it is acceptable w.r.t. some level mapping and some specification.*

The definition of acceptable is more general than that of [1,2] which requires S to be a model of P. Both definitions make the same programs acceptable [11].

Definition 7. *A program P is* **semi-complete** *w.r.t. a specification S if P is complete w.r.t. S for any query Q for which there exists a finite SLD-tree.*

Less formally, the existence of a finite SLD-tree means that P with Q terminates under some selection rule. For a semi-complete program, if a computation for a query Q terminates then all the required by the specification answers for Q have been obtained. Note that a complete program is semi-complete. Also:

Proposition 8 (Completeness). *Let a program P be semi-complete w.r.t. S. The program is complete w.r.t S if*

1. *for each query $A \in S$ there exists a finite SLD-tree, or*
   each $A \in S$ is an instance of a query Q for which a finite SLD-tree exists, or
2. *the program is recurrent, or*
3. *the program is acceptable (w.r.t. a specification S' possibly distinct from S).*

Proving Semi-completeness. We need the following notion.

Definition 9. *A ground atom H is* **covered** *by a clause C w.r.t. a specification S [21] if H is the head of a ground instance $H \leftarrow B_1, \ldots, B_n$ $(n \geq 0)$ of C, such that all the atoms $B_1, \ldots, B_n$ are in S. A ground atom H is* **covered by a program** *P w.r.t. S if it is covered w.r.t. S by some clause $C \in P$.*

For instance, given a specification $S = \{p(s^i(0)) \mid i \geq 0\}$, atom $p(s(0))$ is covered both by $p(s(X)) \leftarrow p(X)$ and by $p(X) \leftarrow p(s(X))$.

Now we present a sufficient condition for semi-completeness. Together with Proposition 8 it provides a sufficient condition for completeness.

Theorem 10 (Semi-completeness). *If all the atoms from a specification S are covered w.r.t. S by a program P then P is semi-complete w.r.t. S.*

Example 11. We show that program SPLIT from Example 5 is complete w.r.t.

$$S_{\text{SPLIT}} = \left\{ \begin{array}{l} s([t_1, \ldots, t_{2n}], [t_1, \cdots, t_{2n-1}], [t_2, \cdots, t_{2n}]), \\ s([t_1, \ldots, t_{2n+1}], [t_1, \cdots, t_{2n+1}], [t_2, \cdots, t_{2n}]) \end{array} \middle| \begin{array}{l} n \geq 0, \\ t_1, \ldots, t_{2n+1} \in \mathcal{HU} \end{array} \right\},$$

where $[t_k, \cdots, t_l]$ denotes the list $[t_k, t_{k+2}, \ldots, t_l]$, for k, l both odd or both even.

Atom $s([\,], [\,], [\,]) \in S_{\mathrm{SPLIT}}$ is covered by clause (1). For $n > 0$, any atom $A = s([t_1, \ldots, t_{2n}], [t_1, \cdots, t_{2n-1}], [t_2, \cdots, t_{2n}])$ is covered by an instance of (2) with a body $B = s([t_2, \ldots, t_{2n}], [t_2, \cdots, t_{2n}], [t_3, \cdots, t_{2n-1}])$. Similarly, for $n \geq 0$ and any atom $A = s([t_1, \ldots, t_{2n+1}], [t_1, \cdots, t_{2n+1}], [t_2, \cdots, t_{2n}])$, the corresponding body is $B = s([t_2, \ldots, t_{2n+1}], [t_2, \cdots, t_{2n}], [t_3, \cdots, t_{2n+1}])$. In both cases, $B \in S_{\mathrm{SPLIT}}$. (To see this, rename each t_i as t'_{i-1}.) So S_{SPLIT} is covered by SPLIT. Thus SPLIT is semi-complete w.r.t. S_{SPLIT}, by Theorem 10.

Now by Proposition 8 the program is complete, as it is recurrent under the level mapping $|s(t, t_1, t_2)| = |t|$, where $|\,[h|t]\,| = 1 + |t|$ and $|f(t_1, \ldots, t_n)| = 0$ (for any ground terms $h, t, t_1, \ldots, t_n$, and any function symbol f distinct from $[\,|\,]$).

By Theorem 4 the program is also correct w.r.t. S_{SPLIT}, as $S_{\mathrm{SPLIT}} \models \mathrm{SPLIT}$. (The details are left to the reader.) Hence $S_{\mathrm{SPLIT}} = \mathcal{M}_{\mathrm{SPLIT}}$.

Note that the sufficient condition of Theorem 10 is equivalent to $S \subseteq T_P(S)$, which implies $S \subseteq \mathrm{gfp}(T_P)$. It is also equivalent to S being a model of ONLY-IF(P) (see e.g. [8] or [13] for a definition).

The notion of semi-completeness is tailored for finite programs. An SLD-tree for a query Q and an infinite program P may be infinite, but with all branches finite. In such case, if the condition of Theorem 10 holds then P is complete for Q [11].

Proving Completeness Directly. Here we present another, declarative, way of proving completeness; a condition is added to Theorem 10 so that completeness is implied directly. This also works for non-terminating programs. However when termination has to be shown anyway, applying Theorem 10 is usually more convenient.

In this section we allow that a level mapping is a *partial* function $|\,|: \mathcal{HB} \hookrightarrow \mathbb{N}$ assigning natural numbers to some atoms.

Definition 12. *A ground atom H is* **recurrently covered** *by a program P w.r.t. a specification S and a level mapping $|\,|: \mathcal{HB} \hookrightarrow \mathbb{N}$ if H is the head of a ground instance $H \leftarrow B_1, \ldots, B_n$ ($n \geq 0$) of a clause of the program, such that $|H|, |B_1|, \ldots |B_n|$ are defined, $B_1, \ldots, B_n \in S$, and $|H| > |B_i|$ for all $i = 1, \ldots, n$.*

For instance, given a specification $S = \{ p(s^i(0)) \mid i \geq 0 \}$, atom $p(s(0))$ is recurrently covered by a program $\{ p(s(X)) \leftarrow p(X). \}$ under a level mapping for which $|p(s^i(0))| = i$. No atom is recurrently covered by $\{ p(X) \leftarrow p(X). \}$. Obviously, if H is recurrently covered by P then it is covered by P. If H is covered by P w.r.t. S and P is recurrent w.r.t. $|\,|$ then H is recurrently covered w.r.t. $S, |\,|$. The same holds for P acceptable w.r.t. an $S' \supseteq S$.

Theorem 13 (Completeness 2). (A reformulation of Theorem 6.1 of [5]). *If, under some level mapping $|\,|: \mathcal{HB} \hookrightarrow \mathbb{N}$, all the atoms from a specification S are recurrently covered by a program P w.r.t. S then P is complete w.r.t. S.*

Example 14. Consider a directed graph E. As a specification for a program describing reachability in E, take $S = S_p \cup S_e$, where

$$S_p = \{\, p(t, u) \mid \text{there is a path from } t \text{ to } u \text{ in } E \,\},$$
$$S_e = \{\, e(t, u) \mid (t, u) \text{ is an edge in } E \,\}.$$

Let P consist of a procedure p: $\{\, p(X, X). \; p(X, Z) \leftarrow e(X, Y), p(Y, Z). \}$ and a procedure e which is a set of unary clauses describing the edges of the graph. Assume the latter is complete w.r.t. S_e. Notice that when E has cycles then infinite SLD-trees cannot be avoided, and completeness of P cannot be shown by Proposition 8.

To apply Theorem 13, let us define a level mapping for the elements of S such that $|e(t, u)| = 0$ and $|p(t, u)|$ is the length of a shortest path in E from t to u (so $|p(t, t)| = 0$). Consider a $p(t, u) \in S$ where $t \neq u$. Let $t = t_0, t_1, \ldots, t_n = u$ be a shortest path from t to u. Then $e(t, t_1), p(t_1, u) \in S$, $|p(t, u)| = n$, $|e(t, t_1)| = 0$, and $|p(t_1, u)| = n - 1$. Thus $p(t, u)$ is recurrently covered by P w.r.t. S and $|\ |$. The same trivially holds for the remaining atoms of S. So P is complete w.r.t. S.

4   Pruning SLD-Trees and Completeness

Pruning some parts of SLD-trees is often used to improve efficiency of programs. It is implemented by using the cut, the if-then-else construct of Prolog, or built-ins, like `once/1`. Pruning preserves the correctness of a logic program, it also preserves termination under a given selection rule, but may violate the program's completeness. We now discuss proving that completeness is preserved.

By a **pruned SLD-tree** for a program P and a query Q we mean a tree with the root Q which is a connected subgraph of an SLD-tree for P and Q. By an *answer* of a pruned SLD-tree we mean the computed answer of a successful SLD-derivation which is a branch of the tree. We will say that a pruned SLD-tree T with root Q is **complete** w.r.t. a specification S if, for any ground $Q\theta$, $S \models Q\theta$ implies that $Q\theta$ is an instance of an answer of T. Informally, such a tree produces all the answers for Q required by S.

We present two approaches for proving completeness of pruned SLD-trees. The first one is based on viewing pruning as skipping certain clauses while building the children of a node. The other deals with a restricted usage of the cut.

4.1   Pruning as Clause Selection

To facilitate reasoning about the answers of pruned SLD-trees, we will now view pruning as applying only certain clauses while constructing the children of a node. So we introduce subsets $\Pi_1, \ldots, \Pi_n$ of P. The intention is that for each node the clauses of one Π_i are used. Programs $\Pi_1, \ldots, \Pi_n$ may be not disjoint.

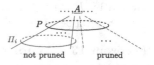

Definition 15. *Given programs $\Pi_1, \ldots, \Pi_n$ ($n > 0$), a* **c-selection rule** *is a function assigning to a query Q' an atom A in Q' and one of the programs $\emptyset, \Pi_1, \ldots, \Pi_n$.*

A **csSLD-tree** *(cs for clause selection) for a query Q and programs $\Pi_1, \ldots, \Pi_n$, via a c-selection rule R, is constructed as an SLD-tree, but for each node its children are constructed using the program selected by the c-selection rule. An answer of a csSLD-tree is defined in the expected way.*

A c-selection rule may choose the empty program, thus making a given node a leaf. Notice that a csSLD-tree for Q and $\Pi_1, \ldots, \Pi_n$ is a pruned SLD-tree for Q and $\bigcup_i \Pi_i$. Conversely, for each pruned SLD-tree T for Q and a (finite) program P there exist $n > 0$, and $\Pi_1, \ldots, \Pi_n \subseteq P$ such that T is a csSLD-tree for Q and $\Pi_1, \ldots, \Pi_n$.

Describing pruning by a c-selection rule is, in a sense, abstract. It does not refer directly to the control constructs in the program. The correspondence between the program and the c-selection rule may be not obvious [11]. A single cut, or if-then-else, may prune children of many nodes in a tree, modifying the behaviour of many procedures of the program. However Examples 19, 20, 21 below suggest that in many cases this difficulty is not substantial.

Example 16. We show that completeness of each of $\Pi_1, \ldots, \Pi_n$ is not sufficient for completeness of a csSLD-tree for $\Pi_1, \ldots, \Pi_n$. Consider a program P:

$$q(X) \leftarrow p(Y, X). \tag{3}$$

$$p(Y, 0). \tag{4}$$

$$p(a, s(X)) \leftarrow p(a, X). \tag{5}$$

$$p(b, s(X)) \leftarrow p(b, X). \tag{6}$$

and programs $\Pi_1 = \{(3), (4), (6)\}$, $\Pi_2 = \{(3), (4), (5)\}$, As a specification for completeness consider $S_0 = \{\, q(s^j(0)) \mid j \geq 0 \,\}$. Each of the programs Π_1, Π_2, P is complete w.r.t. S_0. Assume a c-selection rule R choosing alternatively Π_1, Π_2 along each branch of a tree. Then the csSLD-tree for $q(s^j(0)) \in S_0$ via R (where $j > 2$) has no answers, thus the tree is not complete w.r.t. S_0.

Consider programs $P, \Pi_1, \ldots, \Pi_n$ and specifications $S, S_1, \ldots, S_n$, such that $P \supseteq \bigcup_{i=1}^{n} \Pi_i$ and $S = \bigcup_{i=1}^{n} S_i$. The intention is that each S_i describes which answers are to be produced by using Π_i in the first resolution step. We will call $\Pi_1, \ldots, \Pi_n, S_1, \ldots, S_n$ a **split** (of P and S). Note that $\Pi_1, \ldots, \Pi_n$ or $S_1, \ldots, S_n$ may be not disjoint.

Definition 17. *Let $\mathcal{S} = \Pi_1, \ldots, \Pi_n, S_1, \ldots, S_n$ be a split, and $S = \bigcup S_i$.*

Specification S_i is **suitable** *for an atom A w.r.t. $\mathcal{S}$ when no instance of A is in $S \setminus S_i$. (In other words, when $ground(A) \cap S \subseteq S_i$.) We also say that a program Π_i is* **suitable** *for A w.r.t. $\mathcal{S}$ when S_i is.*

A c-selection rule is **compatible** *with $\mathcal{S}$ if for each non-empty query Q it selects an atom A and a program Π, such that*

- $\Pi \in \{\Pi_1, \ldots, \Pi_n\}$ *is suitable for A w.r.t. S, or*
- *none of $\Pi_1, \ldots, \Pi_n$ is suitable for A w.r.t. S and $\Pi = \emptyset$ (so Q is a leaf).*

A *csSLD-tree for $\Pi_1, \ldots, \Pi_n$ via a c-selection rule compatible with S is said to be **weakly compatible** with S. The tree is **compatible** with S iff for each its nonempty node some Π_i is selected.*

The intuition is that when Π_i is suitable for A then S_i is a fragment of S sufficient to deal with A. It describes all the answers for query A required by S.

The reason of incompleteness of the trees in Example 16 may be understood as selecting a Π_i not suitable for the selected atom. Take $S = \Pi_1, \Pi_2, S_0 \cup S_1'$, $S_0 \cup S_2'$, where $S_1' = \{ p(b, s^i(0)) \mid i \geq 0 \}$ and $S_2' = \{ p(a, s^i(0)) \mid i \geq 0 \}$. In the incomplete trees, Π_1 is selected for an atom $A = p(a, u)$, or Π_2 is selected for an atom $B = p(b, u)$ (where $u \in \mathcal{TU}$). However Π_1 is not suitable for A whenever A has an instance in S (as then $ground(A) \cap S \not\subseteq S_0 \cup S_1'$); similarly for Π_2 and B.

When Π_i is suitable for A then if each atom of S_i is covered by Π_i (w.r.t. S) then using for A only the clauses of Π_i does not impair completeness w.r.t. S:

Theorem 18. *Let $P \supseteq \bigcup_{i=1}^n \Pi_i$ (where $n > 0$) be a program, $S = \bigcup_{i=1}^n S_i$ a specification, and T a csSLD-tree for $\Pi_1, \ldots, \Pi_n$. If*

1. *for each $i = 1, \ldots, n$, all the atoms from S_i are covered by Π_i w.r.t. S, and*
2. *T is compatible with $\Pi_1, \ldots, \Pi_n, S_1, \ldots, S_n$,*
3. *(a) T is finite, or*
   (b) program P is recurrent, or
   (c) P is acceptable (possibly w.r.t. a specification distinct from S), and T is built under the Prolog selection rule

then T is complete w.r.t. S.

We now show three examples of applying this theorem.

Example 19. The following program SAT0 is a simplification of a fragment of the SAT solver of [16] discussed in [9]. Pruning is crucial for the efficiency and usability of the original program.

$$p(P-P, []). \tag{7}$$
$$p(V-P, [B|T]) \leftarrow q(V-P, [B|T]). \tag{8}$$
$$p(V-P, [B|T]) \leftarrow q(B, [V-P|T]). \tag{9}$$
$$q(V-P, \_) \leftarrow V = P. \tag{10}$$
$$q(\_, [A|T]) \leftarrow p(A, T). \tag{11}$$
$$P = P. \tag{12}$$

The program is complete w.r.t. a specification

$$S = \left\{ \begin{array}{l} p(t_0-u_0, [t_1-u_1, \ldots, t_n-u_n]), \\ q(t_0-u_0, [t_1-u_1, \ldots, t_n-u_n]) \end{array} \middle| \begin{array}{l} n \geq 0,\ t_0, \ldots, t_n, u_0, \ldots, u_n \in \mathbb{T}, \\ t_i = u_i \text{ for some } i \in \{0, \ldots, n\} \end{array} \right\} \cup S_=$$

where $\mathbb{T} = \{false, true\} \subseteq \mathcal{HU}$, and $S_= = \{t{=}t \mid t \in \mathcal{HU}\}$. We omit a completeness proof, mentioning only that SAT0 is recurrent w.r.t. a level mapping $|p(t, u)| = 2|u|{+}2$, $|q(t, u)| = 2|u|{+}1$, $|{=}(t, u)| = 0$, where $|u|$ is as in Example 11.

The first case of pruning is due to redundancy within (8), (9); both $\Pi_1 = $ SAT0 $\setminus \{(9)\}$ and $\Pi_2 = $ SAT0 $\setminus \{(8)\}$ are complete w.r.t. S. For any selected atom at most one of (8), (9) is to be used, and the choice is dynamic. As the following reasoning is independent from this choice, we omit further explanations.

So in such pruned SLD-trees the children of each node are constructed using one of programs Π_1, Π_2. Thus they are csSLD-trees for Π_1, Π_2. They are compatible with $\mathcal{S} = \Pi_1, \Pi_2, S, S$ (as Π_1, Π_2 are trivially suitable for any A, due to $S_i = S$ and $S \setminus S_i = \emptyset$ in Definition 17). Each atom of S is covered w.r.t. S both by Π_1 and Π_2. As SAT0 is recurrent, by Theorem 18, each such tree is complete w.r.t. S.

Example 20. We continue with program SAT0 and specification S from the previous example, and add a second case of pruning. When the selected atom is of the form $A = q(s_1, s_2)$ with a ground s_1 then only one of clauses (10), (11) is needed – (10) when s_1 is of the form $t{-}t$, and (11) otherwise. The other clause can be abandoned without losing the completeness w.r.t. S.[1]

Actually, SAT0 is included in a bigger program, say $P = $ SAT0 $\cup \Pi_0$. We skip the details of Π_0, let us only state that P is recurrent, Π_0 does not contain any clause for p or for q, and that P is complete w.r.t. a specification $S' = S \cup S_0$ where S_0 does not contain any p- or q-atom. (Hence each atom of S_0 is covered by Π_0 w.r.t. S'.)

To formally describe the trees for P resulting from both cases of pruning, consider $\mathcal{S} = \Pi_0, \ldots, \Pi_5, S_0, \ldots, S_5$, where

$$\Pi_1 = \{(7), (8)\}, \quad \Pi_2 = \{(7), (9)\}, \quad S_1 = S_2 = S \cap \{p(s, u) \mid s, u \in \mathcal{HU}\},$$
$$\Pi_3 = \{(10)\}, \qquad\qquad\qquad\qquad S_3 = S \cap \{q(t{-}t, s) \mid t, s \in \mathcal{HU}\},$$
$$\Pi_4 = \{(11)\}, \qquad\qquad\qquad\qquad S_4 = S \cap \{q(t{-}u, s) \mid t, u, s \in \mathcal{HU}, t \neq u\},$$
$$\Pi_5 = \{(12)\}, \qquad\qquad\qquad\qquad S_5 = S_=.$$

Each atom from S_i is covered by Π_i w.r.t. S' (for $i = 0, \ldots, 5$). For each q-atom with its first argument ground, Π_3 or Π_4 (or both) is suitable. For each remaining atom from $\mathcal{TB}$, a program from $\Pi_0, \Pi_1, \Pi_2, \Pi_5$ is suitable.

Consider a pruned SLD-tree T for P (employing the two cases of pruning described above). Assume that each q-atom selected in T has its first argument ground. Then T is a csSLD-tree compatible with $\mathcal{S}$. From Theorem 18 it follows that T is complete w.r.t. S'.

The restriction on the selected q-atoms is implemented by means of Prolog delays. This is done in such a way that, for the intended initial queries, floundering is avoided [16] (i.e. an atom is selected in each query). So the obtained

---

[1] The same holds for A of the form $q(s_{11}{-}s_{11}, s_2)$, or $q(s_{11}{-}s_{12}, s_2)$ with non-unifiable s_{11}, s_{12}. The pruning is implemented using the if-then-else construct in Prolog: $q(V - P, [A|T]) :- V = P \rightarrow \mathbf{true}; p(A, T)$. (And the first case of pruning by $p(V - P, [B|T]) :- \mathbf{nonvar}(V) \rightarrow q(V - P, [B|T]); q(B, [V - P|T])$.)

pruned trees are as T above, and the pruning preserves completeness of the program.

Example 21. A Prolog program $\{nop(adam, 0) \leftarrow !.\; nop(eve, 0) \leftarrow !.\; nop(X, 2).\}$ is an example of difficulties and dangers of using the cut in Prolog. Due to the cut, for an atomic query A only the first clause with the head unifiable with A will be used. The program can be seen as logic program $P = \Pi_1 \cup \Pi_2 \cup \Pi_3$ executed with pruning, where (for $i = 1, 2, 3$) Π_i is the i-th clause of the program with the cut removed. The intended meaning is $S = S_1 \cup S_2 \cup S_3$, where $S_1 = \{nop(adam, 0)\}$, $S_2 = \{nop(eve, 0)\}$, and $S_3 = \{nop(t, 2) \in \mathcal{HB} \mid t \notin \{adam, eve\}\}$. Note that all the atoms from S_i are covered by Π_i (for $i = 1, 2, 3$). (We do not discuss here the (in)correctness of the program, but see [11].)

Let S be $\Pi_1, \Pi_2, \Pi_3, S_1, S_2, S_3$. Consider a query $A = nop(t, Y)$ with a ground t. If $t = adam$ then $ground(A) \cap S = S_1$, and only Π_1 is suitable for A w.r.t. S, if $t = eve$ then only Π_2 is. For $t \notin \{adam, eve\}$ the suitable program is Π_3. So for the query A the pruning due to the cuts results in selecting a suitable Π_i, and the obtained csSLD-tree is compatible with S. By Theorem 18 the tree is complete w.r.t. S.

For a query $nop(X, Y)$ or $nop(X, 0)$ only the first clause, i.e. Π_1, is used. However Π_1 is not suitable for the query (w.r.t. S), and the csSLD-tree is not compatible with S. The tree is not complete (w.r.t. S).

4.2   The Cut in the Last Clause

The previous approach is based on a somehow abstract semantics in which pruning is viewed as clause selection. Now we present an approach referring directly to Prolog with the cut. However the usage of the cut is restricted to the last clause of a procedure. The author expects that the general case could be conveniently studied in the context of programs with negation (because if $H \leftarrow A_1, !, A_2$ is followed by $H \leftarrow A_3$ then the latter clause is used only if A_1 fails). We consider LD-resolution, as interaction of the cut with delays introduces additional complications.

We need to reason about the atoms selected in the derivations. So we employ a (non-declarative) approach to reason about LD-derivations, presented in [1]. A specification in this approach, let us call it **call-success specification**, is a pair $pre, post \in \mathcal{TB}$ of sets of atoms, closed under substitution. A program is correct w.r.t. such specification, let us say **c-s-correct**, when in each LD-derivation every selected atom is in pre and each corresponding computed answer is in $post$, provided that the derivation begins with an atomic query from pre. The same holds for non-atomic initial queries, provided that they satisfy a certain condition (are *well asserted* [1]). See [1] or [13] for further explanations, and for a sufficient criterion for c-s-correctness (programs satisfying it are called *well asserted*).

By $vars(E)$ we denote the set of variables occurring in an expression E. For a substitution $\theta = \{X_1/t_1, \ldots, X_n/t_n\}$, let $dom(\theta) = \{X_1, \ldots, X_n\}$, and $rng(\theta) = vars(\{t_1, \ldots, t_n\})$. We begin with generalizing the notion of an atom covered by a clause.

Definition 22. Let S be a specification, and $pre, post$ a call-success specification. A ground atom A is **adjustably covered** by a clause C w.r.t. S and $pre, post$ if A is covered by C and the cut does not occur in C, or the following three conditions hold:

1. C is $H \leftarrow A_1, \ldots, A_{k-1}, !, A_k, \ldots, A_n$,
2. A is covered by $H \leftarrow A_1, \ldots, A_{k-1}$ w.r.t. S,
3. – for any instance $H\rho \in pre$ such that A is an instance of $H\rho$,
   – for any ground instance $(A_1, \ldots, A_{k-1})\rho\eta$ such that $A_1\rho\eta, \ldots, A_{k-1}\rho\eta \in post$,
   – A is covered by $(H \leftarrow A_k, \ldots, A_n)\rho\eta$ w.r.t. S,
   where $dom(\rho) \subseteq vars(H)$, $rng(\rho) \cap vars(C) \subseteq vars(H)$, $dom(\rho) \cap rng(\rho) = \emptyset$, and $dom(\eta) = vars((A_1, \ldots, A_{k-1})\rho)$.

Informally, condition 3 says that A could be produced out of each "related" answer for $A_1, \ldots, A_{k-1}$, and some answer for $A_k, \ldots, A_n$ specified by S. Note that if A is adjustably covered by C w.r.t. $S, pre, post$, where $S \subseteq post$, then A is covered by C w.r.t. S.

Now we are ready to present the sufficient condition for completeness.

Theorem 23 ([12]). *Let S be a specification, $pre, post$ a call-success specification, where $S \subseteq post$. Let T be a pruned LD-tree for a program P and an atomic query Q, where pruning is due to the cut occurring in the last clause(s) of some procedure(s) of P. If*

– T is finite, $Q \in pre$, P is c-s-correct w.r.t. $pre, post$, and
– each $A \in S$ is adjustably covered by a clause of P w.r.t. S and $pre, post$

then T is complete w.r.t. S.

For a non-atomic initial query Q, the condition $Q \in pre$ should be replaced by Q is well asserted w.r.t. $pre, post$ (see [1] for a definition).

We now show two examples of applying this theorem to proving completeness of pruned trees.

Example 24. Consider a program IN:

$$in([\,], L).\qquad\qquad\qquad m(E, [E|L]).$$
$$in([H|T], L) \leftarrow m(H, L), !, in(T, L).\qquad m(E, [H|L]) \leftarrow m(E, L).$$

and specifications

$S = S_m \cup S_{in}$, $pre = pre_m \cup pre_{in}$, $post = post_m \cup post_{in}$, where
$pre_m = \{\, m(u, t) \in \mathcal{TB} \mid t \text{ is a list} \,\}$,
$pre_{in} = \{\, in(u, t) \in \mathcal{HB} \mid u, t \text{ are ground lists} \,\}$,
$post_m = \{\, m(t_i, [t_1, \ldots, t_n]) \in \mathcal{TB} \mid 1 \leq i \leq n \,\}$,
$post_{in} = \{\, in([u_1, \ldots, u_m], [t_1, \ldots, t_n]) \in \mathcal{HB} \mid \{u_1, \ldots, u_m\} \subseteq \{t_1, \ldots, t_n\} \,\}$,
$S_m = post_m \cap \mathcal{HB}$, $S_{in} = post_{in}$.

The program is c-s-correct w.r.t. $pre, post$ (we skip a proof). We show that each atom $A = in(u, t) \in S_{in}$, where $u = [u_1, \ldots, u_m]$, $m > 0$, is adjustably covered by the second clause C of IN. Let C_0 be $in([H|T], L) \leftarrow m(H, L)$. Now A is covered by C_0 w.r.t. S ($A \leftarrow m(u_1, t)$ is a relevant ground instance of C_0).

Take an instance $in([H|T], L)\rho \in pre$ of the head of C. The instance is ground, and the whole $C\rho$ is ground. So in Definition 22, $\rho\eta = \rho$. If A is an instance of (thus equal to) $in([H|T], L)\rho$ then $in(T, L)\rho = in([u_2, \ldots, u_m], t) \in S$ (as $A \in S$). Thus A is covered by $(in([H|T], L) \leftarrow in(T, L))\rho$.

Thus A is adjustably covered by C. It is easy to check that all the remaining atoms of S are covered by IN w.r.t. S, and that IN is recurrent (for $|m(s, t)| = |t|$, $|in(s, t)| = |s| + |t|$, $|t|$ as in Example 11). Thus each LD-tree for IN and a query $Q \in pre$ is finite. By Theorem 23, each such tree pruned due to the cut is complete w.r.t. S. Notice that condition 3 does not hold when non ground arguments of in are allowed in pre_{in}, and that for such queries some answers may be pruned.

Before the next example we introduce a property, which simplifies checking that an atom is adjustably covered by a clause with the cut.

Lemma 25 ([12]). *If condition 3 of Definition 22 holds for an atom $H\rho \in pre$ then it holds for any its instance $H\rho'$ such that A is an instance of $H\rho'$, and ρ' satisfies the requirements of condition 3 (i.e. $dom(\rho') \subseteq vars(H)$, $rng(\rho') \cap vars(C) \subseteq vars(H)$, $dom(\rho') \cap rng(\rho') = \emptyset$).*

Example 26. Consider a program P:

$$p(X, Z) \leftarrow q(X, Y), !, r(Y, Z). \qquad \begin{array}{ll} q(a, a) & r(a, c) \\ q(a, a') & r(a', c) \\ q(b, b) & \end{array}$$

and specifications

$$S = \{\, p(a, c), q(a, a'), q(b, b), r(a, c), r(a', c) \,\},$$
$$post = S \cup \{q(a, a)\},$$
$$pre = \{\, p(a, t) \mid t \in TU \,\} \cup \{\, q(a, t) \mid t \in TU \,\} \cup \{\, r(t, u) \mid t, u \in TU \,\}$$

The program is c-s-correct w.r.t. $pre, post$ (we skip a proof). To check that atom $p(a, c) \in S$ is adjustably covered by the first clause of P, note first that it is covered w.r.t. S by $p(a, c) \leftarrow q(a, a')$. By Lemma 25, it is sufficient to check condition 3 of Definition 22 for $\rho = \{X/a\}$, as $p(X, Z)\rho = p(a, Z)$ is a most general p-atom in pre. If $q(X, Y)\rho\eta \in post$ then $\eta = \{Y/a\}$ or $\eta = \{Y/a'\}$. Hence $r(Y, Z)\rho\eta$ is $r(a, Z)$ or $r(a', Z)$. In both cases, $p(a, c) \leftarrow r(Y\eta, c)$ is a ground instance of $(p(X, Z) \leftarrow r(Y, Z))\rho\eta$ (i.e. of $p(a, Z) \leftarrow r(Y\eta, Z)$) covering $p(a, c)$ w.r.t. S.

The remaining atoms of S are trivially covered by the unary clauses of P. The LD-tree for P and $Q = p(a, Z)$ is finite, hence the LD-tree pruned due to the cut is complete w.r.t. S by Theorem 23.

5   Discussion

Declarativeness. Without declarative ways of reasoning about correctness and completeness of programs, logic programming would not deserve to be called a declarative programming paradigm. The sufficient condition for proving correctness (Theorem 4), that for semi-completeness of Theorem 10, and those for completeness of Proposition 8(2) and Theorem 13 are declarative. Also, the sufficient condition for completeness of pruned trees (Theorem 18), based on clause selection, to a substantial extent abstracts from the operational semantics. On the other hand, the sufficient conditions for program completeness of Propositions 8(1) and 8(3) are not declarative, as they refer to program termination, or depend on the order of atoms in clause bodies.

Declarative completeness proofs employing Proposition 8(2) or Theorem 13 imply termination, or require reasoning similar to that in termination proofs. So proving completeness by means of semi-completeness and termination may be a reasonable compromise between declarative and non-declarative reasoning, as termination has to be shown anyway in most of practical cases.

Granularity of Proofs. Note that the sufficient condition for correctness deals with single clauses, that for semi-completeness – with procedures, and those for completeness take into account a whole program.

Incompleteness Diagnosis. There is a close relation between completeness proving and incompleteness diagnosis [21]. As the reason of incompleteness, a diagnosis algorithm finds an atom from S that is not covered by the program. Thus it finds a reason for violating the sufficient conditions for semi-completeness and completeness of Theorems 10 and 13. So what is diagnosed is lack of semi-completeness. (As the algorithm works with a finite SLD-tree for a program P and a query Q, incompleteness of P for Q implies that P is not semi-complete.)

Approximate Specifications. We found that approximate specifications are crucial in avoiding unnecessary complications in dealing with correctness and completeness of programs (cf. Sect. 2.3, [9,11,13]). For instance, in the main example of [9] (and in its simpler version in Examples 19, and 20) finding an exact specification is not easy, and is unnecessary. The required property of the program is described more conveniently by an approximate specification. Moreover, as this example shows, in program development the semantics of (common predicates in) the consecutive versions of a program may differ. What is unchanged is correctness and completeness w.r.t. an approximate specification.

Approximate Specifications in Program Development. This suggests a generalization of the paradigm of program development by semantics preserving program transformations [19,20]: it is useful and natural to use transformations which only preserve correctness and completeness w.r.t. an approximate specification.

Approximate Specifications in Debugging. In declarative diagnosis [21] the programmer is required to know the exact intended semantics of the program. This is a substantial obstacle to using declarative diagnosis in practice. Instead, an approximate specification can be used, with the specification for correctness (respectively completeness) applied in incorrectness (incompleteness) diagnosis. See [11] for discussion and references.

Interpretations as Specifications. This work uses specifications which are interpretations. (The same kind of specifications is used, among others, in [1], and in declarative diagnosis.) There are however properties which cannot be expressed by such specifications [13]. For instance one cannot express that some instance of an atomic query A should be an answer; one has to specify the actual instance(s). Other approach is needed for such properties, possibly with specifications which are logical theories (where axioms like $\exists X. A$ can be used).

Applications. We want to stress the simplicity and naturalness of the sufficient conditions for correctness (Theorem 4) and semi-completeness (Theorem 10, the condition is a part of each discussed sufficient condition for completeness). Informally, the first one says that the clauses of a program should produce only correct conclusions, given correct premises. The other says that each ground atom that should be produced by P can be produced by a clause of P out of atoms produced by P. The author believes that this is a way a competent programmer reasons about (the declarative semantics of) a logic program.

Paper [9] illustrates practical applicability of the methods presented here. It shows a systematic construction of a non-trivial Prolog program (the SAT solver of [16]). Starting from a formal specification, a definite clause logic program is constructed hand in hand with proofs of its correctness, completeness, and termination under any selection rule. The final Prolog program is obtained by adding control to the logic program (delays and pruning SLD-trees). Adding control preserves correctness and termination. However completeness may be violated by pruning, and by floundering related to delays. By Theorem 18, the program with pruning remains complete.[2] Proving non-floundering is outside of the scope of this work. See [14] for a related analysis algorithm, applicable in this case [17].

The example shows how well "logic" could be separated from "control." The whole reasoning related to correctness and completeness can be done declaratively, abstracting from any operational semantics.

Future Work. A natural continuation is developing completeness proof methods for programs with negation (a first step was made in [13]), maybe also for constraint logic programming and CHR (constraint handling rules). Further examples of proofs are necessary. An interesting task is formalizing and automatizing the proofs, a first step is formalization of specifications. Another issue is overcoming the limitation described in *Interpretations as specifications* above.

---

[2] In [9] a weaker version of Theorem 18 was used, and one case of pruning was discussed informally. A proof covering both cases of pruning is illustrated here in Example 20.

Conclusion. Reasoning about completeness of logic program has been, surprisingly, almost neglected. This paper presents a few sufficient conditions for completeness. As an intermediate step we introduced a notion of semi-completeness. The presented methods are, to a large extent, declarative. Examples suggest that the approach is applicable – maybe at informal level – in practice of Prolog programming. The approach is augmented by two methods of proving completeness in presence of pruning.

References

1. Apt, K.R.: From Logic Programming to Prolog. International Series in Computer Science. Prentice-Hall, Upper Saddle River (1997)
2. Apt, K.R., Pedreschi, D.: Reasoning about termination of pure Prolog programs. Inf. Comput. **106**(1), 109–157 (1993)
3. Bezem, M.: Strong termination of logic programs. J. Log. Program. **15**(1&2), 79–97 (1993)
4. Clark, K.L.: Predicate logic as computational formalism. Technical report 79/59, Imperial College, London (1979)
5. Deransart, P., Małuszyński, J.: A Grammatical View of Logic Programming. The MIT Press, Cambridge (1993)
6. Deville, Y.: Logic Programming: Systematic Program Development. Addison-Wesley, Reading (1990)
7. Deville, Y., Lau, K.-K.: Logic program synthesis. J. Log. Program. **19**(20), 321–350 (1994)
8. Doets, K.: From Logic to Logic Programming. The MIT Press, Cambridge (1994)
9. Drabent, W.: Logic + control: an example. In: Dovier, A., Santos Costa, V. (eds.) Technical Communications of ICLP 2012. LIPIcs, vol. 17, pp. 301–311 (2012). http://drops.dagstuhl.de/opus/volltexte/2012/3631
10. Drabent, W.: Logic + control: an example of program construction, CoRR. abs/1110.4978 (2012). http://arxiv.org/abs/1110.4978.
11. Drabent, W.: Correctness and completeness of logic programs, CoRR. abs/1412.8739 (2014). http://arxiv.org/abs/1412.8739
12. Drabent, W.: On completeness of logic programs, CoRR. abs/1411.3015(2014). http://arxiv.org/abs/1411.3015
13. Drabent, W., Miłkowska, M.: Proving correctness and completeness of normal programs - a declarative approach. Theory Pract. Log. Program. **5**(6), 669–711 (2005)
14. Genaim, S., King, A.: Inferring non-suspension conditions for logic programs with dynamic scheduling. ACM Trans. Comput. Log. **9**(3), 17:1–17:43 (2008)
15. Hogger, C.J.: Introduction to Logic Programming. Academic Press, London (1984)
16. Howe, J.M., King, A.: A pearl on SAT and SMT solving in Prolog. Theor. Comput. Sci. **435**, 43–55 (2012)
17. King, A.: Personal communication, March 2012
18. Kowalski, R.A.: The relation between logic programming and logic specification. In: Hoare, C., Shepherdson, J. (eds.) Mathematical Logic and Programming Languages, pp. 11–27. Prentice-Hall, Upper Saddle River (1985). Also in Phil. Trans. R. Soc. Lond. A, **312**, 345–361(1984)
19. Pettorossi, A., Proietti, M.: Transformation of logic programs: foundations and techniques. J. Log. Program. **19/20**, 261–320 (1994)

20. Pettorossi, A., Proietti, M., Senni, V.: The transformational approach to program development. In: Dovier, A., Pontelli, E. (eds.) GULP. LNCS, vol. 6125, pp. 112–135. Springer, Heidelberg (2010)
21. Shapiro, E.: Algorithmic Program Debugging. The MIT Press, Cambridge (1983)
22. Stärk, R.F.: The theoretical foundations of LPTP (a logic program theorem prover). J. Log. Program. **36**(3), 241–269 (1998)
23. Sterling, L., Shapiro, E.: The Art of Prolog, 2nd edn. The MIT Press, Cambridge (1994)

Polynomial Approximation to Well-Founded Semantics for Logic Programs with Generalized Atoms: Case Studies

Md. Solimul Chowdhury[1]([✉]), Fangfang Liu[2], Wu Chen[3], Arash Karimi[1],
and Jia-Huai You[1]

[1] Department of Computing Science, University of Alberta, Edmonton, Canada
{mdsolimu,arash.karimi,you}@cs.ualberta.ca
[2] College of Computer and Information Science, Shanghai University,
Baoshan, China
ffliu@shu.edu.cn
[3] College of Computer and Information Science, Southwest University,
Chongqing, China
chenwu@swu.edu.cn

Abstract. The well-founded semantics of normal logic programs has two main utilities, one being an efficiently computable semantics with a unique intended model, and the other serving as polynomial time constraint propagation for the computation of answer sets of the same program. When logic programs are generalized to support constraints of various kinds, the semantics is no longer tractable, which makes the second utility doubtful. This paper considers the possibility of tractable but *incomplete methods*, which in general may miss information in the computed result, but never generates wrong conclusions. For this goal, we first formulate a well-founded semantics for logic programs with generalized atoms, which generalizes logic programs with arbitrary aggregates/constraints/dl-atoms. As a case study, we show that the method of removing non-monotone dl-atoms for the well-founded semantics by Eiter et al. actually falls into this category. We also present a case study for logic programs with standard aggregates.

Keywords: Polynomial approximation · Well-founded semantics · Generalized atoms

1 Introduction

Logic programming with negation is a programming paradigm for declarative knowledge representation and problem solving [5]. In a logic program, a problem is represented as a set of rules and the solutions of the problem are determined under a predefined semantics. In more recent years, logic programs have been incorporated with various types of predefined atoms, to enable reasoning with constraints and external knowledge sources. Examples include logic program

© Springer International Publishing Switzerland 2015
M. Proietti and H. Seki (Eds.): LOPSTR 2014, LNCS 8981, pp. 279–296, 2015.
DOI: 10.1007/978-3-319-17822-6_16

with aggregates [1,12,15,21], logic programs with abstract constraint atoms [16,22], dl-programs [8,10], and logic programs with external sources (e.g. HEX-programs [9]).

For a unifying framework for the study of logic programs, following [2], in this paper a constraint atom in its generic form is called *a generalized atom* and logic programs with these atoms are called logic programs with generalized atoms. The main interest of this paper lies in the two case studies of polynomial approximation to the well-founded semantics of these programs.

It is well-known that the well-founded semantics of normal logic programs can be computed in polynomial time [24]. Besides many applications, the mechanism of computing the well-founded semantics under a partial truth value assignment can be employed in the computation of answer sets. This is utilized, for example, in SMODELS [20] by the Expand function (though the term of well-founded semantics was not used explicitly at that time); it is invoked before and after every choice point in search. To serve as constraint propagation, it is essential that this computational process is effective. However, when constraints are incorporated into logic programs, the resulting well-founded semantics becomes intractable, since determining the satisfiability of a generalized atom by all consistent extensions of a partial interpretation is co-NP-complete [1,21].[1] For the utility in constraint propagation in answer set computation, this is a problem.

In this paper, we consider tractable but incomplete methods. That is, we would like to transform a given program P to another one P' in polynomial time, so that

- the well-founded semantics of P' is tractable, and
- the well-founded semantics of P' specifies a subset of conclusions by the well-founded semantics of P (when restricted to the language of P).

Of course, such a transformation should be non-trivial, as the empty program trivially satisfies these properties.

The approach described above is based on a single transformed program P'. In general, however, the computation of an approximation of the well-founded semantics may be carried out by a collection of program components each of which is employed for some specific computations, as long as the overall process takes polynomial time. We will call all of such approaches incomplete methods. But for simplicity, in the following we feel free to describe the effect of an incomplete method by referring to the approach based on a single transformed program P'.

Incomplete methods have practical implications. For example, for computing the well-founded semantics of a logic program with (arbitrary) generalized atoms, one can always compute such an approximation first. For the utility in answer set computation, let us assume that the well-founded semantics of P approximates the answer sets of P, under a suitable definition of answer set. That is, any atom true (resp. false) in the well-founded semantics of P, called *well-founded* (resp.

---

[1] This assumes that the satisfaction of a generalized atom by one interpretation can be determined in polynomial time.

unfounded), remains to hold in any answer set of P. Then, in the computation of answer sets of P, constraint propagation can be performed by computing the well-founded semantics of P' that extends a partial truth value assignment, at any choice point in search for answer sets. The above conditions guarantee that constraint propagation is correct and effective.

It is well-known that an efficient constraint propagation mechanism is essential in all popular search engines, e.g., BCP in Boolean Satisfiability (SAT) [6], the Expand function in Answer Set Programming (ASP) [20], and various consistency techniques in Constraint Programming (CP) [19].

To study incomplete methods, the first question is which well-founded semantics is to be approximated, and how it is defined. In [11], for (disjunctive) logic programs with arbitrary aggregates, the notion of unfounded set is defined but the well-founded semantics itself is not spelled out explicitly. In [10], a well-founded semantics for logic programs with (a certain kind of) monotone dl-atoms is defined, using a similar notion of unfounded set, followed by a comment (cf. Sect. 9.2) that the same approach can be generalized to programs with arbitrary dl-atoms. But the details about this generalization are not provided. In [1], a well-founded semantics for logic programs with monotone and anti-monotone aggregates is presented, also based on a similar notion of unfounded set. Despite these efforts, to our knowledge, there has been no unified approach such that for all of the above classes of logic programs, the well-founded semantics falls into the same theoretical framework. Apparently, a unified approach is important in studying the common properties of logic programs in syntactically different forms. Thus, our first task is to formally define this well-founded semantics for logic programs with generalized atoms, which are first introduced in [2] for the study of answer set semantics for various classes of logic programs. Hence, our work presented here can be viewed as complementing that of [2] for the study of well-founded semantics.

The well-founded semantics for logic programs with arbitrary generalized atoms is not tractable in the general case. We then carry out two case studies on the possibility of polynomial approximation. In the first, we consider dl-programs with arbitrary dl-atoms. In fact, the authors of [10] give a simple rewrite scheme for removing non-monotone dl-atoms from a program, in which every occurrence of the *constraint operator* $\ominus$ is replaced. It is shown that the data complexity for $\ominus$-free dl-programs is tractable. However, we show that the transformation is not faithful - the well-founded semantics of the transformed program in general is not equivalent to the well-founded semantics of the original program. On the other hand, we show that the rewrite scheme is correct. Namely, though it may miss some conclusions, it never generates incorrect ones. This thus provides the first case of polynomial approximation for a practical class of logic programs with generalized atoms.

We then turn our attention to logic programs with aggregates. We adopt the *disjunctive rewrite scheme* of abstract constraint atoms described in [14] for aggregates with standard aggregate functions [21]. We show that disjunctive rewrite is in general an incomplete method - it may miss conclusions but never generates wrong ones for logic programs with (standard) aggregates. For these programs, we formulate two different methods of disjunctive rewrite to construct a polynomial process for the computation of an approximation of the well-founded semantics.

The rest of the paper is organized as follows: Sect. 2 provides syntax and notations. Section 3 defines the well-founded semantics for logic program with generalized atoms and studies its basic properties. Then, we carry out case studies on polynomial approximation for dl-programs in Sect. 4, and for logic programs with aggregates in Sect. 5, followed by related work and discussion in Sect. 6.

2   Preliminaries

2.1   Language

We assume a language $\mathcal{L}$ that includes a countable set of ground atoms Σ (atoms for short). A *literal* is either an atom or its negation. An *interpretation* is a subset of Σ.

Following the spirit of [2], a *generalized atom* α on Σ is a mapping from interpretations to the Boolean truth values $\mathbf{t}$ and $\mathbf{f}$.[2] There is no particular syntax requirement, except that a generalized atom α is associated with a domain, denoted $Dom(\alpha) \subseteq \Sigma$, and the size of α is $|Dom(\alpha)|$.[3] We assume that a generalized atom may only mention atoms in Σ (i.e., it is not nested). An interpretation I *satisfies* α, written $I \models \alpha$, if α maps I to $\mathbf{t}$, I does not satisfy a, written $I \not\models a$, if a maps I to $\mathbf{f}$.

Intuitively, a generalized atom is a constraint, whose semantics is defined externally by how it may be satisfied by sets of atoms.

Since an atom in Σ can be viewed as a special case of a generalized atom, in the sequel, the term *generalized atoms* also includes all atoms in Σ. At times of possible confusion, we may use the term *ordinary atoms* for those in Σ for distinction. As usual, an interpretation I satisfies an ordinary atom a, written $I \models a$, if $a \in I$ (a maps I to $\mathbf{t}$), and I does not satisfy a, written $I \not\models a$, if $a \notin I$ (a maps I to $\mathbf{f}$).

A generalized atom α is *monotone* if for every interpretation I such that $I \models \alpha$, we have $J \models \alpha$ for all interpretations J such that $J \supseteq I$; it is *anti-monotone* if for every interpretation I such that $I \models \alpha$, we have $J \models \alpha$ for all interpretations J such that $J \subseteq I$; it is *non-monotone* if it is not monotone.

Aggregate atoms, abstract constraint atoms, HEX-atoms, dl-atoms, weight constraints, and global constraints in constraint satisfaction can all be modeled by generalized atoms. Below we sketch some examples.

Example 1.

- An *aggregate atom* consists of a mapping from multi-sets to numbers, a comparison operator, and a value. For example, following the notation of

---

[2] Generalized atoms in [2] are essentially conjunctions of generalized atoms defined here.

[3] Domain is needed primarily for complexity measures. It is also indirectly involved in the notion of satisfaction, which gives the semantics of a generalized atom.

[21], $SUM(\{X|p(X)\}) \geq 1$ denotes a (non-ground) aggregate atom: after grounding one can view the set of ground instances of $p(X)$ as the domain of the atom. The semantics of the aggregate is defined by interpretations in which the sum of the arguments in satisfied $p(.)$ is greater than or equal to 1.

- An *abstract constraint atom* (or a *c-atom*) is of the form (D, S), where $D \subseteq \Sigma$ serves as the *domain* and S is a set of subsets of D representing *admissible solutions* of the c-atom [17]. A c-atom (D, S) is satisfied by an interpretation I iff $I \cap D \in S$. For example, the aggregate atom $SUM(\{X|p(X)\}) \geq 1$ above with domain $D = \{p(1), p(-1), p(2)\}$ can be represented by a c-atom (D, S) where $S = \{\{p(1)\}, \{p(2)\}, \{p(-1), p(2)\}, \{p(1), p(2)\}, \{p(1), p(-1), p(2)\}\}$.
Note that a c-atom contains an *internal* specification of how it may be satisfied by sets of atoms, while the satisfiability of a generalized atom is defined externally. In both cases, the meaning of such an atom is defined by how it is satisfied by sets of atoms, in spite of the syntactic differences in appearance.
- Global constraints in Constraint Satisfaction can be represented by generalized atoms, by giving a name and domain of the constraint. Such a constraint is considered "built-in", i.e., its meaning is pre-defined. For example, a *allDiff* global constraint modeling the pigeon hole problem, can be specified by a generalized atom with the domain that consists of atoms each of which represents a pigeon taking a hole. Suppose there are two pigeons x and y and two holes $\{1, 2\}$. It can be represented by a generalized atom, say g, such that $Dom(g) = \{x(1), x(2), y(1), y(2)\}$, where, for example, $x(1)$ means pigeon x is in hole 1; g is satisfied by interpretations such that every pigeon is in a unique hole and no hole can hold up more than one pigeon.

2.2 Logic Programs with Generalized Atoms

A *logic program with generalized atoms* (or sometimes just called a *program*) is a finite set of rules of the form: $a \leftarrow \beta_1, \ldots, \beta_m, \neg\beta_{m+1}, \ldots, \neg\beta_n$, where $m, n \geq 0$, a is an ordinary atom and β_i $(1 \leq i \leq n)$ are generalized atoms. A rule is *normal* if all β_i are ordinary atoms, and a program is *normal* if all rules in it are normal. A program is called *a logic program with monotone and anti-monotone generalized atoms*, if every generalized atom in P is monotone or anti-monotone.

For a rule r of the above form, the head of the rule is denoted by $H(r) = a$ and the body of the rule by $B(r) = \{\beta_1, \ldots, \beta_m, \neg\beta_{m+1}, \ldots, \neg\beta_n\}$. Also, we define $Pos(r) = \{\beta_1, \ldots, \beta_m\}$ and $Neg(r) = \{\beta_{m+1}, \ldots, \beta_n\}$ to denote positive atoms and negative atoms of $B(r)$ respectively. We may use sets as conjunctions. A *generalized literal* is either a generalized atom α, or its negation, $\neg\alpha$. Note that, without confusion, ordinary literals are (special cases of) generalized literals.

Let I be an interpretation, β a generalized atom and r a rule. Recall that if β maps I to t we say that I satisfies β and write $I \models \beta$. We define that $I \models \neg\beta$ if β maps I to f, and $I \models B(r)$ if $I \models l$ for all $l \in B(r)$. I is a *model* of a set of rules P if I satisfies every $r \in P$.

Let I be an interpretation. By $\bar{I} = \{a \mid a \in \Sigma \backslash I\}$, we denote the set of atoms exclusive of I.

Well-founded semantics are typically defined by building a *partial interpreta-tion*. Let S be a set of literals. We define $S^+ = \{a \mid a \in S\}$, $S^- = \{a \mid \neg a \in S\}$, and $\neg.S = \{\neg a \mid a \in S\}$. S is *consistent* if $S^+ \cap S^- = \emptyset$. A *partial interpreta-tion* S is a consistent subset of $\Sigma \cup \neg.\Sigma$. Any atom not appearing in S is said to be *undefined*. A *consistent extension* of S is an interpretation I such that $S \subseteq I \cup \neg.\bar{I}$. Note that a consistent extension here is a (two-valued) interpreta-tion, i.e., all atoms are assigned a truth value.[4] In the sequel, we restrict Σ to the set of atoms appearing in P. This is typically called the *Herbrand base* of P and is denoted by HB_P.

2.3   Well-Founded Semantics for Normal Logic Programs

To place our work in perspective, we briefly review the well-founded semantics for normal logic programs, which can be defined alternatively in several ways, one of which is based on the notion of unfounded set, which we adopt here.

Let P be a normal logic program and S a partial interpretation. A set $U \subseteq \Sigma$ is *an unfounded set* of P w.r.t. S, if for every $a \in U$ and every rule $r \in P$ with $H(r) = a$, either (i) $\neg b \in S \cup \neg.U$ for some $b \in Pos(r)$, or (ii) $b \in S$ for some $b \in Neg(r)$. The greatest unfounded set of P w.r.t. S always exists, which is denoted by $U_P(S)$.

Intuitively, unfounded atoms are those that can be safely assumed to be false without affecting the evaluation of the rules under the given interpretation.

We then define two operators

- $T_P(S) = \{H(r) \mid r \in P, Pos(r) \cup \neg.Neg(r) \subseteq S\}$
- $W_P(S) = T_P(S) \cup \neg.U_P(S)$

The operator W_P is monotone, and thus has a least fixpoint, which defines the well-founded semantics of P.

3   Well-Founded Semantics for Logic Programs with Generalized Atoms

In the definition of the well-founded semantics for normal logic programs, when an atom a is in a partial interpretation S, it is clear that a remains to be satisfied by all consistent extensions of S. However, this is not the case in general for a non-monotone generalized atom. We thus extend the notion of truth (resp. falsity) to the notion of *persistent truth* (resp. *persistent falsity*) under a partial interpretation.

Definition 1. *Let α be a generalized atom and S a partial interpretation.*

- *if α is an ordinary atom, it is* persistently true *(resp. persistently false) under S if $\alpha \in S^+$ (resp. $\alpha \in S^-$);*

---

[4] This is to be consistent with the notion of an extension of a partial interpretation introduced in [11].

- *Otherwise, α is* persistently true *(resp.* persistently false*) under S if for all consistent extensions I of S, $I \models \alpha$ (resp. $I \not\models \alpha$).*
- *$\neg\alpha$ is* persistently true *under S if α is* persistently false *under S.*
- *$\neg\alpha$ is* persistently false *under S if α is* persistently true *under S.*

Intuitively, a generalized atom α is persistently true (resp. persistently false) relative to a partial interpretation S, iff α is true (resp. false) relative to all consistent extensions of S, i.e., the truth or falsity of α remains unaffected when all undefined atoms w.r.t. S get assigned in any possible way.

The definition above naturally extends to conjunctions of generalized literals. The following definition can be seen as a paraphrase of the notion of unfounded set for logic programs with aggregates [11].

Definition 2 (Unfounded Set). *Let P be a logic program with generalized atoms and S a partial interpretation. A set $U \subseteq HB_P$ is an* unfounded set *of P relative to S if for each $r \in P$ with $H(r) \in U$, some generalized literal in $B(r)$ is persistently false w.r.t. $(S \setminus U) \cup \neg.U$.*

The definition says that an atom a is in an unfounded set U, relative to S, because, for every rule with a in the head, at least one body literal is false in all consistent extensions of $(S \setminus U) \cup \neg.U$.

Definition 3 (Unfounded-Free Interpretation). *Let P be a logic program with generalized atoms and S a partial interpretation. S is* unfounded-free*, if $S \cap U = \emptyset$, for each unfounded set U of P relative to S.*

The following lemma has been proved for logic programs with aggregates [11]. The same proof can be adopted for programs with generalized atoms.

Lemma 1. *Let P be a logic program with generalized atoms and S an unfounded-free interpretation. (i) Unfounded sets of P relative to S are closed under union. (ii) The greatest unfounded set of P relative to S exists, which is the union of all unfounded sets of P relative to S.*

We now define the operators that are needed for the definition of well-founded semantics.

Definition 4. *Let P be a logic program with generalized atoms and S an unfounded free partial interpretation. Define the operators T_P, U_P, and W_P as follows:*

(i) $T_P(S) = \{H(r) \mid r \in P, B(r)$ is persistently true under $S\}$.
(ii) $U_P(S)$ is the greatest unfounded set of P relative to S.
(iii) $W_P(S) = T_P(S) \cup \neg.U_P(S)$.

As a notation, we define $W_P^0 = \emptyset$, and $W_P^{i+1} = W_P(W_P^i)$, for all $i \geq 0$. Note that W_P^0 is an unfounded-free interpretation and so is every W_P^i, for $i \geq 0$. Thus in every step the greatest unfounded set is computed relative to an unfounded-free set.

Lemma 2. *The operators T_P, U_P, and W_P are all monotone.*

Note that the operator W_P is well-defined, i.e., it is a mapping on unfounded-free partial interpretations, which, along with the subset relation, forms a complete lattice. Since W_P is monotone, it follows from the Knaster-Tarski fixpoint theorem [23] that the least fixpoint of W_P exists.

Definition 5 (Well-Founded Semantics). *Let P be a logic program with generalized atoms. The well-founded semantics (WFS) of P, denoted WFS(P), is defined as the least fixpoint of the operator W_P, denoted $lfp(W_P)$. An atom $a \in \Sigma$ is well-founded (resp. unfounded) relative to P iff a (resp. $\neg a$) is in $lfp(W_P)$.*

Observe that, the only difference in the operators defined here from those defined for normal logic programs is in the evaluation of body literals - being true (resp. false) has been replaced by being persistently true (resp. persistently false). It thus follows that the well-founded semantics for logic programs with generalized atoms is a generalization of the well-founded semantics for normal logic programs, and it treats monotone, anti-monotone, and non-monotone generalized atoms in a uniform manner.

Example 2. Consider program P below, where generalized atoms are aggregates.

$$r_1 : p(-1).$$
$$r_3 : p(3) \leftarrow SUM(\{X \mid p(X)\}) > -4.$$
$$r_2 : p(-2) \leftarrow SUM(\{X \mid p(X)\}) \leq 2.$$
$$r_4 : p(-4) \leftarrow SUM(\{X \mid p(X)\}) \leq 0.$$

The aggregates under SUM are self-explaining, e.g., $SUM(\{X \mid p(X)\}) \leq 2$ means that the sum of X for satisfied atoms $p(X)$ is less than or equal to 2. We start with $W_P^0 = \emptyset$, and then $W_P^1 = \{p(-1)\}$. Observe that the body of r_2 is persistently true under W_P^1. We then have $W_P^2 = \{p(-1), p(-2)\}$, and $W_P^3 = \{p(-1), p(-2), p(-4)\}$. Now the body of r_3 is persistently false under W_P^3. So, $W_P^4 = \{p(-1), p(-2), p(-4), \neg p(3)\}$ and, WFS(P) $= lfp(W_P) = W_P^4$.

3.1 Complexity

Here, let us assume that for a generalized atom α, $Dom(\alpha)$ is finite and the relation $I \models \alpha$ (resp. $I \models \neg\alpha$) can be determined in polynomial time in the size of $Dom(\alpha)$. This is the case for practical aggregates in logic programming, reasoning with Horn clauses, and satisfiability testing of the combined complexity for the *DL-Lite* family of description logics [3].

In general, the problem of computing the WFS of a program is intractable since determining whether a generalized atom is persistently true or persistently false under a partial interpretation is in general intractable.

Proposition 1. *Let α be a generalized atom. Checking whether α is persistently true relevant to a partial interpretation S is Co-NP-complete.*

However, the WFS for logic programs with monotone and anti-monotone generalized atoms is tractable. Let α be a generalized atom and S a partial interpretation. To check whether α is persistently true under S, if α is monotone we test $S^+ \models \alpha$, and if α is anti-monotone we test $\Sigma \backslash S^- \models \alpha$. On the other hand, if α is monotone, then α is persistently false under S iff $\Sigma \backslash S^- \not\models \alpha$, and if α is anti-monotone, then α is persistently false under S iff $S^+ \not\models \alpha$.

As the number of distinct atoms is at most the size of P, and the greatest unfounded set can be generated incrementally in polynomial time, the following proposition can be proved in a way similar to the claim for polynomial time computation of WFS of logic programs with monotone and anti-monotone aggregates.

Proposition 2. *Let P be a logic program with monotone and anti-monotone generalized atoms. Then, $lfp(W_P)$ can be computed in polynomial time.*

4   Polynomial Approximation for DL-Programs

In this section we will introduce dl-programs [8,10], show how these programs can be viewed as instances of logic programs with generalized atoms, and present a polynomial approximation to the well-founded semantics of these programs.

4.1   Description Logic Program

We assume that the reader has some familiarity with description logics (DLs) [4], which are decidable fragments of first order logic.

A *dl-program* is a combined knowledge base $KB = (L, P)$, where L is a *DL knowledge base*, which is a collection of axioms in the underlying DL, and P is a *rule base*, which is a finite set of rules of this form: $h \leftarrow A_1, \ldots, A_m, \neg B_1, \ldots, \neg B_n$, where h is an atom, and A_i and B_i are atoms or *dl-atoms*,[5] which are of the form

$$DL[S_1 op_1 p_1, \cdots, S_m op_m p_m; Q](\mathbf{t}) \tag{1}$$

in which S_i is a concept or role from the vocabulary of L, $op_i \in \{\uplus, \cup, \cap\}$, p_i is a predicate symbol only appearing in P whose arity matches that of S_i, and $Q(\mathbf{t})$ is called a *dl-query*, where $\mathbf{t}$ is a list of constants and Q is a concept, a role, or a concept inclusion axiom, built from the vocabulary of L.

Intuitively, $S_i \uplus p_i$ extends S_i by the extension of p_i, and $S_i p_i$ analogously extends $\neg S_i$; the expression $S \cap p_i$ instead constrains S_i to p_i. It is clear that the operator $\cap$ (which we call the *constraint operator*) may cause a dl-program to be non-monotone; a dl-atom which is free of the constraint operator is monotone.

In this paper, we assume that a dl-program is ground, and define the *Herbrand base* of P, denoted HB_P, to be the set of all ground atoms $p(\mathbf{t})$, where p appears in P and $\mathbf{t}$ is a tuple of constants. Interpretations are subsets of HB_P.

Definition 6. *Let $KB = (L, P)$ be a dl-program and $I \subseteq HB_P$ an interpretation. We define the satisfaction relation under L, denoted $\models_L$, as follows:*

---

[5] For simplicity, we assume that equality does not appear in rules.

1. $I \models_L \top$ and $I \not\models_L \bot$.
2. For any atom $a \in HB_P$, $I \models_L a$ if $a \in I$.
3. For any (ground) dl-atom $A = DL[S_1op_1p_1, \cdots, S_mop_mp_m; Q](\mathbf{c})$ occurring in P, $I \models_L A$ if $L \cup \bigcup_{i=1}^m A_i \models Q(\mathbf{c})$, where

$$A_i = \begin{cases} \{S_i(\mathbf{e}) \mid p_i(\mathbf{e}) \in I\}, & \text{if } op_i = \uplus; \\ \{\neg S_i(\mathbf{e}) \mid p_i(\mathbf{e}) \in I\}, & \text{if } op_i = \uplus; \\ \{\neg S_i(\mathbf{e}) \mid p_i(\mathbf{e}) \notin I\}, & \text{if } op_i = \cap. \end{cases}$$

Intuitively, dl-programs can be seen as instances of logic programs with generalized atoms, if each dl-atom is mapped to a generalized atom while preserving its semantics. To make this precise, let us define a mapping ξ from dl-programs to programs with generalized atoms.

Definition 7. Let $KB = (L, P)$ be a dl-program, and $\xi(KB)$ denote the set of rules obtained from KB in the following way: for each $r \in P$, we have $r' \in \xi(KB)$ such that r' is r except that each dl-atom g appearing in r is replaced by a generalized atom α_g such that

– $Dom(\alpha_g)$ is the set of atoms appearing in P, and
– We identify Σ with HB_P, and define the semantics of α_g as: for all interpretations I, $I \models_L g$ iff $I \models \alpha_g$.

That is, ξ is a transformation that preserves satisfiability under $\models_L$.

Definition 8. Let $KB = (L, P)$ be a dl-program. The well-founded semantics (WFS) of KB is defined in terms of the well-founded semantics of $\xi(KB)$.

Example 3. Consider a dl-program $KB = (\emptyset, P)$, where P consists of

$$\begin{aligned} r_1: & \quad p(a) \leftarrow \neg DL[S_1 \cap q, S_2 \uplus r; \neg S_1 \sqcap S_2](a). \\ r_2: & \quad q(a) \leftarrow DL[S \uplus q; S](a). \\ r_3: & \quad r(a) \leftarrow DL[S \cap q; \neg S](a). \end{aligned}$$

Let $P' = \xi(KB)$. We have $W_{P'}^0 = \emptyset$. Next, consider r_1 and the only dl-atom in it. Let $I = \{r(a)\}$, which is a consistent extension of $W_{P'}^0$. By $S_1 \cap q$ in the dl-atom, that $q(a) \notin I$ leads to $\neg S_1(a)$; similarly $r(a) \in I$ leads to $S_2(a)$. Thus, the query $Q[a] = \neg S_1(a) \sqcap S_2(a)$ is satisfied by $\{\neg S_1(a), S_2(a)\}$. Hence $DL[S_1 \cap q, S_2 \uplus r; \neg S_1 \sqcap S_2](a)$ evaluates to true under I and its negation to false. As $\neg DL[S_1 \cap q, S_2 \uplus r; \neg S_1 \sqcap S_2](a)$ is not persistently true, we do not derive $p(a)$. But $\{q(a)\}$ is an unfounded set relative to $\emptyset$, since the dl-atom in the body of rule r_2 is persistently false relative to all consistent extensions of $W_{P'}^0 \cup \{\neg q(a)\}$. The reader can verify that $W_{P'}^1 = \{\neg q(a)\}$, $W_{P'}^2 = \{\neg q(a), r(a)\}$, and $W_{P'}^3 = \{\neg q(a), \neg p(a), r(a)\}$, which is the least fixpoint of $W_{P'}$.

Due to the one-to-one correspondence between the satisfaction under $\models_L$ and that under $\models$, in the following, we may refer to the WFS of a dl-program directly without applying the mapping ξ explicitly. The WFS of a dl-program KB is thus denoted by $lfp(W_{KB})$. We can then apply the notion of persistent truth and falsity directly to dl-programs. In this way, it can be shown easily that the well-founded semantics defined in [10] for dl-programs with dl-atoms that are free of the constraint operator coincides with the well-founded semantics of the corresponding logic program with generalized atoms.

4.2 Removing Non-monotone Dl-Atoms as a Polynomial Approximation

Clearly, non-monotone dl-atoms are the result of applying the constraint operator $⩀$, as the satisfiability of such atoms depends on truth value of propositional atoms exclusive of a given interpretation. In [10], the authors suggest a polynomial time rewrite to remove the constraint operator.

Definition 9 *(Transformation π). Let $KB = (L, P)$ be a dl-program. $\pi(KB)$ is a dl-program obtained from KB by*

1. *replacing each occurrence of $S_i ⩀ p_i$ with $S_i ⊎ \bar{p}_i$ ($\bar{p}_i$ is the complement of p_i), and*
2. *for each atom $p_i(\mathbf{t})$ that occurs in the head of some rule in P, add the following rule*

$$\bar{p}_i(\mathbf{t}) \leftarrow \neg DL[S_i' ⊎ p_i; S_i'](\mathbf{t})$$

where S_i' is a fresh concept or role name.

Note that π is a polynomial transformation.

Since the transformation π does not affect DL knowledge base L in KB, we may write $\pi(P)$ to denote the set of all transformed rules, and $\pi(r)$ to denote the set of transformed rules for $r \in P$. By the transformation π, any dl-program $KB = (L, P)$ can be rewritten to a dl-program $\pi(KB)$ free of the constraint operator, and thus $\pi(KB)$ only contains monotone dl-atoms (then for such a dl-atom α, $\neg\alpha$ is anti-monotone). By Proposition 2, we know that, if query answering with the underlying DL is tractable, then the WFS of $\pi(KB)$ can be computed in polynomial time.

The question is whether the transformation π is faithful. That is, whether it is the case that for all dl-programs KB, the WFS of KB is equivalent to the WFS of $\pi(KB)$, barring freshly added concept/role names in $\pi(KB)$. The following example shows that the answer to this question is no.

Example 4. Consider a single-rule dl-program $KB = (\emptyset, P)$,[6] where P consists of a single rule

$$p(a) \leftarrow DL[S ⊎ p, S ⩀ p; \neg S](a).$$

The WFS of KB is $\{p(a)\}$. On the other hand, $\pi(P)$ consists of two rules

$$p(a) \leftarrow DL[S ⊎ p, S ⊎ \bar{p}; \neg S](a). \qquad \bar{p}(a) \leftarrow \neg DL[S' ⊎ p; S'](a).$$

Clearly, the WFS of $\pi(KB)$ is $\emptyset$, as neither $p(a)$ nor $\bar{p}(a)$ is derivable under $\emptyset$, and neither is unfounded relative to $\emptyset$.

---

[6] This example was originally provided by Yisong Wang.

However, we can show that the transformation π is correct, i.e., given a dl-program KB, when restricted to the language of KB, all the well-founded (resp. unfounded) atoms relative to $\pi(KB)$ are well-founded (resp. unfounded) relative to KB.

Let $KB = (L, P)$ be a dl-program and $\tau(KB)$ its signature. We denote the WFS of $\pi(KB)$, restricted to $\tau(KB)$, by $\text{WFS}(\pi(KB))|_{\tau(KB)}$ (similarly, $lfp(W_{\pi(KB)})|_{\tau(KB)}$). For simplicity, we just write τ for $\tau(KB)$.

We now give the main theorem of this section.

Theorem 1. *Let* $KB = (L, P)$ *be a dl-program. Then* $\text{WFS}(\pi(KB))|_{\tau} \subseteq \text{WFS}(KB)$.

5 Polynomial Approximation for Logic Programs with Aggregates

Aggregates can be viewed as special cases of generalized atoms. In this case, logic programs with aggregates are instances of logic programs with generalized atoms. In this section, we show an incomplete method for approximating the well-founded semantics for these programs.

5.1 Syntax and Semantics of Logic Programs with Aggregates

Following [21], an *aggregate* (or *aggregate atom*) is a constraint on a set of atoms taking the form

$$aggr(\{X \mid p(X)\}) \text{ op } Result \tag{2}$$

where *aggr* is an *aggregate function*. The standard aggregate functions are those in $\{\text{SUM, COUNT, AVG, MAX, MIN}\}$. The set $\{X|p(X)\}$ is called an *intensional set*, where p is a predicate, and X is a variable which takes value from a set $D(X) = \{a_1, \ldots, a_n\}$, called the *variable domain*. The relational operator *op* is from $\{=, \neq, <, >, \leq, \geq\}$ and *Result* is a numeric constant.

The *domain* of an aggregate A, denoted $Dom(A)$, is the set of atoms $\{p(a) \mid a \in D(X)\}$. The size of an aggregate is $|Dom(A)|$.

For an aggregate A, the intensional set $\{X|p(X)\}$, the variable domain $D(X)$, and the domain of an aggregate $Dom(A)$ can also be a multiset which may contain duplicate members. Since multiple occurrences of an atom in a multiset can be represented by distinct symbols whose equivalence can be reinforced by adding simple normal rules, in the following, we only discuss the case of domain being a set.

Let I be an interpretation. I is a *model* of (*satisfies*) an aggregate A, denoted $I \models A$, if $aggr(\{a \mid p(a) \in I \cap Dom(A)\}) \text{ op } Result$ holds, otherwise I is not a model of (does not satisfy) A, denoted $I \not\models A$.

For instance, consider the aggregate $A = SUM(\{X|p(X)\}) \geq 2$, where $D(X) = \{-1, 1, 1, 2\}$. For the sets $I_1 = \{p(2)\}$ and $I_2 = \{p(-1), p(1)\}$, we have $I_1 \models A$ and $I_2 \not\models A$.

A *logic program with aggregates* (or an *aggregate logic program*) is a finite set of rules of this form: $h \leftarrow A_1, \ldots, A_k, \neg B_1, \ldots, \neg B_m, G_1, \ldots, G_n$, where h, A_i and B_j are atoms and G_i are aggregates.

We then can define a mapping from from aggregate programs to programs with generalized atoms, in exactly the same way as in Definition 7. Namely, the domain of an aggregate A is the domain of the corresponding generalized atom g_A and the satisfiability of g_A is identical to that of A.

5.2   Disjunctive Rewrite as a Polynomial Approximation

To optimize programs with constraint atoms, in [14], replacement techniques are studied, where a complex constraint may be decomposed into simpler ones. In one replacement scheme, the authors propose to rewrite a program with disjunctive encoding for c-atoms under the answer set semantics. The idea is to encode a complex c-atom by a disjunction of simpler c-atoms. We apply this idea to aggregates.

A *disjunctive encoding* of an aggregate A is a disjunction of aggregates A_i ($1 \leq i \leq m$), denoted by $d(A_1, \ldots, A_m)$, such that for any interpretation I, $I \models A$ iff $I \models A_i$ ($1 \leq i \leq m$). That is, disjunctive encoding preserves satisfaction.

In [21], the authors show that the determination of persistent truth of an aggregate atom involving SUM/AVG and $\neq$ at the same time is intractable, while determining the same for all other aggregate atoms is tractable. Now, by definition, an aggregate atom A is persistently false under S iff the complement of A is persistently true under S. As a result, determining *persistent falsity* of SUM/AVG involving the $=$ operator is also intractable. Thus the goal of disjunctive rewrite for aggregate logic programs is to transform away aggregates of the form $f(.) \neq c$ for computing well-founded atoms and $f(.) = c$ for computing unfounded atoms, where $f \in \{SUM, AVG\}$.

Definition 10. (Disjunctive Rewrite). *Let P be a logic program with aggregates. The disjunctive rewrite of P produces two programs, one for polynomial time computation of well-founded atoms of P, denoted as P_w and the other for polynomial time computation of unfounded atoms of P, denoted as P_u. We define P_w as:*

For each occurrence of aggregate atom of the form $f(.) \neq c$ in P, where $f \in \{SUM, AVG\}$, we replace that atom with a unique symbol α and add the following two rules: $\alpha \leftarrow f(.) > c$ and $\alpha \leftarrow f(.) < c$.

and define P_u as:

For each occurrence of aggregate atom of the form $f(.) = c$ in P, where $f \in \{SUM, AVG\}$, we replace that atom with the conjunction of two aggregates, $f(.) \leq c$ and $f(.) \geq c$.

By an abuse of notation, let us denote the pair of programs P_w and P_u by $P_{(w,u)}$. Now, we revise the definition of the operator W_P of Definition 4 as follows:

$$W_{P_{(w,u)}}(S) = T_{P_w}(S) \cup \neg.U_{P_u}(S).$$

It can be shown that the operator $W_{P_{(w,u)}}$ is monotone, thus its least fixpoint can be computed iteratively. Again, let us denote by WFS($P_{(w,u)}$) the least fixpoint of the operator $W_{P_{(w,u)}}$.

The following example shows that the disjunctive rewrite is an incomplete method.

Example 5. Consider the following aggregate logic program P

$$p(2) \leftarrow SUM(\{X|p(X)\}) \neq -1. \qquad p(-3) \leftarrow p(2). \qquad p(1) \leftarrow.$$

where $HB_P = \{p(1), p(2), p(-3)\}$. WFS($P$) is computed as follows: $W_P^0 = \emptyset$, $W_P^1 = \{p(1)\}$, $W_P^2 = \{p(1), p(2)\}$, and $W_P^3 = \{p(1), p(2), p(-3)\}$, which is the least fixpoint of W_P.

By disjunctive rewrite, we have P_w below

$$p(2) \leftarrow \alpha. \qquad p(-3) \leftarrow p(2). \qquad p(1) \leftarrow.$$
$$\alpha \leftarrow SUM(\{X : p(X)\}) > -1. \qquad \alpha \leftarrow SUM(\{X : p(X)\}) < -1.$$

As P contains no aggregate atom of the form $f(.) = c$, we have $P_u = P$. Then, we have $T_{P_w}(\emptyset) = \{p(1)\}$ and $U_{P_u}(\emptyset) = \emptyset$, thus $W_{P_{(w,u)}}(\emptyset) = \{p(1)\}$. It can be verified easily that this is a fixpoint of $W_{P_{(w,u)}}$, i.e., WFS($P_{(w,u)}$) = $\{p(1)\}$.

It can be seen that disjunctive rewrite produces stronger constraints. To show this, let us extend the notion of persistent truth and falsity to disjunction of literals in a natural way. Then, we can see that, given a disjunction of aggregates, say $(f(.) < c) \vee (f(.) > c)$, and a partial interpretation S, the fact that $f(.) < c$ is persistently true under S or $f(.) > c$ is persistently true under S implies that $(f(.) < c) \vee (f(.) > c)$ is persistently true under S, but the converse does not hold in general - that $(f(.) < c) \vee (f(.) > c)$ is satisfied by all consistent extensions of S does not imply that $f(.) < c$ is persistently true under S or $f(.) > c$ is persistently true under S, because it may be due to that some consistent extensions satisfy $f(.) < c$ and others satisfy $f(.) > c$.

Similarly, for the computation of unfounded set, that $f(.) = c$ is persistently false iff $f(.) \neq c$ is persistently true if either $f(.) > c$ is persistently true or $f(.) < c$ is persistently true iff either $f(.) \leq c$ is persistently false or $f(.) \geq c$ is persistently false. The converse for the if statement above does not hold because that $f(.) = c$ is persistently false may be due to the fact that for some consistent extensions $f(.) < c$ holds and for the others $f(.) > c$ holds.

The above arguments actually give a proof sketch of the following theorem.

Theorem 2. *Let P be a logic program with aggregates. Then, WFS($P_{(w,u)}$)$|_{\tau(P)} \subseteq$ WFS(P), where $\tau(P)$ denotes the original language of P.*

6   Related Work and Discussion

The well-founded semantics defined in this paper for logic programs with generalized atoms is based on essentially the same notion of unfounded set formulated in [11]. By the work of [18], for logic programs with aggregates, this well-founded semantics is known to approximate answer sets based on the notion of *conditional satisfaction* [22]. This well-founded semantics is different from that of [25], which approximates *answer sets by reduct* [22]. The WFS of [25] is weaker than the WFS defined here, but without any reduction on complexity.

In [13], the authors present a well-founded semantics for hybrid Minimal Knowledge and Negation as Failure (MKNF) knowledge bases, which is a local closed world extension of the MKNF DL knowledge base. The well-founded semantics defined in [13] is shown to be tractable, if the chosen DL fragment is tractable. As shown in Proposition 1 of this paper, even if we assume the entailment relation $I \models \phi$ is tractable, for interpretation I and generalized atom ϕ, computing the WFS is still not. This is inevitable since classic formulas under the scope of negation are anti-monotone while generalized atoms may be neither monotone nor anti-monotone. The precise relationship between the MKNF WFS and the WFS defined here requires further study.

If we assume that the domain of a generalized atom is bounded [2] the well-founded semantics can be computed in polynomial time. This assumption is reasonable only for generalized atoms with small domains.

For improving propagation efficiency for HEX-programs, a decision criterion is introduced in [7] to allow to decide if further check is necessary (with the external sources) to complete the computation of the Unfounded Set (UFS) of the guessing program Q obtained from a given HEX-program P w.r.t. an interpretation A. The decision criterion is as follows: is there any atom dependency cycle that exists in P, which contains external edges (e-cycle)? Following this decision criterion, the authors devise a program decomposition technique which decomposes a given HEX-program into two types of components - one type of component is with e-cycles and other type of component does not have e-cycles. UFS checking is needed only for the components which do have e-cycles. Thus this technique avoids UFS checking when it is not necessary. This work however does not prevent complexity jump in constraint propagation. The decision criterion reduces computational cost linearly. In our case studies, we focus on subtle aspects of computation that cause complexity jump, which may be avoided by incomplete methods. We establish the links between such incomplete methods with the well-founded semantics of the underlying logic program.

We wonder whether the idea of incomplete methods can be pursued for HEX-programs in general. If yes, it will be interesting to study characterizations of the type of information that may be captured, or lost, by such an approximation.

Many practical logic programs use weight constraints, which are essentially the *SUM* aggregates. It would be interesting to see whether the methods proposed in this paper are applicable to some of these programs, and if yes, how the propagators for weight constraints of the currently available ASP systems (such as clasp[7]) can benefit from disjunctive rewrite presented in this paper.

A   Appendix: Proof of Theorem 1

Proof. We prove the claim by induction on the construction of $lfp(W_{\pi(KB)})$ and $lfp(W_{KB})$.

(a) Base: $W^0_{\pi(KB)}|_\tau = W^0_{KB} = \emptyset$.

(b) Step: Assume, for all $k \geq 0$, $W^k_{\pi(KB)}|_\tau \subseteq W^k_{KB}$, and prove $W^{k+1}_{\pi(KB)}|_\tau \subseteq W^{k+1}_{KB}$. By definition, we know

$$W^{k+1}_{\pi(KB)}|_\tau = T_{\pi(KB)}(W^k_{\pi(KB)})|_\tau \cup \neg U_{\pi(KB)}(W^k_{\pi(KB)})|_\tau \tag{3}$$

$$W^{k+1}_{KB} = T_{KB}(W^k_{KB}) \cup \neg . U_{KB}(W^k_{KB}) \tag{4}$$

To prove $W^{k+1}_{\pi(KB)}|_\tau \subseteq W^{k+1}_{KB}$, it is sufficient to prove both of

$$T_{\pi(KB)}(W^k_{\pi(KB)})|_\tau \subseteq T_{KB}(W^k_{KB}) \tag{5}$$

$$U_{\pi(KB)}(W^k_{\pi(KB)})|_\tau \subseteq U_{KB}(W^k_{KB}) \tag{6}$$

Below, let us assume that at most one dl-atom may appear in a rule. The proof can be generalized to arbitrary rules by the same argument, for the transformation of one dl-atom at a time.

(i) We first prove (5). Let $a \in T_{\pi(KB)}(W^k_{\pi(KB)})|_\tau$. By definition, $\exists r' \in \pi(P)$ such that $B(r')$ is persistently true under $W^k_{\pi(KB)}$. WLOG, assume for some $r \in P$, $\pi(r) = \{r', r''\}$, as illustrated in (7) below, in which a dl-atom appears positively, which is replaced by rules in (7) of $\pi(r)$.

$$r : a \leftarrow \ldots, DL[\ldots, S_j \cap p_j, \ldots; Q](e), \ldots \tag{7}$$
$$r' : a \leftarrow \ldots, DL[\ldots, S_j \uplus \bar{p}_j, \ldots; Q](e), \ldots$$
$$r'' : \bar{p}_j(\mathbf{e}) \leftarrow \neg DL[S'_j \uplus p_j; S'](\mathbf{e}) \tag{8}$$

Let D denote the dl-atom in (7), and D' the corresponding dl-atom in (7). If the operator $\cap$ does not occur in D, then trivially $B(r)$ is persistently true under W^k_{KB}, and thus $a \in T_{KB}(W^k_{KB})$.

Otherwise, since $B(r')$ is persistently true under $W^k_{\pi(KB)}$, we have D' is persistently true under $W^k_{\pi(KB)}$, and by the fact that D' is monotone, we have $W^k_{\pi(KB)} \models_L D'$. The atom $\bar{p}_j(\mathbf{e})$ may or may not play a role in the entailment $L \cup \bigcup_{i=1}^m D'_i \models Q(\mathbf{e})$ (cf. Definition (6)). The proof is trivial if it does not. Otherwise, $\bar{p}_j(\mathbf{e})$ is well-founded already w.r.t. $W^k_{\pi(KB)}$, and by the last rule in (7), $p_j(\mathbf{e})$ is unfounded w.r.t. $W^{k'}_{\pi(KB)}$, for some $k' < k$. By induction

---

hypothesis, we know $W^{k'}_{\pi(KB)}|_\tau \subseteq W^{k'}_{KB}$, thus $p_j(\mathbf{e})$ is unfounded w.r.t. W^k_{KB}. It follows D is persistently true under W^k_{KB}, and by the assumption that D is the only dl-atom in (7) and that $B(r')$ is persistently true under $W^k_{\pi(KB)}$, we have $B(r)$ is persistently true under W^k_{KB}. Thus, $a \in T_{KB}(W^k_{KB})$.

If D appears negatively in rule body, the proof is similar because, given a partial interpretation S, $\neg D$ is persistently true under S iff D is persistently false under S, and we just need to swap *well-founded* and *unfounded* in the argument above.

(ii) To prove (6), assume that $a \in U_{\pi(KB)}(W^k_{\pi(KB)})|_\tau$ and we show $a \in U_{KB}(W^k_{KB})$. Consider the case of (7). WLOG, assume that r is the only rule in P with a in the head. If the fact $a \in U_{\pi(KB)}(W^k_{\pi(KB)})$ is independent of $\bar{p}_j(\mathbf{e})$, the proof is trivial. Otherwise, that $a \in U_{\pi(KB)}(W^k_{\pi(KB)})$ is because D' is persistently false under $\neg.U' \cup W^k_{\pi(KB)}$, where U' is the greatest unfounded set relative to $W^k_{\pi(KB)}$. This implies that $\bar{p}_j(\mathbf{e})$ must be unfounded w.r.t. $W^k_{\pi(KB)}$, and it follows that $p_j(\mathbf{e})$ is well-founded already w.r.t. $W^{k'}_{\pi(KB)}$, for some $k' < k$. Then, by induction hypothesis, we have that $p_j(\mathbf{e})$ is well-founded w.r.t. W^k_{KB}, which implies that $B(r)$ is persistently false under $\neg.U \cup W^k_{KB}$, where U is the greatest unfounded set w.r.t. W^k_{KB}. It follows $a \in U_{KB}(W^k_{KB})$. The proof for the case where a dl-atom appears negatively in rule body is similar.

Hence, the proof is completed. □

References

1. Alviano, M., Calimeri, F., Faber, W., Leone, N., Perri, S.: Unfounded sets and well-founded semantics of answer set programs with aggregates. J. Artif. Intell. Res. **42**, 487–527 (2011)
2. Alviano, M., Faber, W.: The complexity boundary of answer set programming with generalized atoms under the FLP semantics. In: Cabalar, P., Son, T.C. (eds.) LPNMR 2013. LNCS, vol. 8148, pp. 67–72. Springer, Heidelberg (2013)
3. Artale, A., Calvanese, D., Kontchakov, R., Zakharyaschev, M.: The DL-Lite family and relations. J. Artif. Intell. Res. **36**, 1–69 (2009)
4. Baader, F., Calvanese, D., McGuinness, D., Nardi, D., Patel-Schneider, P.F. (eds.): The Description Logic Handbook: Theory, Implementation and Applications. Cambridge University Press, Cambridge (2003)
5. Baral, C.: Knowledge Representation, Reasoning and Declarative Problem Solving. Cambridge University Press, New York (2003)
6. Bordeaux, L., Hamadi, Y., Zhang, L.: Propositional satisfiability and constraint programming: a comparative survey. ACM Comput. Surv. **38**(4), 1–54 (2006)
7. Eiter, T., Fink, M., Krennwallner, T., Redl, C., Schüller, P.: Efficient hex-program evaluation based on unfounded sets. J. Artif. Intell. Res. (JAIR) **49**, 269–321 (2014)
8. Eiter, T., Ianni, G., Lukasiewicz, T., Schindlauer, R., Tompits, H.: Combining answer set programming with description logics for the semantic web. Artif. Intell. **172**(12–13), 1495–1539 (2008)
9. Eiter, T., Ianni, G., Schindlauer, R., Tompits, H.: A uniform integration of higher-order reasoning and external evaluations in answer-set programming. In: IJCAI, pp. 90–96 (2005)

10. Eiter, T., Lukasiewicz, T., Ianni, G., Schindlauer, R.: Well-founded semantics for description logic programs in the semantic web. ACM Trans. Comput. Log. **12**(2) (2011), Article 3
11. Faber, W.: Unfounded sets for disjunctive logic programs with arbitrary aggregates. In: Baral, C., Greco, G., Leone, N., Terracina, G. (eds.) LPNMR 2005. LNCS (LNAI), vol. 3662, pp. 40–52. Springer, Heidelberg (2005)
12. Faber, W., Pfeifer, G., Leone, N., Dell'Armi, T., Ielpa, G.: Design and implementation of aggregate functions in the DLV system. TPLP **8**(5–6), 545–580 (2008)
13. Knorr, M., Júlio Alferes, J., Hitzler, P.: Local closed world reasoning with description logics under the well-founded semantics. Artif. Intell. **175**(9–10), 1528–1554 (2011)
14. Liu, G., Goebel, R., Janhunen, T., Niemelä, I., You, J.-H.: Strong equivalence of logic programs with abstract constraint atoms. In: Delgrande, J.P., Faber, W. (eds.) LPNMR 2011. LNCS, vol. 6645, pp. 161–173. Springer, Heidelberg (2011)
15. Liu, G., You, J.-H.: Lparse programs revisited: semantics and representation of aggregates. In: Garcia de la Banda, M., Pontelli, E. (eds.) ICLP 2008. LNCS, vol. 5366, pp. 347–361. Springer, Heidelberg (2008)
16. Liu, L., Pontelli, E., Cao Son, T., Truszczyński, M.: Logic programs with abstract constraint atoms. Artif. Intell. **174**, 295–315 (2010)
17. Marek, V.W., Truszczynski, M: Logic programs with abstract constraint atoms. In: Proceedings of AAAI-04, pp. 86–91 (2004)
18. Pelov, N., Denecker, M., Bruynooghe, M.: Well-founded and stable semantics of logic programs with aggregates. Theory Pract. Log. Program. **7**, 301–353 (2007)
19. Rossi, F., Van Beek, P., Walsh, T. (eds.): Global constraints. Handbook of Constraint Programming. Elsevier, New York (2006)
20. Simons, P., Niemelä, I., Soininen, T.: Extending and implementing the stable model semantics. Artif. Intell. **138**(1–2), 181–234 (2002)
21. Cao Son, T., Pontelli, E.: A constructive semantic characterization of aggregates in answer set programming. TPLP **7**(3), 355–375 (2007)
22. Pontelli, E., Huy Tu, P.: Answer sets for logic programs with arbitrary abstract constraint atoms. J. Artif. Intell. Res. **29**, 353–389 (2007)
23. Tarski, A.: A lattice-theoretical fixpoint theorem and its applications. Pac. J. Math. **5**(2), 285–309 (1955)
24. Van Gelder, A., Ross, K.A., Schlipf, J.S.: The well-founded semantics for general logic programs. J. ACM **38**(3), 620–650 (1991)
25. Wang, Y., Lin, F., Zhang, M., You, J.-H.: A well-founded semantics for basic logic programs with arbitrary abstract constraint atoms. In: AAAI (2012)

Program Transformation and
Optimization

Declarative Compilation for Constraint Logic Programming

Emilio Jesús Gallego Arias[1], James Lipton[2(✉)], and Julio Mariño[3]

[1] University of Pennsylvania, Philadelphia, USA
emilioga@cis.upenn.edu
[2] Wesleyan University, Middletown, USA
jlipton@wesleyan.edu
[3] Universidad Politécnica de Madrid, Madrid, Spain
jmarino@fi.upm.es

Abstract. We present a new declarative compilation of logic programs with constraints into variable-free relational theories which are then executed by rewriting. This translation provides an algebraic formulation of the abstract syntax of logic programs. Management of logic variables, unification, and renaming apart is completely elided in favor of algebraic manipulation of variable-free relation expressions. We prove the translation is sound, and the rewriting system complete with respect to traditional SLD semantics.

Keywords: Logic programming · Constraint programming · Relation algebra · Rewriting · Semantics

1 Introduction

Logic programming is a paradigm based on proof search and directly programming with logical theories. This is done to achieve *declarative transparency*: guaranteeing that execution respects the mathematical meaning of the program. The power that such a paradigm offers comes at a cost for formal language research and implementation. Management of logic variables, unification, renaming variables apart and proof search are cumbersome to handle formally. Consequently, it is often the case that the formal definition of these aspects is left outside the semantics of programs, complicating reasoning about them and the introduction of new declarative features.

We address this problem here by proposing a new compilation framework – based on ideas of Tarski [21] and Freyd [9] – that encodes logic programming syntax into a variable-free algebraic formalism: relation algebra. Relation algebras are pure equational theories of structures containing the operations of composition, intersection and convolution. An important class of relation algebras is the so-called *distributive relation algebras with quasi-projections*, which also incorporate union and projections.

We present the translation of constraint logic programs to such algebras in 3 steps. First, for a CLP program P with signature Σ, we define its associated

© Springer International Publishing Switzerland 2015
M. Proietti and H. Seki (Eds.): LOPSTR 2014, LNCS 8981, pp. 299–316, 2015.
DOI: 10.1007/978-3-319-17822-6_17

relation algebra $\mathbf{QRA}_\Sigma$, which provides both the target object language for program translation and formal axiomatization of constraints and logic variables. Second, we introduce a constraint compilation procedure that maps constraints to variable-free relation terms in $\mathbf{QRA}_\Sigma$. Third, a program translation procedure compiles constraint logic programs to an equational theory over $\mathbf{QRA}_\Sigma$.

The *key feature* of the semantics and translation is its variable-free nature. Programs that contain logical variables are represented as ground terms in our setting, thus all reasoning and execution is reduced to algebraic equality, allowing the use of rewriting. The resulting system is sound and complete with respect to SLD resolution. Our compilation provides a solution to the following problems:

- Underspecification of abstract syntax and logic variable management in logic programs: solved by the the inclusion of metalogical operations directly into the compilation process.
- Interdependence of compilation and execution strategies: solved by making target code completely orthogonal to execution.
- Lack of transparency in compilation (for subsequent optimization and abstract interpretation): solved by making target code a low-level yet *fully declarative* translation of the original program.

Variable Elimination and Relation Composition. We illustrate the spirit of translation, and in particular the variable elimination procedure, by considering a simple case, namely the transitive closure of a graph:

```
edge(a,b).            connected(X,X).
edge(b,c).            connected(X,Y) :- edge(X,Z), connected(Z,Y).
edge(a,e).
edge(e,f).
```

In this carefully chosen example, the elimination of variables and the translation to binary relation symbols is immediate:

$$\mathbf{edge} = (a,b) \cup (b,c) \cup (a,e) \cup (a,e) \cup (e,f)$$
$$\mathbf{connected} = \mathbf{id} \cup \mathbf{edge}; \mathbf{connected}$$

The key feature of the resulting term is the composition $\mathbf{edge}; \mathbf{connected}$. The logical variable Z is eliminated by the composition of relations allowing the use of variable free object code. A query $\mathbf{connected}(a, X)$ is then modeled by the relation $\mathbf{connected} \cap (a,a)\mathbf{1}$ where $\mathbf{1}$ is the (maximal) universal relation. Computation can proceed by rewriting the query using a suitable orientation of the relation algebra equations and unfolding pertinent recursive definitions.

Handling actual arbitrary constraint logic programs is more involved. First, we use sequences and projection relations to handle predicates involving an arbitrary number of arguments and an unbounded number of logic variables; second, we formalize constraints in a relational way.

Projections and permutations algebraically encode all the operations of logical variables – disjunctive and conjunctive clauses are handled with the help of the standard relational operators $\cap$, $\cup$.

Constraint Logic Programming Conventions. We refer the reader to [16] for basic definitions of logic programming over Horn clauses, and [12] for background on the syntax and semantics of constraint logic programming. In this paper we fix a signature Σ, a set of terms $\mathcal{T}_\Sigma(\mathcal{X})$, and a subset $\mathcal{C}$ of all first-order formulas over Σ closed under conjunction and existential quantification to be the set of *constraint formulas* as well as a Σ-structure $\mathcal{D}$, called the *constraint domain*. Constraint logic programs are sets of Horn clauses. We use vector notation extensively in the paper, to abbreviate Horn clauses with constraints $p \leftarrow q_1, \ldots, q_n$, where p is an atomic formula and q_i may be an atomic formula or a constraint. For instance, in our vector notation, a clause is written $p(\boldsymbol{t}[\boldsymbol{x}]) \leftarrow \boldsymbol{q}(\boldsymbol{u}[\boldsymbol{x}, \boldsymbol{y}])$, where the boldface symbols indicate vectors of variables $\boldsymbol{x}, \boldsymbol{y}$, terms $\boldsymbol{t}, \boldsymbol{u}$ (depending on variables $\boldsymbol{x}$, etc...) and predicates $\boldsymbol{q}$ (depending on terms $\boldsymbol{u}$).

2 Relation Algebras and Signatures

In this section, we define $\mathbf{QRA}_\Sigma$, a relation algebra in the style of [9,21] formalizing a CLP signature Σ and a constraint domain $\mathcal{D}$. We define its language, its equational theory and semantics.

2.1 Relational Language and Theory

The relation language R_Σ is built from a set $\mathsf{R}_\mathcal{C}$ of relation constants for constant symbols a set $\mathsf{R}_\mathcal{F}$ of relation constants for function symbols from Σ, and a set of relation constants for primitive predicates $\mathsf{R}_{\mathcal{CP}}$, as well as a fixed set of relation constants and operators detailed below. Let us begin with $\mathsf{R}_\mathcal{C}$. Each constant symbol $a \in \mathcal{C}_\Sigma$ defines a constant symbol $(a, a) \in \mathsf{R}_\mathcal{C}$, each function symbol $f \in \mathcal{F}_\Sigma$ defines a constant symbol R_f in $\mathsf{R}_\mathcal{F}$. Each predicate symbol $r \in \mathcal{CP}_\Sigma$ defines a constant symbol r in $\mathsf{R}_{\mathcal{CP}}$. We write R_Σ for the full relation language:

$$\mathsf{R}_\mathcal{C} = \{(a, a) \mid a \in \mathcal{C}_\Sigma\} \quad \mathsf{R}_\mathcal{F} = \{\mathsf{R}_f \mid f \in \mathcal{F}_\Sigma,\} \quad \mathsf{R}_{\mathcal{CP}} = \{r \mid r \in \mathcal{CP}_\Sigma\}$$
$$\mathsf{R}_{atom} ::= \mathsf{R}_\mathcal{C} \mid \mathsf{R}_\mathcal{F} \mid \mathsf{R}_{\mathcal{CP}} \mid id \mid di \mid \mathbf{1} \mid \mathbf{0} \mid hd \mid tl$$
$$\mathsf{R}_\Sigma ::= \mathsf{R}_{atom} \mid \mathsf{R}_\Sigma^\circ \mid \mathsf{R}_\Sigma \cup \mathsf{R}_\Sigma \mid \mathsf{R}_\Sigma \cap \mathsf{R}_\Sigma \mid \mathsf{R}_\Sigma \mathsf{R}_\Sigma$$

The constants $\mathbf{1}, \mathbf{0}, id, di$ respectively denote the universal relation (whose standard semantics is the set of all ordered pairs on a certain set), the empty relation, the identity (diagonal) relation, and identity's complement. Juxtaposition RR represents relation composition (often written R;R) and R° is the inverse of R. We write hd and tl for the head and tail relations. The projection of an n-tuple onto its i-th element is written P_i and defined as $P_1 = hd, P_2 = tl; hd, \ldots, P_n = tl^{n-1}; hd$.

$\mathbf{QRA}_\Sigma$ (Fig. 1) is the standard theory of distributive relation algebras, plus Tarski's quasiprojections [21], and equations axiomatizing the new relations of R_Σ. Note that products and their projections are axiomatized in a relational, variable-free manner.

$$R \cap R = R \qquad R \cap S = S \cap R \qquad R \cap (S \cap T) = (R \cap S) \cap T$$
$$R \cup R = R \qquad R \cup S = S \cup R \qquad R \cup (S \cup T) = (R \cup S) \cup T$$
$$R \, id = R \qquad R0 = 0 \qquad 0 \subseteq R \subseteq 1$$
$$R \cup (S \cap R) = R = (R \cup S) \cap R$$
$$R(S \cup T) = RS \cup RT \qquad (S \cup T)R = SR \cup TR$$
$$R \cap (S \cup T) = (R \cap S) \cup (R \cap T)$$
$$(R \cup S)^\circ = R^\circ \cup S^\circ \qquad (R \cap S)^\circ = S^\circ \cap R^\circ$$
$$R^{\circ\circ} = R \qquad (RS)^\circ = S^\circ R^\circ$$
$$R(S \cap T) \subseteq RS \cap RT \qquad RS \cap T \subseteq (R \cap TS^\circ)S$$
$$id \cup di = 1 \qquad id \cap di = 0$$

$$hd(hd)^\circ \cap tl(tl)^\circ \subseteq id \qquad (hd)^\circ hd \subseteq id, \ (tl)^\circ tl \subseteq id \qquad (hd)^\circ tl = 1$$
$$1(c,c)1 = 1 \qquad (c,c) \subseteq id$$

Fig. 1. QRA$_\Sigma$

2.2 Semantics

Let Σ be a constraint signature and $\mathcal{D}$ a Σ-structure. Write $t^{\mathcal{D}}$ for the interpretation of a term $t \in \mathcal{T}_\Sigma$. We define $\mathcal{D}^\dagger$ to be the union of $\mathcal{D}^0 = \{\langle\rangle\}$ (the empty sequence), $\mathcal{D}$ and $\mathcal{D}$-finite products, for example: $\mathcal{D}^2, \mathcal{D}^2 \times \mathcal{D}, \mathcal{D} \times \mathcal{D}^2, \ldots$ We write $\langle a_1, \ldots, a_n \rangle$ for members of the n-fold product associating to the right, that is to say, $\langle a_1, \langle a_2, \ldots, \langle a_{n-1}, a_n \rangle \cdots \rangle\rangle$. Furthermore, we assume right-association of products when parentheses are absent. Note that the 1 element sequence does not exist in the domain, so we write $\langle a \rangle$ for a as a convenience.

Let $\mathsf{R}_\mathcal{D} = \mathcal{D}^\dagger \times \mathcal{D}^\dagger$. We make the power set of $\mathsf{R}_\mathcal{D}$ into a model of the relation calculus by interpreting atomic relation terms in a certain canonical way, and the operators in their standard set-theoretic interpretation. We interpret hd and tl as projections in the model.

Definition 1. *Given a structure $\mathcal{D}$ a relational $\mathcal{D}$-**interpretation** is a mapping $[\![\_]\!]^{\mathcal{D}^\dagger}$ of relational terms into $\mathsf{R}_\mathcal{D}$ satisfying the identities in Fig. 2. The function α used in this table and elsewhere in this paper refers to the arity of its argument, whether a relation or function symbol from the underlying signature.*

Theorem 1. *Equational reasoning in $\mathbf{QRA}_\Sigma$ is sound for any interpretation:*

$$\mathbf{QRA}_\Sigma \vdash R = S \implies [\![R]\!]^{\mathcal{D}^\dagger} = [\![S]\!]^{\mathcal{D}^\dagger}$$

3 Program Translation

We define constraint and program translation to relation terms. To this end, we define a function $\dot{K}$ from constraint formulas with – possibly free – logic variables to a variable-free relational term. $\dot{K}$ is the core of the variable elimination mechanism and will appear throughout the rest of the paper.

$$[\![\mathbf{1}]\!]^{\mathcal{D}^\dagger} = \mathsf{R}_A \qquad\qquad [\![tl]\!]^{\mathcal{D}^\dagger} = \{(\langle a,b\rangle, b) \mid a,b \in \mathcal{D}^\dagger\}$$
$$[\![\mathbf{0}]\!]^{\mathcal{D}^\dagger} = \emptyset \qquad\qquad\qquad [\![R^\circ]\!]^{\mathcal{D}^\dagger} = ([\![R]\!]^{\mathcal{D}^\dagger})^\circ$$
$$[\![id]\!]^{\mathcal{D}^\dagger} = \{(u,u) \mid u \in \mathcal{D}^\dagger\} \qquad [\![R \cup S]\!]^{\mathcal{D}^\dagger} = [\![R]\!]^{\mathcal{D}^\dagger} \cup [\![S]\!]^{\mathcal{D}^\dagger}$$
$$[\![di]\!]^{\mathcal{D}^\dagger} = \{(u,v) \mid u \neq v\} \qquad\quad [\![R \cap S]\!]^{\mathcal{D}^\dagger} = [\![R]\!]^{\mathcal{D}^\dagger} \cap [\![S]\!]^{\mathcal{D}^\dagger}$$
$$[\![hd]\!]^{\mathcal{D}^\dagger} = \{(\langle a,b\rangle, a) \mid a,b \in \mathcal{D}^\dagger\} \quad [\![(c,c)]\!]^{\mathcal{D}^\dagger} = \{(c^{\mathcal{D}}, c^{\mathcal{D}})\}$$
$$[\![RS]\!]^{\mathcal{D}^\dagger} = [\![R]\!]^{\mathcal{D}^\dagger}; [\![S]\!]^{\mathcal{D}^\dagger}$$
$$[\![R_f]\!]^{\mathcal{D}^\dagger} = \{(x, yu) \mid x = f^{\mathcal{D}}(a_1, \ldots, a_n) \wedge y = \langle a_1, \ldots, a_n\rangle, a_i \in \mathcal{D}, u \in \mathcal{D}^\dagger, n = \alpha(f)\}$$
$$[\![r]\!]^{\mathcal{D}^\dagger} = \{(xu, xu) \mid x = \langle a_1, \ldots, a_n\rangle \wedge r^{\mathcal{D}}(a_1, \ldots, a_n), a_i \in \mathcal{D}, u \in \mathcal{D}^\dagger, n = \alpha(r)\}$$

Fig. 2. Standard interpretation of binary relations.

The reader should keep in mind that there are two kinds of predicate symbols in a constraint logic program: *constraint predicates* r which are translated by the function $\acute{K}$ above to relation terms r, and *defined* or program predicates.

We translate defined predicates – and CLP programs – to equations $\overline{p} \doteq R$, where $\overline{p}$ will be drawn from a set of definitional variables standing for program predicate names p, and R is a relation term. The set of definitional equations can be both seen as an executable specification, by understanding it in terms of the rewriting rules given in this paper; or as a declarative one, by unfolding the definitions and using the standard set-theoretic interpretation of binary relations.

3.1   Constraint Translation

We fix a canonical list $x_1, \ldots, x_n$ of variables occurring in all terms, so as to translate them to variable-free relations in a systematic way. There is no loss of generality as later, we transform programs into this canonical form.

Definition 2 (Term Translation). *Define a translation function* $K : \mathcal{T}_\Sigma(\mathcal{X}) \to \mathsf{R}_\Sigma$ *from first-order terms to relation expressions as follows:*

$$K(c) = (c,c)\mathbf{1}$$
$$K(x_i) = P_i^\circ$$
$$K(f(t_1, \ldots, t_n)) = \mathsf{R}_f; \bigcap_{i \leq n} P_i; K(t_i)$$

This translation is extended to vectors of terms as follows $K(\langle t_1, \ldots, t_n\rangle) = \bigcap_{i \leq n} P_i; K(t_i)$.

The semantics of the relational translation of a term is the set of all of the instances of that term, paired with the corresponding instances of its variables. For instance, the term $x_1 + s(s(x_2))$ is translated to the relation $+; (P_1; P_1^\circ \cap P_2; \mathsf{s}; \mathsf{s}; P_2^\circ)$.

Lemma 1. *Let* $t[\mathbf{x}]$ *be a term of* $\mathcal{T}_\Sigma(\mathcal{X})$ *whose free variables are among those in the sequence* $\mathbf{x} = x_1, \ldots, x_m$. *Then, for any sequences* $\mathbf{a} = a_1, \ldots, a_m \in \mathcal{D}^\dagger, \mathbf{u} \in \mathcal{D}^\dagger$ *and any* $b \in \mathcal{D}$ *we have*

$$(b, \mathbf{au}) \in [\![K(t[\mathbf{x}])]\!]^{\mathcal{D}^\dagger} \iff b = t^{\mathcal{D}}[\mathbf{a}/\mathbf{x}]$$

We will translate constraints over m variables to partially coreflexive relations over the elements that satisfy them. A binary relation R is *coreflexive* if it is contained in the identity relation, and it is *i-coreflexive* if its i-th projection is contained in the *identity relation*: $P_i^\circ ; R; P_i \subseteq id$. Thus, for a variable x_i *free* in a constraint, the translation will be i-coreflexive.

We now formally define two *partial identity relation expressions* I_m, Q_i for the translation of existentially quantified formulas, in such a way that if a constraint $\varphi[x]$ over m variables is translated to an m-coreflexive relation, the formula $\exists x_i.\ \varphi[x]$ corresponds to a coreflexive relation in all the positions but the i-th one, as x_i is no longer free. In this sense Q_i may be seen as a *hiding* relation.

Definition 3. *The partial identity relation expressions I_m, Q_i for $m, i > 0$ are defined as:*

$$I_m := \bigcap_{1 \leq i \leq m} P_i(P_i)^\circ \qquad Q_i = I_{i-1} \cap J_{i+1} \qquad J_i = tl^i; (tl^\circ)^i$$

I_m is the identity on sequences up to the first m elements. Q_i is the identity on all but the i-th element, with the i-th position relating arbitrary pairs of elements.

Definition 4 (Constraint Translation). *The $\dot{K} : \mathcal{L}_D \to R_\Sigma$ translation function for constraint formulas is:*

$$\dot{K}(p(t_1, \ldots, t_n)) = (\bigcap_{i \leq n} K(t_i)^\circ; P_i^\circ); \mathsf{p}; (\bigcap_{i \leq n} P_i; K(t_i))$$
$$\dot{K}(\varphi \wedge \theta) \qquad = \dot{K}(\varphi) \cap \dot{K}(\theta)$$
$$\dot{K}(\exists x_i.\ \varphi) \qquad = Q_i; \dot{K}(\varphi); Q_i$$

As an example, the translation of the constraint $\exists x_1, x_2.s(x_1) \leq x_2$ is

$$Q_1; Q_2; (P_1^\circ; s^\circ; P_1 \cap P_2^\circ; P_2); \leq; (P_1; s; P_1^\circ \cap P_2; P_2^\circ); Q_1; Q_2$$

Lemma 2. *Let $\varphi[x]$ be a constraint formula with free variables among $x = x_1, \ldots, x_m$. Then, for any sequences $a = a_1, \ldots, a_m$, u and u' of members of $\mathcal{D}$*

$$(au, au') \in [\![\dot{K}(\varphi[x])]\!]^{\mathcal{D}^\dagger} \iff \mathcal{D} \models \varphi[a/x]$$

3.2   Translation of Constraint Logic Programs

To motivate the technical definitions below, we illustrate the program translation procedure with an example. Assume a language with constant 0, a unary function symbol s, constraint predicate $=$ and program predicate *add*. We can write the traditional Horn clause definition of Peano addition:

```
add(0,X,X).
add(s(X),Y,s(Z)) :- add(X,Y,Z).
```

This program is first *purified:* the variables in the head of the clauses defining each predicate are chosen to be a sequence of fresh variables x_1, x_2, x_3, with all bindings stated as equations in the tail.

$$add(x_1, x_2, x_3) \longleftarrow x_1 = 0, x_2 = x_3.$$
$$add(x_1, x_2, x_3) \longleftarrow \exists x_4, x_5 x_1 = s(x_4), x_3 = s(x_5), add(x_4, x_2, x_5))$$

The clauses are combined into a single definition similar to the Clark completion of a program. We also use the variable permutation π sending $x_1, x_2, x_3, x_4, x_5 \mapsto x_4, x_2, x_5, x_1, x_3$ to rewrite the occurrence of the predicate add in the tail so that its arguments coincide with those in the head:

$$add(x_1, x_2, x_3) \leftrightarrow (x_1 = 0, x_2 = x_3)$$
$$\vee \exists x_4, x_5, x_1 = s(x_4), x_3 = s(x_5), w_\pi \, add(x_1, x_2, x_3).$$

Now we apply relational translation $\dot{K}$ defined above to all relation equations, and eliminate the existential quantifier using the *partial identity operator* I_3 defined above. We represent the permutation π using the relation expression W_π that simulates its behavior in a variable-free manner and replace the predicate add with a corresponding *relation variable* $\overline{add}$. (A formal definition of W_π and its connection with function w_π is given below, see Definition 7 and Lemma 4.)

$$\overline{add} \stackrel{\circ}{=} \dot{K}(x_1 = o \wedge x_2 = x_3) \cup I_3((\dot{K}(x_1 = s(x_4) \wedge x_3 = s(x_5)) \cap W_\pi \, \overline{add} \, W_\pi^\circ)))$$

Now we give a description of the general translation procedure. We first process programs to their complete database form as defined in [6], which given the executable nature of our semantics reflects the choice to work within the minimal semantics. The main difference in our processing of a program P to its completed form P' is that a strict policy on variable naming is enforced, so that the resulting completed form is suitable for translation to relational terms.

Definition 5 (General Purified Form for Clauses). *For a clause $p(t[y]) \leftarrow q(v[y])$, let $h = \alpha(p)$, $y = |y|$, $v = |v|$, and $m = h + y + v$. Assume vectors:*

$$\begin{aligned}
x &= x_h x_t = x_h x_y x_v = x_1, \ldots, x_h, x_{h+1}, \ldots, x_{h+y}, x_{h+y+1}, \ldots, x_m \\
x_h && = x_1, \ldots, x_h \\
x_t & = x_y x_v = & x_{h+1}, \ldots, x_{h+y}, x_{h+y+1}, \ldots, x_m \\
x_y && = & x_{h+1}, \ldots, x_{h+y} \\
x_v && = & x_{h+y+1}, \ldots, x_m
\end{aligned}$$

the clause's GPF form is:

$$p(x_h) \leftarrow \exists^{h\uparrow}.((x_h = t[x_y] \wedge x_v = v[x_y]), q(x_v))$$

$\exists^{n\uparrow}$ denotes existential closure with respect to all variables whose index is greater than n. x_h and x_t stand for head and tail variables. A program is in GPF form iff every one of its clauses is. After the GPF step, we perform Clark's completion.

Definition 6 (Completion of a Predicate). *We define Clark's completed form for a predicate p with clauses $cl_1, \ldots, cl_n$ in GPF form:*

$$\left.\begin{array}{l} p(\boldsymbol{x}_h) \leftarrow_{cl_1} tl_1 \\ \cdots \\ p(\boldsymbol{x}_h) \leftarrow_{cl_n} tl_k \end{array}\right\} \xRightarrow{\ Clark's\ comp.\ } p(\boldsymbol{x}_h) \leftrightarrow tl_1 \vee \cdots \vee tl_k$$

The above definition easily extends to programs. Completed forms are translated to relations by using $\dot{K}$ for the constraints, mapping conjunction to $\cap$ and $\vee$ to $\cup$. Existential quantification, recursive definitions and parameter passing are handled in a special way which we proceed to detail next.

Existential Quantification: Binding Local Variables. Variables *local* to the tail of a clause are existentially quantified. For technical reasons — simpler rewrite rules — we use the *partial identity* relation I_n, rather than the Q_n relation defined in the previous sections. I_n acts as an existential quantifier for all variables of index greater than a given number.

Lemma 3. *Let $\boldsymbol{a} = a_1, \ldots, a_n \in \mathcal{D}$, $\boldsymbol{x} = x_1, \ldots, x_n$, let φ be a constraint over m free variables, with $m > n$, $\boldsymbol{y}$ a vector of length k such that $n + k = m$, and $\boldsymbol{u}, \boldsymbol{v} \in \mathcal{D}^\dagger$, then:*

$$(\boldsymbol{au}, \boldsymbol{av}) \in [\![I_n; \dot{K}(\varphi[\boldsymbol{xy}]); I_n]\!]^{\mathcal{D}^\dagger} \iff \mathcal{D} \models (\exists^{n\uparrow}.\varphi[\boldsymbol{xy}])[\boldsymbol{a}/\boldsymbol{x}]$$

Recursive Predicate Definitions. We shall handle recursive predicate definitions by extending the relational language with a set of definitional symbols $\overline{p}, \overline{q}, \overline{r}, \ldots$ for predicates. Then, a recursive predicate $\overline{p}$ is translated to a definitional equation $\overline{p} \overset{\circ}{=} R(\overline{p}_1, \ldots, \overline{p}_n)$, spelled out in Definition 8 where the notation $R(\overline{p}_1, \ldots, \overline{p}_n)$ indicates that relation R resulting from the translation may depend on predicate symbols $\overline{p}_1, \ldots, \overline{p}_n$. Note that R is monotone in $\overline{p}_1, \ldots, \overline{p}_n$. Consequently, using a straightforward fixed point construction we can extend the interpretation $[\![\_]\!]^{\mathcal{D}^\dagger}$ to satisfy $[\![\overline{p}]\!]^{\mathcal{D}^\dagger} = [\![R(\overline{p}_1, \ldots, \overline{p}_n)]\!]^{\mathcal{D}^\dagger}$, thus preserving soundness when we adjoin the definitional equations to $\mathbf{QRA}_\Sigma$. The details are given in Subsect. 3.3, below.

Parameter Passing. The information about the order of parameters in each pure atomic formula $p(x_{i_1}, \ldots, x_{i_r})$ is captured using permutations. Given a permutation $\pi : \{1..n\} \to \{1..n\}$, the function w_π on formulas and terms is defined in the standard way by its action over variables. We write W_π for the corresponding relation:

Definition 7 (Switching Relations). *Let $\pi : \{1..n\} \to \{1..n\}$ be a permutation. The switching relation expression W_π, associated to π is:*

$$W_\pi = \bigcap_{j=1}^{n} P_{\pi(j)}(P_j)^\circ.$$

```
male(terach). male(haran). male(isaac). male(lot).

female(sarah). female(milcah). female(yiscah).

father(terach,haran). father(haran,lot). ↩
    father(haran,milcah).

mother(sarah,isaac).

parent(X,Y) ← father(X,Y).
parent(X,Y) ← mother(X,Y).

sibling(S1,S2) ← S1≠S2, parent(Par,S1), parent(Par,S2).

brother(Brother,Sib) ← male(Brother), sibling(Brother,Sib).
```

Fig. 3. Biblical family relations in prolog.

Lemma 4. *Fix a permutation π and its corresponding w_π and W_π. Then:*

$$[\![\dot{K}(w_\pi(p(x_1,\ldots,x_n)))]\!] = [\![W_\pi\dot{K}(p)W_\pi^\circ]\!]$$

The Translation Function. Now we may define the translation for defined predicates.

Definition 8 (Relational Translation of Predicates). *Let $h, p(\boldsymbol{x}_h)$ be as in Definition 5. The translation function Tr from completed predicates to relational equations is defined by:*

$$
\begin{aligned}
Tr(p(\boldsymbol{x}_h) &\leftrightarrow cl_1 \vee \cdots \vee cl_k) = (\overline{p} \stackrel{\circ}{=} Tr_{cl}(cl_1) \cup \cdots \cup Tr_{cl}(cl_k)) \\
Tr_{cl}(\exists^{h\uparrow}.\boldsymbol{p}) &= I_h; (Tr_l(p_1) \cap \cdots \cap Tr_l(p_n)); I_h \\
Tr_l(\varphi) &= \dot{K}(\varphi) \qquad\qquad\qquad \varphi \text{ a constraint} \\
Tr_l(p_i(\boldsymbol{x}_i)) &= W_\pi; \overline{p_i}; W_\pi^\circ \quad \text{such that } \pi(x_1,\ldots,x_{\alpha(p_i)}) = \boldsymbol{x}_i
\end{aligned}
$$

where $\boldsymbol{x}_i$ is the original sequence of variables in p_i in the Clark completion of the program, and π a permutation that transforms the ordered sequence of length $\alpha(p)$ starting at x_1 to $\boldsymbol{x}_i$.

We will sometimes write $I_n(R)$ for $I_n R I_n$ and $W_\pi(R)$ for $W_\pi R W_i^\circ$.

Example 1. Figure 3 shows a fragment of a constraint logic program to represent a family relations database [20].

Consider the translation of the program predicates mother, parent, sibling and brother. We write the program in general purified form:

$$
\begin{aligned}
mother(x_1,x_2) &\Longleftrightarrow (x_1 = sarah) \wedge (x_2 = isaac) \\
parent(x_1,x_2) &\Longleftrightarrow father(x_1,x_2) \vee mother(x_1,x_2) \\
sibling(x_1,x_2) &\Longleftrightarrow \exists x_3.\ x_1 \neq x_2 \wedge parent(x_3,x_1) \wedge parent(x_3,x_2) \\
brother(x_1,x_2) &\Longleftrightarrow male(x_1) \wedge sibling(x_1,x_2)
\end{aligned}
$$

Letting σ_1 and σ_2 be the permutations $\langle 1, 2, 3 \rangle \longrightarrow \langle 2, 3, 1 \rangle$ and $\langle 1, 2, 3 \rangle \longrightarrow \langle 3, 2, 1 \rangle$ respectively we obtain

$$\overline{mother} = \dot{K}(x_1 = sarah) \cap \dot{K}(x_2 = isaac)$$
$$\overline{parent} = \overline{father} \cup \overline{mother}$$
$$\overline{sibling} = \dot{K}(x_1 \neq x_2) \cap I_2[W_{\sigma_1} \overline{parent} W_{\sigma_1}^o \cap W_{\sigma_2} \overline{parent} W_{\sigma_2}^o] I_2$$
$$\overline{brother} = \overline{male} \cap \overline{sibling}$$

The query $brother(X, milcah)$ leads to the rewriting of the term $\dot{K}(x_2 = milcah) \cap \overline{brother}$ to $\dot{K}(x_2 = milcah) \cap \dot{K}(x_1 = lot)$.

3.3   The Least Relational Interpretation Satisfying Definitional Equations

Let P be a program and $\overline{p}_1, \ldots, \overline{p}_n$ be a sequence of *relation variables*, one for each predicate symbol p_i in the language of P. We define the extended relation calculus $R_\Sigma(\overline{p}_1, \ldots, \overline{p}_n)$ to be the set of terms generated by $\overline{p}_1, \ldots, \overline{p}_n$ and the terms of R_Σ. More formally

$$R_{atom} \quad ::= \overline{p}_1 \mid \cdots \mid \overline{p}_n \mid R_C \mid R_F \mid R_{CP} \mid id \mid di \mid 1 \mid 0 \mid hd \mid tl$$
$$R_\Sigma(\overline{p}_1, \ldots, \overline{p}_n) ::= R_{atom} \mid R_\Sigma{}^\circ \mid R_\Sigma \cup R_\Sigma \mid R_\Sigma \cap R_\Sigma \mid R_\Sigma R_\Sigma$$

Observe that the relational translation of Definition 8 maps programs to sets of definitional equations $\overline{p}_i \doteq R_i(\overline{p}_1, \ldots, \overline{p}_n)$ over $R_\Sigma(\overline{p}_1, \ldots, \overline{p}_n)$. Let $\mathcal{F}$ be the set of all n such definitional equations.

Given a structure $\mathcal{D}$ we now lift the definition of $\mathcal{D}$-*interpretation* given in Definition 1 to the extended relation calculus. An extended interpretation $[\![\ ]\!] : R_\Sigma(\overline{p}_1, \ldots, \overline{p}_n) \longrightarrow R_\mathcal{D}$ is a function satisfying the identities in Fig. 2 as well as mapping each relation variable $\overline{p}_i$ to an arbitrary member $[\![\overline{p}_i]\!]$ of $R_\mathcal{D}$. Given a structure $\mathcal{D}$ for the language of a program, its action is completely determined by its values at the $\overline{p}_i$. Note that the set $\mathcal{I}$ of all such interpretations forms a CPO, a complete partial order with a least element, under pointwise operations. That is to say, any *directed* set $\{[\![\ ]\!]_d : d \in \Lambda\}$ of interpretations has a supremum $\bigvee_{d \in \Lambda} [\![\ ]\!]_d$ given by $T \mapsto \bigcup_{d \in \Lambda} [\![T]\!]_d$. The directedness assumption is necessary. For example, to show that a pointwise supremum of interpretations $\bigvee_{d \in \Lambda} [\![\ ]\!]_d$ preserves composition (one of the 13 identities of Fig. 2), we must show that for any relation terms R and S we have $\bigcup_{d \in \Lambda} [\![RS]\!]_d = \bigcup_{d \in \Lambda} [\![R]\!]_d; \bigcup_{d \in \Lambda} [\![S]\!]_d$. However the right hand side of this identity is equal to $\bigcup_{d, e \in \Lambda \times \Lambda} [\![R]\!]_d; [\![S]\!]_e$. But since the family of interpretations is directed, for every pair d, e of indices in Λ there is an $m \in \Lambda$ with $[\![\ ]\!]_d, [\![\ ]\!]_e \leq [\![\ ]\!]_m$, hence $\bigcup_{d, e \in \Lambda \times \Lambda} [\![R]\!]_d; [\![S]\!]_e \leq \bigcup_{m \in \Lambda} [\![R]\!]_m [\![S]\!]_m$. The reverse inequality is immediate and we obtain $\bigcup_{d \in \Lambda} [\![R]\!]_d; \bigcup_{d \in \Lambda} [\![S]\!]_d = \bigcup_{d \in \Lambda} [\![RS]\!]_d$.

The least element of the collection $\mathcal{I}$ is the interpretation $[\![\ ]\!]_0$ given by $[\![\overline{p}_i]\!]_0 = \emptyset$ for all i ($1 \leq i \leq n$).

In the remainder of this section, the word *interpretation* will refer to an extended $\mathcal{D}$-interpretation.

Lemma 5. *Let* $[\![\,]\!]$ *and* $[\![\,]\!]'$ *be interpretations. If for all* i $[\![\overline{p}_i]\!] \subseteq [\![\overline{p}_i]\!]'$ *then* $[\![\,]\!] \leq [\![\,]\!]'$.

Proof. By induction on the structure of extended relations. For all relational constants c we have $[\![c]\!] = [\![c]\!]'$ We will consider one of the inductive cases, namely that of composition. Suppose $[\![R]\!] \subseteq [\![R]\!]'$ and $[\![S]\!] \subseteq [\![S]\!]'$. Then we must show that $[\![RS]\!] \subseteq [\![RS]\!]'$. But this follows immediately by a set-theoretic argument, since $(x,u) \in [\![R]\!]$ and $(u,y) \in [\![S]\!]$ imply, by inductive hypothesis, that $(x,u) \in [\![R]\!]'$ and $(u,y) \in [\![S]\!]'$. It can also be proved using the axioms of **QRA**$_\Sigma$ by showing that $A \cup A' = A'$ and $B \cup B' = B'$ imply $AB \cup A'B' = A'B'$. We leave the remaining cases to the reader.

We will now define a operator $\Phi_{\mathcal{F}}$ from interpretations to interpretations, show it continuous and define the *interpretation generated by* $\mathcal{F}$ as its least fixed point. This interpretation will be the least extension of a given relational $\mathcal{D}$-interpretation satisfying the equations in $\mathcal{F}$.

Definition 9. *Let* P *be a program, with predicate symbols* $\{p_1, \ldots, p_n\}$. *Fix a structure* $\mathcal{D}$ *for the language of* P. *Let* $\mathcal{F}$ *be the set of definitional equations* $\{\overline{p}_i \overset{\circ}{=} R_i(\overline{p}_1, \ldots, \overline{p}_n) : i \in \mathbb{N}\}$ *produced by the translation* Tr *of* P *of Definition 8. Let* $\mathcal{I}$ *be the set of extended* $\mathcal{D}$-*interpretations, with poset structure induced pointwise. Then we define the operator* $\Phi_{\mathcal{F}} : \mathcal{I} \longrightarrow \mathcal{I}$ *as follows*

$$\Phi_{\mathcal{F}}([\![\,]\!])(\overline{p}_i) = [\![R_i(\overline{p}_1, \ldots, \overline{p}_n)]\!].$$

Theorem 2. $\Phi_{\mathcal{F}}$ *is a continuous operator, that is to say it preserves suprema of directed sets.*

Proof. Let $\{[\![\,]\!]_d : d \in \Lambda\}$ be a directed set of interpretations. By Lemma 5 it suffices to show that for all p_i

$$\Phi_{\mathcal{F}}(\bigvee_{d \in \Lambda} [\![\,]\!]_d)(\overline{p}_i) = (\bigvee_{d \in \Lambda} \Phi_{\mathcal{F}}([\![\,]\!]_d))(\overline{p}_i).$$

Let $[\![\,]\!]^* = \bigvee_{d \in \Lambda} [\![\,]\!]_d$. Then $\Phi_{\mathcal{F}}(\bigvee_{d \in \Lambda} [\![\,]\!]_d)(\overline{p}_i) = [\![R_i(\overline{p}_1, \ldots, \overline{p}_n)]\!]^*$, which in turn is the union $\bigcup_{d \in \Lambda} [\![R_i(\overline{p}_1, \ldots, \overline{p}_n)]\!]_d$. But this is equal to $\bigcup_{d \in \Lambda} \Phi_{\mathcal{F}}([\![\,]\!]_d)(\overline{p}_i)$. Therefore $\Phi_{\mathcal{F}}(\bigvee_{d \in \Lambda} [\![\,]\!]_d) = \bigvee_{d \in \Lambda} \Phi_{\mathcal{F}}([\![\,]\!]_d)$.

By Kleene's fixed point theorem $\Phi_{\mathcal{F}}$ has a least fixed point $[\![\,]\!]^\dagger$ in $\mathcal{I}$. This fixed point is, in fact, the union of all $\Phi_{\mathcal{F}}^{(n)}([\![\,]\!]_0), (n \in \mathbb{N})$. By virtue of its being fixed by $\Phi_{\mathcal{F}}$ we have $[\![\overline{p}_i]\!]^\dagger = [\![R_i(\overline{p}_1, \ldots, \overline{p}_n)]\!]^\dagger$. That is to say, all equations in $\mathcal{F}$ are true in $[\![\,]\!]^\dagger$, which is the least interpretation with this property under the pointwise order.

4   A Rewriting System for Resolution

In this section, we develop a rewriting system for proof search based on the equational theory **QRA**$_\Sigma$, which will be proven equivalent to the traditional

$$m_1 \;\; : I_m(\dot{K}(\psi)) \qquad\quad \overset{P}{\longmapsto} \dot{K}(\exists^{m\uparrow}.\psi) \qquad \text{Hiding meta-reduction}$$
$$m_1* : I_m(0) \qquad\qquad\quad \overset{P}{\longmapsto} 0$$
$$m_2 \;\; : W_\pi(\dot{K}(\psi)) \qquad\quad \overset{P}{\longmapsto} \dot{K}(w_\pi(\psi)) \qquad \text{Permutation meta-reduction}$$
$$m_2* : W_\pi(0) \qquad\qquad\quad\; \overset{P}{\longmapsto} 0$$
$$m_3 \;\; : \dot{K}(\psi_1) \cap \dot{K}(\psi_2) \overset{P}{\longmapsto} \dot{K}(\psi_1 \wedge \psi_2) \quad \mathcal{D} \models \psi_1 \wedge \psi_2$$
$$m_3 \;\; : \dot{K}(\psi_1) \cap \dot{K}(\psi_2) \overset{P}{\longmapsto} 0 \qquad\qquad \mathcal{D} \not\models \psi_1 \wedge \psi_2$$
$$m_4 \;\; : \dot{K}(\psi) \cap \bar{q} \qquad\quad \overset{P}{\longmapsto} \dot{K}(\psi) \cap (\Theta) \quad \text{where } \bar{q} \doteq \Theta \in Tr(P)$$

Fig. 4. Constraint meta-reductions

operational semantics for CLP. In Sect. 5 we will show that answers obtained by resolution correspond to answers yielded by our rewriting system and conversely.

The use of ground terms permits the use of rewriting, overcoming the practical and theoretical difficulties that the existence of logic variables causes in equational reasoning. Additionally, we may speak of *executable* semantics: we use the same function to compile and interpret CLP programs in the relational denotation.

For practical reasons, we don't rewrite over the full relational language, but we will use a more compact representation of the relations resulting from the translation.[1]

Formally, the signature of our rewriting system is given by the following term-forming operations over the sort $\mathcal{T}_R$: $\mathsf{I} : (\mathbb{N} \times \mathcal{T}_R) \to \mathcal{T}_R$, $\mathsf{W} : (\mathsf{Perm} \times \mathcal{T}_R) \to \mathcal{T}_R$, $\mathsf{K} : \mathcal{L}_\mathcal{D} \to \mathcal{T}_R$, $\cup : (\mathcal{T}_R \times \mathcal{T}_R) \to \mathcal{T}_R$ and $\cap : (\mathcal{T}_R \times \mathcal{T}_R) \to \mathcal{T}_R$. Thus, for instance, the relation $I_n; R; I_n$ is formally represented in the rewriting system as $\mathsf{I}(n, \mathsf{R})$, provided R can be represented in it. In practice we make use of the conventional relational notation I_n, W_π when no confusion can arise.

4.1 Meta-Reductions

We formalize the interface between the rewrite system and the constraint solver as meta-reductions (Fig. 4). Every meta-reduction uses the constraint solver in a black-box manner to perform constraint manipulation and satisfiability checking.

Lemma 6. *All meta-reductions are sound: if* $m_i : l \overset{P}{\longmapsto} r$ *then* $[\![l]\!]^{\mathcal{D}^\dagger} = [\![r]\!]^{\mathcal{D}^\dagger}$.

4.2 A Rewriting System for SLD Resolution

We present a rewriting system for proof search in Fig. 5. We prove local confluence. Later we will prove that a query rewrites to a term in the canonical form $\dot{K}(\psi) \cup R$ iff the leftmost branch of the associated SLD-tree of the program is finite.

---

[1] There is no problem in defining the rewriting system using the general relational signature, but we would need considerably more rules for no gain.

$$p_1 : \mathbf{0} \cup R \qquad\qquad \overset{P}{\longmapsto} R$$
$$p_2 : \mathbf{0} \cap R \qquad\qquad \overset{P}{\longmapsto} \mathbf{0}$$
$$p_3 : W_\pi(R \cup S) \qquad \overset{P}{\longmapsto} W_\pi(R) \cup W_\pi(S)$$
$$p_4 : I_n(R \cup S) \qquad \overset{P}{\longmapsto} I_n(R) \cup I_n(S)$$
$$p_5 : (R \cup S) \cap T \qquad \overset{P}{\longmapsto} (R \cap T) \cup (S \cap T)$$
$$p_6 : \dot{K}(\psi) \cap (R \cup S) \qquad \overset{P}{\longmapsto} (\dot{K}(\psi) \cap R) \cup (\dot{K}(\psi) \cap S)$$
$$p_7 : \dot{K}(\psi) \cap (R \cap W_\pi(\overline{q_i})) \overset{P}{\longmapsto} (\dot{K}(\psi) \cap R) \cap W_\pi(\overline{q_i})$$
$$p_8 : \dot{K}(\psi) \cap W_\pi(\overline{q}) \qquad \overset{P}{\longmapsto} W_\pi^\circ(W_\pi(\dot{K}(\psi)) \cap \overline{q})$$
$$p_9 : \dot{K}(\psi) \cap I_m(R) \qquad \overset{P}{\longmapsto} I_m(I_m(\dot{K}(\psi)) \cap R) \cap \dot{K}(\psi)$$

Fig. 5. Rewriting system for SLD.

Lemma 7. $\overset{P}{\longmapsto}$ *is sound: if* $p_i : l \overset{P}{\longmapsto} r$ *then* $[\![l]\!]^{\mathcal{D}^\dagger} = [\![r]\!]^{\mathcal{D}^\dagger}$.

Lemma 8. *If we give higher priority to* p_7 *over* p_8, $\overset{P}{\longmapsto}$ *is locally confluent.*

A *left outermost strategy* gives priority to p_7 over p_8.

5   Operational Equivalence

We prove that our rewriting system over relational terms simulates "traditional" SLD proof search specified as a transition-based operational semantics (i.e. [7, 12]). For reasons of space, we give a high-level overview of the proof. The full details can be found in the online technical report.

Recall a *resolvent* is a sequence of atoms or constraints $\boldsymbol{p}$. We write $\square$ for the empty resolvent. We assume given a constraint domain $\mathcal{D}$ and its satisfaction relation $\mathcal{D} \models \varphi$. A *program state* is an ordered pair $\langle \boldsymbol{p} \,|\, \varphi \rangle$ where $\boldsymbol{p}$ is a resolvent and φ is a constraint (called the *constraint store*). The notation $cl : p(\boldsymbol{u}[\boldsymbol{y}]) \leftarrow q(\boldsymbol{v}[\boldsymbol{z}])$ indicates that $p(\boldsymbol{u}[\boldsymbol{y}]) \leftarrow q(\boldsymbol{v}[\boldsymbol{z}])$ is a program clause with label cl. Then, the standard operational semantics for SLD resolution can be defined as the following transition system over program states:

Definition 10 (Standard SLD Semantics).

$$\langle \varphi, \boldsymbol{p} \,|\, \psi \rangle \quad \overset{cs}{\longrightarrow}_l \quad \langle \boldsymbol{p} \,|\, \psi \wedge \varphi \rangle \qquad \text{iff } \mathcal{D} \models \psi \wedge \varphi$$
$$\langle p(\boldsymbol{t}[\boldsymbol{x}]), \boldsymbol{p} \,|\, \varphi \rangle \overset{res_{cl}}{\longrightarrow}_l \langle q(\boldsymbol{v}[\sigma(\boldsymbol{z})]), \boldsymbol{p} \,|\, \varphi \wedge (\boldsymbol{u}[\sigma(\boldsymbol{y})] = \boldsymbol{t}[\boldsymbol{x}]) \rangle$$
$$\text{where: } cl : p(\boldsymbol{u}[\boldsymbol{y}]) \leftarrow q(\boldsymbol{v}[\boldsymbol{z}])$$
$$\mathcal{D} \models \varphi \wedge (\boldsymbol{u}[\sigma(\boldsymbol{y})] = \boldsymbol{t}[\boldsymbol{x}])$$
$$\sigma \text{ a renaming apart for } \boldsymbol{y}, \boldsymbol{z}, \boldsymbol{x}$$

Taking the previous system as a reference, the proof proceeds in two steps: we first define a new transition system that internalizes renaming apart and proof search, and we prove it equivalent to the standard one.

Second, we show a simulation relation between the fully internalized transition system and a transition system defined over relations, which is implemented by the rewriting system of Sect. 4.

With these two equivalences in place, the main theorem is:

Theorem 3. *The rewriting system of Fig. 5 implements the transition system of Definition 10. Formally, for every transition* $(r_1, r_2) \in (\to_l)^*$,

$$\exists n.(Tr(r_1), Tr(r_2)) \in (\overset{P}{\longmapsto})^n$$

and

$$\forall r_3.(Tr(r_1), r_3) \in (\overset{P}{\longmapsto})^n \Rightarrow Tr(r_2) = r_3$$

Thus, given a program P, relational rewriting of translation will return an answer constraint $K(\varphi)$ iff SLD resolution from P reaches a program state $\langle \Box \mid \varphi' \rangle$, with $\varphi \iff \varphi'$.

In the next section, we briefly describe the main intermediate system used in the proof.

5.1   The Resolution Transition System

The crucial part of the SLD-simulation proof is the definition of a new extended transition system over program states that will internalize both renaming apart and the proof-search tree. It is an intermediate system between relation rewriting and traditional proof search.

The first step towards the new system is the definition of an extended notion of state. In the standard system of Definition 10, a state is a resolvent plus a constraint store. Our extended notion of state includes:

- A notion of *scope*, which is captured by a natural number which can be understood as the number of global variables of the state.
- A notion of *substate*, which includes information about parameter passing in the form of a *permutation*.
- A notion of clause *selection*, and
- a notion of *failure* and *parallel state*, which represents failures in the search tree and alternatives.

Such states are enough to capture all the meta-theory of constraint logic programming except recursion, which operates meta-logically by replacing predicate symbols by their definitions. Formally:

Definition 11. *The set $\mathcal{PS}$ of resolution states is inductively defined as:*

- $\langle fail \rangle$.
- $\langle \boldsymbol{p} | \varphi \rangle_n$, *where* $p_i \equiv P_i(\boldsymbol{x}_i)$ *is an atom, or a constraint* $p_i \equiv \psi$, $\boldsymbol{x}_i$ *a vector of variables,* φ *a constraint store and* n *a natural number.*
- $\langle {}^\pi PS, \boldsymbol{p} | \varphi \rangle_n$, *where* PS *is a resolution state, and* π *a permutation.*
- $\langle {}^\pi \blacktriangleright PS, \boldsymbol{p} | \varphi \rangle_n$, *the "select state". It represents the state just before selecting a clause to proceed with proof search.*
- $(PS_1 \mid PS_2)$. *The bar is parallel composition, capturing choice in the proof search tree.*

$$\langle \psi, \boldsymbol{p} \,|\, \varphi \rangle_n \xrightarrow{\text{constraint}}_p \langle \boldsymbol{p} \,|\, \varphi \wedge \psi \rangle_n$$

$$\langle \psi, \boldsymbol{p} \,|\, \varphi \rangle_n \xrightarrow{\text{fail}}_p \langle \text{fail} \rangle$$

> if $\varphi \wedge \psi$ is not satisfiable

$$\langle p(\boldsymbol{x}), \boldsymbol{p} \,|\, \varphi \rangle_n \xrightarrow{\text{call}}_p \langle^\pi \blacktriangleright (\langle q_1 \,|\, \top \rangle_h \,\| \ldots \| \langle q_k \,|\, \top \rangle_h), \boldsymbol{p} \,|\, \varphi \rangle_n$$

> if $p(\boldsymbol{x}_h) \leftarrow \exists^{h\uparrow}.(q_1 \vee \ldots \vee q_k) \in P'$, $\pi(\boldsymbol{x}) = \boldsymbol{x}_h$

$$\langle^\pi \blacktriangleright (\langle q \,|\, \psi \rangle_h \,\| PS), \boldsymbol{p} \,|\, \varphi \rangle_n \xrightarrow{\text{select}}_p (\langle^\pi \langle q \,|\, \psi \wedge \Delta_h^\pi(\varphi) \rangle_h, \boldsymbol{p} \,|\, \varphi \rangle_n \,\| \langle^\pi \blacktriangleright PS, \boldsymbol{p} \,|\, \varphi \rangle_n)$$

$$\langle^\pi \langle \square \,|\, \psi \rangle_h, \boldsymbol{p} \,|\, \varphi \rangle_n \xrightarrow{\text{return}}_p \langle \boldsymbol{p} \,|\, \nabla_h^\pi(\psi, \varphi) \rangle_n$$

$$\langle^\pi \langle \text{fail} \rangle, \boldsymbol{p} \,|\, \varphi \rangle_n \xrightarrow{\text{return}}_p \langle \text{fail} \rangle$$

$$\langle^\pi PS, \boldsymbol{p} \,|\, \varphi \rangle_n \xrightarrow{\text{sub}}_p \langle^\pi PS', \boldsymbol{p} \,|\, \varphi \rangle_n$$

> if $PS \neq \langle \square \,|\, \psi \rangle_n$, $PS \neq \langle \text{fail} \rangle$, and $PS \rightarrow_p PS'$

$$(\langle \text{fail} \rangle \,\| PS) \xrightarrow{\text{backtrack}}_p PS$$

$$(PS_1 \,\| PS_2) \xrightarrow{\text{seq}}_p (PS_1' \,\| PS_2)$$

> if $PS \neq \langle \text{fail} \rangle$, and $PS_1 \rightarrow_p PS_1'$

(We omit the case in *select* where the left side has no *PS* component which happens when the number of clauses for a given predicate is one ($k = 1$))

Fig. 6. Resolution transition system

The *resolution transition system* $\rightarrow_P \subseteq (\mathcal{PS} \times \mathcal{PS})$ is shown in Fig. 6. The two first transitions deal with the case where a constraint is first in the resolvent, failing or adding it to the constraint store in case it is satisfiable.

When the head of the resolvent is a defined predicate, the *call* transition will replace it by its definition, properly encapsulated by a select state equipped with the permutation capturing argument order.

The *select* transition performs two tasks: first, it modifies the current constraint store adding the appropriate permutation and scoping (n, π); second, it selects the first clause for proof search.

The *return* transitions will either propagate failure or undo the permutation and scoping performed at call time.

sub, *backtrack*, and *seq* are structural transitions with a straightforward interpretation from a proof search perspective.

Then, we have the following lemma:

Lemma 9. *For all queries* $\langle \boldsymbol{p} | \varphi \rangle_n$, *the first successful* $\rightarrow_l$ *derivation using a SLD strategy uniquely corresponds to a* $\rightarrow_p$ *derivation:*

$$\langle \boldsymbol{p} | \varphi \rangle_n \rightarrow_l \ldots \rightarrow_l \langle \square \,|\, \varphi' \rangle_n \iff \langle \boldsymbol{p} | \varphi \rangle_n \rightarrow_p \ldots \rightarrow_p (\langle \square \,|\, \varphi' \rangle_n \,\| PS)$$

for some resolution state PS.

Corollary 1. *The transition systems of Definition 10 and Fig. 6 are answer-equivalent: for any query they return the same answer constraint.*

With this lemma in place, the proof of Theorem 3 is completed by showing a simulation between the resolution system and a transition system induced by relation rewriting.

6   Related and Future Work

Previous Work: The paper is the continuation of previous work in [4,10,15] considerably extended to include constraint logic programming, which requires a different translation procedure and a different rewriting system.

In particular, the presence of constraints in this paper permits a different translation of the Clark completion of a program and plays a crucial role in the proof of completeness, which was missing in earlier work. The operational semantics is also new.

Related Work: A number of solutions have been proposed to the syntactic specification problem. There is an extensive literature treating abstract syntax of logic programming (and other programming paradigms) using encodings in higher-order logic and the lambda calculus [18], which has been very successful in formalizing the treatment of substitution, unification and renaming of variables, although it provides no special framework for the management and progressive instantiation of logic variables, and no treatment of constraints. Our approach is essentially orthogonal to this, since it relies on the complete elimination of variables, substitution, renaming and, in particular, existentially quantified variables. Our reduction of management of logic variables to variable free rewriting is new, and provides a complete solution to their formal treatment.

An interesting approach to syntax specification is the use of nominal logic [5,22] in logic programming, another, the formalization of logic programming in categorical logic [1,2,8,13,19] which provides a mathematical framework for the treatment of variables, as well as for derivations [14]. None of the cited work gives a solution that simultaneously includes logic variables, constraints and proof search strategies however.

Bellia and Occhiuto [3] have defined a new calculus, the C-expression calculus, to eliminate variables in logic programming. We believe our translation into the well-understood and scalable formalism of relations is more applicable to extensions of logic programming. Furthermore the authors do not consider constraints.

Future Work: A complementary approach to this work is the use of category theory, in particular the Freyd's theory of *tabular allegories* [9] which extends the relation calculus to an abstract category of relations providing native facilities for generation of fresh variables and a categorical treatment of monads. A first attempt in this direction has been published by the authors in [11]. It would be interesting to extend the translation to hereditarily Harrop or higher order logic [17] by using a stronger relational formalism, such as Division and Power Allegories. Also, the framework would yield important benefits if it was extended to include relation and set constraints explicitly.

7   Conclusion

We have developed a declarative relational framework for the compilation of Constraint Logic programming that eliminates logic variables and gives an algebraic treatment of program syntax. We have proved operational equivalence to the classical approach. Our framework has several significant advantages.

Programs can be analyzed, transformed and optimized entirely within this framework. Execution is carried out by rewriting over relational terms. In these two ways, specification and implementation are brought much closer together than in the traditional logic programming formalism.

References

1. Amato, G., Lipton, J., McGrail, R.: On the algebraic structure of declarative programming languages. Theor. Comput. Sci. **410**(46), 4626–4671 (2009), http://www.sciencedirect.com/science/article/B6V1G-4WV15VS-7/2/5475111b9 a9642244a208e9bd1fcd46a (abstract Interpretation and Logic Programming: In honor of professor Giorgio Levi)
2. Asperti, A., Martini, S.: Projections instead of variables: a category theoretic interpretation of logic programs. In: ICLP, pp. 337–352 (1989)
3. Bellia, M., Occhiuto, M.E.: C-expressions: a variable-free calculus for equational logic programming. Theor. Comput. Sci. **107**(2), 209–252 (1993)
4. Broome, P., Lipton, J.: Combinatory logic programming: computing in relation calculi. In: ILPS'94: Proceedings of the 1994 International Symposium on Logic programming, pp. 269–285. MIT Press, Cambridge (1994)
5. Cheney, J., Urban, C.: Alpha prolog: a logic programming language with names, binding, and alpha-equivalence (2004)
6. Clark, K.L.: Negation as failure. In: Gallaire, H., Minker, J. (eds.) Logic and Data Bases, pp. 293–322. Plenum Press (1977)
7. Comini, M., Levi, G., Meo, M.C.: A theory of observables for logic programs. Inf. Comput. **169**(1), 23–80 (2001)
8. Finkelstein, S.E., Freyd, P.J., Lipton, J.: A new framework for declarative programming. Theor. Comput. Sci. **300**(1–3), 91–160 (2003)
9. Freyd, P., Scedrov, A.: Categories, Allegories. North Holland Publishing Company, Amsterdam (1991)
10. Gallego Arias, E.J., Lipton, J., Mariño, J., Nogueira, P.: First-order unification using variable-free relational algebra. Log. J. IGPL **19**(6), 790–820 (2011). http:// jigpal.oxfordjournals.org/content/19/6/790.abstract
11. Gallego Arias, E.J., Lipton, J.: Logic programming in tabular allegories. In: Dovier, A., Costa, V.S. (eds.) Technical Communications of the 28th International Conference on Logic Programming, ICLP 2012, September 4–8, 2012, Budapest, Hungary. LIPIcs, vol. 17, pp. 334–347. Schloss Dagstuhl—Leibniz-Zentrum fuer Informatik (2012)
12. Jaffar, J., Maher, M.J.: Constraint logic programming: a survey. J. Log. Program. **19/20**, 503–581 (1994). http://citeseer.ist.psu.edu/jaffar94constraint.html
13. Kinoshita, Y., Power, A.J.: A fibrational semantics for logic programs. In: Dyckhoff, R., Herre, H., Schroeder-Heister, P. (eds.) ELP. LNCS, vol. 1050, pp. 177–191. Springer, Heidelberg (1996)
14. Komendantskaya, E., Power, J.: Coalgebraic derivations in logic programming. In: Bezem, M. (ed.) CSL. LIPIcs, vol. 12, pp. 352–366. Schloss Dagstuhl—Leibniz-Zentrum fuer Informatik (2011)

15. Lipton, J., Chapman, E.: Some notes on logic programming with a relational machine. In: Jaoua, A., Kempf, P., Schmidt, G. (eds.) Using Relational Methods in Computer Science, pp. 1–34. Technical report Nr. 1998-03, Fakultät für Informatik, Universität der Bundeswehr München, July 1998
16. Lloyd, J.W.: Foundations of Logic Programming. Springer, New York (1984)
17. Miller, D., Nadathur, G., Pfenning, F., Scedrov, A.: Uniform proofs as a foundation for logic programming. Ann. Pure Appl. Log. 51(1–2), 125–157 (1991)
18. Pfenning, F., Elliot, C.: Higher-order abstract syntax. In: PLDI'88: Proceedings of the ACM SIGPLAN 1988 Conference on Programming Language Design and Implementation, pp. 199–208. ACM, New York (1988)
19. Rydeheard, D.E., Burstall, R.M.: A categorical unification algorithm. In: Proceedings of a Tutorial and Workshop on Category Theory and Computer Programming, pp. 493–505. Springer, New York (1986)
20. Sterling, L., Shapiro, E.: The Art of Prolog. The MIT Press, Cambridge (1986)
21. Tarski, A., Givant, S.: A Formalization of Set Theory Without Variables, Colloquium Publications, vol. 41. American Mathematical Society, Providence (1987)
22. Urban, C., Pitts, A.M., Gabbay, M.J.: Nominal unification. Theor. Comput. Sci. 323(1–3), 473–497 (2004)

Pre-indexed Terms for Prolog

J.F. Morales[1](✉) and M. Hermenegildo[1,2]

[1] IMDEA Software Institute, Madrid, Spain
josef.morales@imdea.org
[2] School of Computer Science, Technical University
of Madrid, Madrid, Spain

Abstract. Indexing of terms and clauses is a well-known technique used in Prolog implementations (as well as automated theorem provers) to speed up search. In this paper we show how the same mechanism can be used to implement efficient reversible mappings between different term representations, which we call *pre-indexings*. Based on user-provided term descriptions, these mappings allow us to use more efficient data encodings internally, such as prefix trees. We show that for some classes of programs, we can drastically improve the efficiency by applying such mappings at selected program points.

1 Introduction

Terms are the most important data type for languages and systems based on first-order logic, such as (constraint) logic programming languages or resolution-based automated theorem provers. Terms are inductively defined as variables, atoms, numbers, and compound terms (or structures) comprised by a functor and a sequence of terms[1]. Two main representations for Prolog terms have been proposed. Early Prolog systems, such as the Marseille and DEC-10 implementations, used *structure sharing* [2], while the WAM [1,15] –and consequently most modern Prolog implementations– use *structure copying*. In structure sharing, terms are represented as a pair of pointers, one for the structure skeleton, which is shared among several instances, and another for the binding environment, which determines a particular instantiation. In contrast, structure copying makes a copy of the structure for each newly created term. The encoding of terms in memory resembles tree-like data structures.

In order to speed up resolution, sophisticated term indexing has been implemented both in Prolog [1,7] and automated theorem provers [6]. By using specialized data structures (such as, e.g., tries), indexing achieves sub-linear complexity

---

Research supported in part by projects EU FP7 318337 *ENTRA*, Spanish MINECO TIN2012-39391 *StrongSoft* and TIN2008-05624 *DOVES*, and Comunidad de Madrid ICE-2731 *NGREENS Software*. We would also like to thank Rémy Haemmerlé and the anonymous reviewers for providing valuable comments and suggestions.

[1] Additionally, many Prolog systems implement an extension mechanism for variable domains using *attributed variables*.

M. Proietti and H. Seki (Eds.): LOPSTR 2014, LNCS 8981, pp. 317–331, 2015.
DOI: 10.1007/978-3-319-17822-6_18

in clause selection. Similar techniques are used to efficiently store predicate solutions in tabling [13]. This efficient indexing is typically also supported in dynamic predicates, i.e., for predicates whose facts or clauses can be changed dynamically during program execution. This results in a mechanism that is often very attractive for storing and manipulating program data: indexed dynamic predicates offer the benefits of efficient key-value data structures while hiding the implementation details from the user program.

Modulo some issues like variable sharing, there is thus a duality in programming style between *explicitly* encoding data as terms or encoding data *implicitly* as tuples in dynamic predicates, in order to exploit the built-in indexing provided by this representation. For example, the set $\{1, 2, 3, \ldots, n\}$ is represented naturally as the term [1,2,3,...,n] (equivalent to a linked list). However, depending on the lifetime and operations to be performed on the data, binary trees, some other map-like structure, or dynamic predicates may be preferable. Which representation is most efficient or convenient is very application-dependent and it would be desirable to be able to explore the relative merits of the alternative representations with minimal changes in the program. Unfortunately, in practice such changes in representation typically mean significant modifications, which propagate throughout the whole program. Even worse, it is also frequent to find code where, after changes motivated by such performance considerations, the data is represented in the end in a quite unnatural way.

The goal of this paper is to study the merits of term indexing *during term creation* rather than at clause selection time. We exploit the fact that data has frequently a fixed skeleton structure, and introduce a mapping in order to index and share that part. This mapping is derived from program declarations specifying term encoding (called *rtypes*, for r*epresentation types*) and annotations defining the program points where *pre-indexing* of terms is to be performed. This is done on top of structure copying, so that no large changes are required in a typical Prolog runtime system. Moreover, the approach does not require large changes in program structure, which makes *rtypes* easily interchangeable.

We have implemented a prototype as a Ciao [4] package that deals with *rtype* declarations as well as with some additional syntactic sugar that we provide for marking pre-indexing points.

2   Background

We follow the definitions and naming conventions for *term indexing* of [3,6]. Given a set of terms $\mathcal{L}$ (the *indexed terms*), a binary relation R over terms (the *retrieval condition*), and a term t (the *query term*), we want to identify the subset $\mathcal{M} \subseteq \mathcal{L}$ consisting of all the terms l such that $R(l, t)$ holds (i.e., such that l is R-compatible with t). We are interested in the following retrieval conditions R (where σ is a substitution):

- $unif(l, t) \Leftrightarrow \exists \sigma \; l\sigma = t\sigma$ (unification)
- $inst(l, t) \Leftrightarrow \exists \sigma \; l = t\sigma$ (instance check)

- $gen(l, t) \Leftrightarrow \exists \sigma\ l\sigma = t$ (generalization check)
- $variant(l, t) \Leftrightarrow \exists \sigma\ l\sigma = t$ and σ is a renaming substitution   (variant check)

Example 1. Given $\mathcal{L} = \{h(f(A)), h(f(B, C)), h(g(D))\}$, $t = h(f(1))$, and $R = $ *unif*, then $\mathcal{M} = \{h(f(A))\}$.

The objective of *term indexing* is to implement fast retrieval of candidate terms. This is done by processing the indexed set $\mathcal{L}$ into specialized data structures (*index construction*) and modifying this index when terms are inserted or deleted from $\mathcal{L}$ (*index maintenance*).

When the retrieval condition makes use of the function symbols in the query and indexed terms, it is called *function symbol based indexing*.

In Prolog, indexing finds the set of program clauses such that their heads unify with a given literal in the goal. In tabled logic programming, this is also interesting for detecting if a new goal is a variant or subsumed by a previously evaluated subgoal [5,12].

Limitations of Indexing. Depending on the part of the terms that is indexed and the supporting data structure, the worst case cost of indexing is proportional to the size of the term. When computing hash keys, the whole term needs to be traversed (e.g., computing the key for `h(f(A))` requires walking over h and f). This may be prohibitively costly, not only in the maintenance of the indices, but also in the lookup. As a compromise many systems rely only on first argument, first level indexing (with constant hash table lookup, relying on linear search for the selected clauses). However, when the application needs stronger, multi-level indexing, lookup costs are repeated many times for each clause selection operation.

3 Pre-indexing

The goal of pre-indexing is to move lookup costs to term building time. The idea that we propose herein is to use a bijective mapping between the standard and the pre-indexed representations of terms, at selected program points. The fact that terms can be partially instantiated brings in a practical problem, since binding a variable may affect many precomputed indices (e.g., precomputed indices for `H=h(X)`, `G=g(X)` may need a change after `X=1`). Our solution to this problem is to restrict the mapping to terms of a specific form, based on *instantiation types*, defined as (possibly recursive) unary predicates. For convenience, the user-defined instantiation types are extended with the native definitions **any** (that represents any term or variable) and **nv** (that represents any **nonvar** term).

Definition 1 (Instantiation Type Check). *We say that t is an instance of an* instantiation type τ *(defined as a unary predicate), written as* $check_\tau(t)$, *if there exists a term l in the answers of τ and $gen(l, t)$ (or $inst(t, l)$).*

For conciseness, we will describe the restricted form of instantiation types used herein using a specialized syntax "`:- rtype` *Name* `--->` $Cons_1$ `;` $\ldots$ `;` $Cons_n$", where each $Cons_i$ is a term constructor. A term constructor is composed of a functor name and a number of arguments, where each argument is another *rtype* name. E.g.,:[2]

```
:- rtype lst ---> [] ; [any|lst]
```

The rule above thus corresponds to the predicate:

```
lst([]).
lst([_|Xs]) :- lst(Xs).
```

Example 2. According to the definition above for `lst`, the terms `[1,2,3]` and `[_,2]` belong to `lst` while `[1|_]` does not. If `nv` were used instead of `any` in the definition above then `[_,2]` would also not belong to `lst`.

Type-based Pre-indexing. The idea behind pre-indexing is to maintain specialized indexing structures for each *rtype* (which in this work is done based on user annotations). We denote as inhabitants of *rtype* τ the set of the most general terms (w.r.t. *gen* relation) that are instances of τ. The indexing structure will keep track of the *rtype* inhabitants constructed during the execution dynamically, assigning a unique identifier (the pre-index key) to each representant (modulo variants). E.g., for `lst` we could assign $\{[]\mapsto k_0,\ [\_]\mapsto k_1, [\_,\_]\mapsto k_2,\ldots\}$ (that is, k_i for each list of length i). Note that special `any` does not define a concrete term constructor and is not pre-indexed, while `nv` represents all possible term constructors with `any` as arguments.

For every term t so that $check_\tau(t)$, then exists l in the inhabitants of τ such that $gen(l,t)$. That is, there exists a substitution σ such that $t = l\sigma$. The pre-indexing of a term replaces t by a simpler term using the inhabitant key k and the substitution σ. Since k is unique for each inhabitant this translation has inverse. The translation between pre-indexed and non-pre-indexed forms is defined in terms of a *pre-indexing casting*.

Definition 2 (Pre-Indexing Cast). *A pre-indexing cast of type τ is a bijective mapping with the set of terms defined by $check_\tau$ as domain, denoted by $\#\tau$, with the following properties:*

1. *for every term t so that $check_\tau(t)$ (which defines the domain of the mapping), and substitution σ, then $\#\tau(t\sigma) = \#\tau(t)\sigma$ (σ-commutative)*
2. *the main functor of $\#\tau(t)$ encodes the (indexed) structure of the arguments (so that it uniquely identifies the rtype inhabitant).*

E.g., for `[1,2,3]` and `lst` the pre-indexed term would be $k_1(1,2,3)$.

Informally, the first property ensures that pre-indexing casts can be selectively introduced in a program (whose terms are instantiated enough) without altering the (substitution) semantics. Moreover, the meaning of many built-ins is also preserved after pre-indexing, as expressed in the following theorem.

---

[2] Despite the syntax being similar to that described in [10], note that the semantics is not equivalent.

Theorem 1 (Built-in Homomorphism). *Given $check_\tau(x)$ and $check_\tau(y)$, then $unif(x, y) \Leftrightarrow unif(\#\tau(x), \#\tau(y))$ (equivalently for gen, inst, variant, and other built-ins like* ==/2, ground/1*).*

Proof. $unif(x, y) \Leftrightarrow$ [def. of unif] $\exists \sigma \; x\sigma = y\sigma$. Since $\#\tau$ is bijective, then $\#\tau(x\sigma) = \#\tau(y\sigma) \Leftrightarrow$ [σ-commutative] $\#\tau(x)\sigma = \#\tau(y)\sigma$. Given the def. of *unif*, it follows that $unif(\#\tau(x), \#\tau(y))$. The proofs for other built-ins are similar.

In this work we do not require the semantics of built-ins like @< (i.e., *term ordering*) to be preserved, but if desired this can be achieved by selecting carefully the order of keys in the pre-indexed term. Similarly, functor arity in principle will not be preserved since ground arguments that are part of the *rtype* structure are allowed to be removed.

3.1 Building Pre-Indexed Terms

We are interested in building terms directly into their pre-indexed form. To achieve this we take inspiration from WAM compilation. Complex terms in variable-term unifications are decomposed into simple variable-structure unifications $X = f(A_1, \ldots, A_n)$ where all the A_i are variables. In WAM bytecode, this is further decomposed into a put_str f/n (or get_str f/n) instruction followed by a sequence of unify_arg A_i. These instructions can be expressed as follows:

```
put_str(X,F/N,S0,S1),    % | F/N |
unify_arg(A1,S1,S2)      % | F/N | A1 |
...
unify_arg(An,Sn,S)       % | F/N | A1 | ... | An |
```

where the S_i represent each intermediate heap state, which is illustrated in the comments on the right.

Assume that each argument A_i can be decomposed into its indexed part $A_i k$ and its value part $A_i v$ (which may omit information present in the key). *Pre-indexing* builds terms that encode $A_i k$ into the main functor by incremental updates:

```
g_put_str(X,F/N,S0,S1),  % | F/N |
g_unify_arg(A1,S1,S2)    % | F/N<A1k> | A1v |
...
g_unify_arg(An,Sn,S)     % | F/N<A1k,...,Ank> | A1v | ... | Anv |
```

The *rtype* constructor annotations (that we will see in Sect. 3.2) indicate how the functor and arguments are indexed.

Cost Analysis. Building and unifying pre-indexed terms have impact both on performance and memory usage. First, regarding time, although pre-indexing operations can be slower, clause selection becomes faster, as it avoids repetitive lookups on the fixed structure of terms. In the best case, $O(n)$ lookups (where n

is the size of the term) become $O(1)$. Other operations like unification are sped-up (e.g., earlier failure if keys are different). Second, pre-indexing has an impact on memory usage. Exploiting the data structure allows more compact representations, e.g., `bitpair(bool,bool)` can be assigned an integer as key (without storage costs). In other cases, the supporting *index structures* may effectively share the common part of terms (at the cost of maintaining those structures).

3.2 Pre-Indexing Methods

Pre-indexing is enabled in an *rtype* by annotating each constructor with modifiers that specify the *indexing method*. Currently we support compact trie-like representations and packed integer encodings.

Trie representation is specified with the `index(Args)` modifier, which indicates the order in which arguments are walked in the decision-tree. The process is similar to term creation in the heap, but instead of moving a heap pointer, we combine it with walking through a trie of nodes. Keys are retrieved from the term part that corresponds to the *rtype* structure.

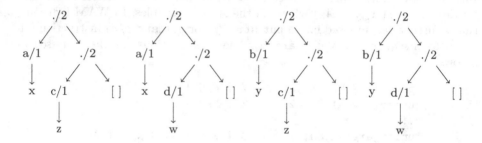

Fig. 1. Example terms for pre-indexing

For example, let us consider the input set of terms $[a(x), c(z)]$, $[a(x), d(w)]$, $[b(y), c(z)]$, $[b(y), d(w)]$, where a, b, c, d are function symbols and x, y, z, w are variable symbols. The heap representation is shown in Fig. 1.[3] We will compare different *rtype* definitions for representing these terms.

As mentioned before, **nv** represents the *rtype* for any **nonvar** term (where its main functor is taking part in pre-indexing). The declaration:

```
:- rtype lst ---> [] ; [nv|lst]:::index([0,1,2]).
```

specifies that the lookup order for $[\_|\_]$ is (a) the constructor name (*./2*), (b) the first argument (not a pre-indexed term, but takes part in pre-indexing), and (c) the second argument (pre-indexed). The resulting trie is in Fig. 2. In the figure, each node number represents a position in the trie. Singly circled nodes are temporary nodes, doubly circled nodes are final nodes. Final nodes encode

---

[3] Remember that `[1,2]` = `.(1,.(2,[]))`.

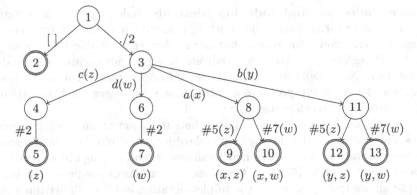

Fig. 2. Index for example terms (*rtype* `1st` `--->` `[]` ; `[nv|1st]:::index([0,1,2])`)

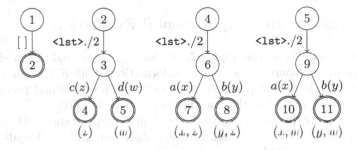

Fig. 3. Index for example terms (*rtype* `1st` `--->` `[]` ; `[nv|1st]:::index([2,0,1])`)

terms. The initial node (*#1*) is unique for each *rtype*. Labels between nodes indicate the lookup input. They can be constructor names (e.g., `./2`), nv terms (e.g., *b(y)*), or other pre-indexed `1st` (e.g., *#2* for *[]*, or *#5(z)* for *[c(z)]*). The arguments are placeholders for the non-indexed information. That is, a term *[a(g),c(h)]* would be encoded as *#9(g,h)*.

Trie indexing also supports *anchoring* on non-root nodes. Consider this declaration:

```
:- rtype 1st ---> [] ; [nv|1st]:::index([2,0,1]).
```

Figure 3 shows the resulting trie (which has been separated into different subtrees for the sake of clarity). For `./2`, the lookup now starts from the second argument, then the constructor name, and finally the first argument. The main difference w.r.t. the previous indexing method is that the beginning node is another pre-indexed term. This may lead to more optimal memory layouts and need fewer lookup operations. Note that constructor names in the edges from initial nodes need to be prefixed with the name of the *rtype*. This is necessary to avoid ambiguities, since the initial node is no longer unique.

Garbage Collection and Indexing Methods. Indexing structures require special treatment for garbage collection[4]. In principle, it would not be necessary to keep in a trie nodes for terms that are no longer reachable (e.g., from the heap, WAM registers, or dynamic predicates), except for caching to speed-up node creation. Node removal may make use of lookup order. That is, if a key at a temporary level n corresponds to an atom that is no longer reachable, then all nodes above n can be safely discarded.

Anchoring on non-root nodes allows the simulation of interesting memory layouts. For example, a simple way to encode objects in Prolog is by introducing a *new object* operation that creates new fresh atoms, and storing object attributes with a dynamic `objattr(ObjId, AttrName, AttrValue)` predicate. Anchoring on `ObjId` allows fast deletion (at the implementation level) of all attributes of a specific object when it becomes unreachable.

4   Applications and Experimental Evaluation

To show the feasibility of the approach, we have implemented the pre-indexing transformations as source-to-source transformations within the Ciao system. This is done within a Ciao package which defines the syntax and processes the *rtype* declarations as well as the marking of pre-indexing points.

As examples, we show algorithmically efficient implementations of the Lempel-Ziv-Welch (LZW) lossless data compression algorithm and the Floyd-Warshall algorithm for finding the shortest paths in a weighted graph, as well as some considerations regarding supporting module system implementation. In the following code, `forall/2` is defined as `\+ (Cond, \+ Goal)`.

4.1   Lempel-Ziv-Welch Compression

Lempel-Ziv-Welch (LZW) [16] is a lossless data compression algorithm. It encodes an input string by building an indexed dictionary D of words and writing a list of dictionary indices, as follows:

1- $D := \{w \mid w \text{ has length } 1\}$ (all strings of length one).
2- Remove from input the longest prefix that matches some word W in D, and emit its dictionary index.
3- Read new character C, $D := D \cup \mathsf{concat}(W, C)$, go to step 2; otherwise, stop.

A simple Prolog implementation is shown in Figs. 4 and 5. Our implementation uses a *dynamic* predicate `dict/2` to store words and corresponding numeric indices (for output). Step 1 is implemented in the `build_dict/1` predicate.

---

[4] Automatic garbage collection of indexing structures is not supported in the current implementation.

```
1   compress(Cs, Result) :-              % Compress Cs
2       build_dict(256),                 % Build the dictionary
3       compress_(Cs, #1st([]), Result).
4
5   compress_([], W, [I]) :-             % Empty, output code for W
6       dict(W,I).
7   compress_([C|Cs], W, Result) :-     % Compress C
8       WC = #1st([C|^W]),
9       ( dict(WC,_) ->                  % WC is in dictionary
10          W2 = WC,
11          Result = Result0
12      ; dict(W,I),                     % WC not in dictionary
13          Result = [I|Result0],        % Output the code for W
14          insert(WC),                  % Add WC to the dictionary
15          W2 = #1st([C])
16      ),
17      compress_(Cs, W2, Result0).
```

Fig. 4. LZW Compression: main code.

Steps 2 and 3 are implemented in the compress_/3 predicate[5]. For encoding words we use lists. We are only interested in adding new characters and word matching. For that, list construction and unification are good enough. We keep words in reverse order so that appending a character is done in constant time. For constant-time matching, we use an *rtype* for pre-indexing lists. The implementation is straighforward. Note that we add a character to a word in WC = #1st([C|^W]) (Line 8). The annotation (whose syntax is implemented as a user-defined Prolog operator) is used by the compiler to generate the pre-indexed version of term construction. In this case, it indicates that words are pre-indexed using the 1st *rtype* and that W is already pre-indexed (indicated by the escape ^ prefix). Thus we can effectively obtain optimal algorithmic complexity.

Performance Evaluation. We have encoded three files of different format and size (two HTML files and a Ciao bytecode object) and measured the performance of alternative indexing and pre-indexing options. The experimental results for the algorithm implementation are shown in Table 1[6]. The columns under *indexing* show the execution time in seconds for different indexing methods:

---

[5] We use updates in the dynamic program database as an instrumental example for showing the benefits of preindexing from an operational point of view. It is well known that this style of programming is often not desirable. The illustrated benefits of preindexing can be easily translated to more declarative styles (like declaring and composing effects in the type system) or more elaborate evaluation strategies (such as tabling, that uses memoization techniques).

[6] Despite the simplicity of the implementation, we obtain compression rates similar to gzip.

```
1    % Mapping between words and dictionary index
2    :- data dict/2.
3
4    % NOTE: #lst can be changed or removed, ^ escapes cast
5    % Anchors to 2nd arg in constructor
6    :- rtype lst ---> [] ; [int|lst]:::index([2,0,1]).
7
8    build_dict(Size) :-                    % Initial dictionary
9        assertz(dictsize(Size)),
10       Size1 is Size - 1,
11       forall(between(0, Size1, I),       % Single code entry for I
12           assertz(dict(#lst([I]), I))).
13
14   insert(W) :-                           % Add W to the dictionary
15       retract(dictsize(Size)), Size1 is Size + 1, assertz(dictsize(Size1)),
16       assertz(dict(W, Size)).
```

Fig. 5. LZW Compression: auxiliary code and *rtype* definition for words.

Table 1. Performance of LZW compression (in seconds) by indexing method.

	Data size		Indexing (time)		
	Original	Result	None	Clause	Term
Data1	1326	732	0.074	0.025	0.015
Data2	83101	20340	49.350	1.231	0.458
Data3	149117	18859	93.178	2.566	0.524

none indicates that no indexing is used (except for the default first argument, first level indexing); *clause* performs multi-level indexing on dict/2; *term* uses pre-indexed terms.

Clearly, disabling indexing performs badly as the number of entries in the dictionary grows, since it requires one linear (w.r.t. the dictionary size) lookup operation for each input code. Clause indexing reduces lookup complexity and shows a much improved performance. Still, the cost has a linear factor w.r.t. the word size. Term pre-indexing is the faster implementation, since the linear factor has disappeared (each word is uniquely represented by a trie node).

4.2   Floyd-Warshall

The Floyd-Warshall algorithm computes the shortest paths problem in a weighted graph in $O(n^3)$ time, where n is the number of vertices. Let $G = (V, E)$ be a weighted directed graph, $V = v_1, \ldots, v_n$ the set of vertices, $E \subseteq V^2$, and $w_{i,j}$ the weight associated to edge (v_i, v_j) (where $w_{i,j} = \infty$ if $(v_i, v_j) \notin E$ and $w_{i,i} = 0$). The algorithm is based on incrementally updating an estimate on the shortest path between each pair of vertices until the result is optimal. Figure 6 shows a

```
1   floyd_warshall :-
2       % Initialize distance between all vertices to infinity
3       forall((vertex(I), vertex(J)), assertz(dist(I,J,1000000))),
4       % Set the distance from V to V to 0
5       forall(vertex(V), set_dist(V,V,0)),
6       forall(weight(U,V,W), set_dist(U,V,W)),
7       forall((vertex(K), vertex(I), vertex(J)),
8           (dist(I,K,D1),
9            dist(K,J,D2),
10           D12 is D1 + D2,
11           mindist(I,J,D12))).
12
13  mindist(I,J,D) :- dist(I,J,OldD), ( D < OldD -> set_dist(I,J,D) ; true ).
14
15  set_dist(U,V,W) :- retract(dist(U,V,_)), assertz(dist(U,V,W)).
```

Fig. 6. Floyd-Warshall code

simple Prolog implementation. The code uses a dynamic predicate dist/3 to store the computed minimal distance between each pair of vertices. For each vertex k, the distance between each (i, j) is updated with the minimum distance calculated so far.

Performance Evaluation. The performance of our Floyd-Warshall implementation for different sizes of graphs is shown in Fig. 7. We consider three indexing methods for the dist/3 predicate: *def* uses the default first order argument indexing, *t12* computes the vertex pair key using two-level indices, *p12* uses a packed integer representation (obtaining a single integer representation for the pair of vertices, which is used as key), and *p12a* combines *p12* with a specialized array to store the dist/3 clauses.

The execution times are consistent with the expected algoritmic complexity, except for *def*. The linear relative factor with the rest of methods indicates that the complexity without proper indexing is $O(n^4)$. On the other hand, the plots also show that specialized computation of keys and data storage (*p12* and *p12a*) outperforms more generic encoding solutions (*t12*).

4.3   Module System Implementations

Module systems add the notion of modules (as separate namespaces) to predicates or terms, together with visibility and encapsulation rules. This adds a significantly complex layer on top of the program database (whether implemented in C or in Prolog meta-logic as hidden tables, as in Ciao [4]). Nevertheless, almost no changes are required in the underlying emulator machinery or program semantics. Modular terms and goals can be perfectly represented as M:T terms and a program transformation can systematically introduce M from the context. However, this would include a noticeable overhead. To solve this issue,

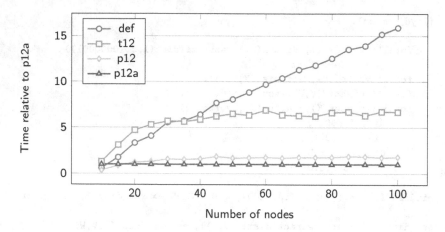

Fig. 7. Execution time for Floyd-Warshall

Ciao reserves special atom names for module-qualified terms (currently, only predicates).

We can see this optimization as a particular case of pre-indexing, where the last step in module resolution (which maps to the internal representation) is a pre-indexing cast for an mpred *rtype*:

$$:- \text{ rtype mpred } ---> \text{ nv:nv } ::: \text{ index}([1,0,2]).$$

For example, given a module M = lists and goal G = append(X,Y,Z), the pre-indexed term MG = #mpred(M:G) can be represented as 'lists:append' (X,Y,Z),[7] where the first functor encodes both the module and the predicate name. To enable meta-programming, when MG is provided, both M and G can be recovered.

Internally, another rewrite step replaces predicate symbols by actual pointers in the bytecode, which removes yet another indirection step. This indicates that it would be simple to reuse pre-indexing machinery for module system implementations, e.g., to enhance modules with hierarchies or provide better tools for meta-programming. In principle, pre-indexing would bring the advantages of efficient low-level code with the flexibility of Prolog-level meta representation of modules. Moreover, anchoring on M mimics a memory layout where predicate tables are stored as key-value tables inside module data structures.

5   Related Work

There has been much previous work on improving indexing for Prolog and logic programming. Certain applications involving large data sets need any- and multi-argument indexing. In [7] an alternative to static generation of multi-argument indexing is presented. The approach presented uses dynamic schemes

---

[7] Note that the identifier does not need any symbolic description in practice.

for demand-driven indexing of Prolog clauses. In [14] a new extension to Prolog indexing is proposed. User-defined indexing allows the programmer to index both instantiated and constrained variables. It is used for range queries and spatial queries, and allows orders of magnitude speedups on non-trivial datasets.

Also related is ground-hashing for tabling, studied in [17]. This technique avoids storing the same ground term more than once in the table area, based on computation of hash codes. The approach proposed adds an extra cell to every compound term to memoize the hash code and avoid the extra linear time factor.

Our work relates indexing techniques (which deal with fast lookup of terms in collections) with term representation and encoding (which clearly benefits from specialization). Both problems are related with optimal data structure implementation. Prolog code is very often used for prototyping and then translated to (low-level) imperative languages (such as C or C++) if scalability problems arise. This is however a symptom that the emulator and runtime are using suboptimal data structures which add unnecessary complexity factors. Many specialized data structures exist in the literature, with no clear winner in all cases. If they can be directly implemented in Prolog, they are often less efficient than their low-level counterparts (e.g., due to data immutability). Without proper abstraction they obscure the program to the point where a low-level implementation may not be more complex. On the other hand, adding them to the underlying Prolog machines is not trivial. Even supporting more than one term representation may have prohibitive costs (e.g., efficient implementations require a low number of tags, small code that fits in the instruction cache, etc.). Our work aims at reusing the indexing machinery when possible and specializing indexing for particular programs.

The need for the right indexing data structures to get optimal complexity is also discussed in [11] in the context of CHR. In [9] an improved term encoding for indexed ground terms that avoids the costs of additional hash-tables is presented. This offers similar results to anchoring in pre-indexing. Reusing the indexing machinery is also studied in [8], which shows term flattening and specialization transformations.

6   Conclusions and Future Work

Traditionally, Prolog systems index terms during clause selection (in the best case, reducing a linear search to constant time). Despite that, index lookup is proportional to the size of the term. In this paper we have proposed a mixed approach where indexing is precomputed during term creation. To do that, we define a notion of instantiation types and annotated constructors that specify the indexing mode. The advantage of this approach is that lookups become sub-linear. We have shown experimentally that this approach improves clause indexing and that it has other applications, for example for module system implementation.

These results suggest that it may be interesting to explore lower-level indexing primitives beyond clause indexing. This work is also connected with structure sharing. In general, pre-indexing annotations allow the optimization of simple Prolog programs with scalability problems due to data representation.

As future work, there are some open lines. First, we plan to polish the current implementation, which is mostly based on program rewriting and lacks garbage collection of indexing tables. We expect major performance gains by optimizing some operations at the WAM or C level. Second, we want to extend our repertoire of indexing methods and supporting data structures. Finally, *rtype* declarations and annotations could be discovered and introduced automatically via program analysis or profiling (with heuristics based on cost models).

References

1. Ait-Kaci, H.: Warren's Abstract Machine, A Tutorial Reconstruction. MIT Press, Cambridge (1991)
2. Boyer, R., More, J.: The sharing of structure in theorem-proving programs. Mach. Intell. **7**, 101–116 (1972)
3. Graf, P. (ed.): Term Indexing. LNCS, vol. 1053. Springer, Heidelberg (1996)
4. Hermenegildo, M.V., Bueno, F., Carro, M., López, P., Mera, E., Morales, J., Puebla, G.: An overview of ciao and its design philosophy. Theory Pract. Logic Program. **12**(1–2), 219–252 (2012). http://arxiv.org/abs/1102.5497
5. Johnson, E., Ramakrishnan, C.R., Ramakrishnan, I.V., Rao, P.: A space efficient engine for subsumption-based tabled evaluation of logic programs. In: Middeldorp, A., Sato, T. (eds.) FLOPS 1999. LNCS, vol. 1722, pp. 284–299. Springer, (1999)
6. Ramakrishnan, I.V., Sekar, R.C., Voronkov, A.: Term indexing. In: Robinson, J.A., Voronkov, A. (eds.) Handbook of Automated Reasoning, pp. 1853–1964 Elsevier and MIT Press (2001)
7. Santos Costa, V., Sagonas, K., Lopes, R.: Demand-driven indexing of prolog clauses. In: Dahl, V., Niemelä, I. (eds.) ICLP 2007. LNCS, vol. 4670, pp. 395–409. Springer, Heidelberg (2007)
8. Sarna-Starosta, B., Schrijvers, T.: Transformation-based indexing techniques for Constraint Handling Rules. In: CHR, RISC Report Series 08–10, pp.3–18. University of Linz, Austria (2008)
9. Sarna-Starosta, B., Schrijvers, T.: Attributed data for CHR indexing. In: Hill, P.M., Warren, D.S. (eds.) ICLP 2009. LNCS, vol. 5649, pp. 357–371. Springer, Heidelberg (2009)
10. Schrijvers, T., Santos Costa, V., Wielemaker, J., Demoen, B.: Towards typed prolog. In: de la Banda, M.G., Pontelli, E. (eds.) ICLP 2008. LNCS, vol. 5366, pp. 693–697. Springer, Heidelberg (2008)
11. Sneyers, J., Schrijvers, T., Demoen, B.: The computational power and complexity of constraint handling rules. ACM Trans. Program. Lang. Syst. **31**(2), 8:1–8:42 (2009)
12. Swift, T., Warren, D.S.: Tabling with answer subsumption: implementation, applications and performance. In: Janhunen, T., Niemelä, I. (eds.) JELIA 2010. LNCS, vol. 6341, pp. 300–312. Springer, Heidelberg (2010)
13. Swift, T., Warren, D.S.: XSB: extending prolog with tabled logic programming. TPLP **12**(1–2), 157–187 (2012)
14. Vaz, D., Costa, V.S., Ferreira, M.: User defined indexing. In: Hill, P.M., Warren, D.S. (eds.) ICLP 2009. LNCS, vol. 5649, pp. 372–386. Springer, Heidelberg (2009)

15. Warren, D.H.D.: An Abstract Prolog Instruction Set. Technical Report 309, Artificial Intelligence Center, SRI International, 333 Ravenswood Ave, Menlo Park CA 94025 (1983)
16. Welch, T.A.: A technique for high-performance data compression. IEEE Comput. **17**(6), 8–19 (1984)
17. Zhou, N.F., Have, C.T.: Efficient tabling of structured data with enhanced hash-consing. TPLP **12**(4–5), 547–563 (2012)

Author Index

Printed in the United States
By Bookmasters